Algebra
Problem Strings

Pamela Weber Harris and **Kara Louise Imm**

with

B. Michelle Rinehart
Susan M. Simmons

Discovering
Mathematics

Discovering Algebra
Discovering Geometry
Discovering Advanced Algebra

For more information on the **Discovering Mathematics** series,
visit k12.kendallhunt.com

Kendall Hunt
publishing company

The graphs were produced in Desmos®.
Cover image © Shutterstock, Inc.

Kendall Hunt
publishing company

www.kendallhunt.com
Send all inquiries to:
4050 Westmark Drive
Dubuque, IA 52004-1840

Copyright © 2017 by Kendall Hunt Publishing Company

ISBN: 978-1-5249-4396-7 (Coil bound)
978-1-5249-2319-8 (Perfect bound)

Printed in the United States of America

1 2 3 4 5 6 7 8 9 10 22 21 20 19 18 17

Contents

*Lesson numbers correspond to lesson numbers in *Discovering Algebra*, 3rd edition.

Acknowledgments

We gratefully acknowledge the influence of the work in the following: *Young Mathematicians at Work*, *Contexts for Learning*, *Math In Context*, the Freudenthal Institute, NumberStrings.com, Dr. Rachel Lambert of Chapman University, Andrew Stadel, Clotheslinemath.com, Texas Instruments' Teachers Teaching with Technology, *Functions Modeling Change*, Frank Demana and Bert Waits and their *Precalculus Mathematics: A Graphing Approach*, and the *Math Vision Project*.

Special thanks from Pam to:

- My husband, Daniel Harris, for supporting me in all of my crazy ventures. And for cooking dinners, cleaning house, and reminding me to sleep.

- Cameron Harris for his thoughtful editing of the Introduction. Thanks for helping me put on paper what I mean.

- Matthew Harris for the graphics production. His attention to detail and great advice is much appreciated.

- Craig Harris who willingly acts the student and pushes back, "No one would do that."

- Abigail Harris, my "I can't do it unless I understand why" girl, for pushing me to understand why like no one else.

- Debra Plowman for supporting me to produce the *Focus on Algebra* series of professional development workshops, my first foray into high school problem strings.

- Kim Montague for answering the calls, "Kim, this isn't your content area, but is this a problem string? No? Then how can I make it one? How do *you* think about this concept/model/strategy? Really? That's a thing? I'm going to write a string so I can think that way too!"

- Kathy Hale for providing me a venue to present to teachers and build my own content knowledge at the same time.

- Scott Hendrickson for getting me started. He is the master teacher who has integrity in his teaching—it all fits together.

- Michelle Rinehart for running with my half-baked ideas and making them shine. You found the clever twist so many times that makes a string *fun!*

- Sue Simmons for writing and editing and catching so many subtleties. And keeping me straight in so many areas of my life. You're a joy professionally and personally.

- Jerry Murdock, Ellen Kamischke, Eric Kamischke as the original task sequencers for giving me the first and best glimpse of what it means to teach real math. When I described problem strings to them as a lesson structure, they replied, "Isn't that just good teaching?" Yes, yes it is.

Special thanks from Kara to:

- Cathy Fosnot and Maarten Dolk for naming and living so many important ideas about the teaching and learning of mathematics, and for believing that I, too, could engender the kind of vibrant mathematical communities that you envisioned. And also for my first introduction to number strings!

- Janan Hamm, Nicole Shields, Carla Neufeldt Abatie, Sharon Steiner, Rachel Carr, Peggy Tsue, Ada Okun, Nancy Gonzalez, and Monica Mendoza (the Math in the City "east" and "west" crews) for kid-centered collaborations full of love and laughter.

- Hundreds of New York City teachers who we work with at Math in the City, but especially to Marcelle Good and Geoffrey Enriquez for co-developing early strings together.

- The Math for America (MfA-NY) community for their excitement about strings for high school algebra—"Finally, a whole book for us."

- Barbara Dubitsky (Bank Street College) who reminded me of the centrality of strong relationships with kids in the work of teaching.
- The powerful classroom communities of New York, NY; St. Louis, MO; Fall River, MA; Washington, DC; Buffalo, NY, Geneseo, NY, Binghamton, NY, Grand Cayman Islands, and Gothenburg, Sweden who anchor my ideas in the real lives of real (and lovely) humans.

Deep gratitude from both of us to:

- Tim Pope at Kendall Hunt for so many things, including the idea and support for this project and suggestions for statistics and probability problem strings.
- Kelly Fagan at Kendall Hunt for a fabulous job of editing in a tight time line.
- The myriads of teachers and students who have played with mathematics with us, trusted that problem strings were as powerful and exciting as we claimed, and given us invaluable feedback along the way.

Introduction

Teaching and Learning Mathematics

We believe that teaching mathematics is about mentoring mathematicians. This means helping students to mathematize experiences both in and out of school. To mathematize, according to Dutch mathematician Hans Freudenthal, means to structure, model, and interpret one's "lived world" mathematically. That is, to look for connections, make conjectures, seek to generalize, and justify reasoning. As teachers of algebra, our job is not to distribute to students a set of prescribed rules and facts, but to help them to structure and schematize, thus creating mathematical relationships in their minds.

To help students construct mathematics in this way, we draw from a variety of lesson structures, including, but not limited to: inquiry, investigations followed by math discussions, and mini-lessons. The important mini-lesson structure that we highlight in this book is called *problem strings*. I, (Pamela Weber Harris) developed and introduced problem strings in my book *Building Powerful Numeracy for Middle and High School Students* (2011). But long before this, the idea of mental math routines focused on students' strategies originated in the work of Cathy Twomey Fosnot and Maarten Dolk, when they introduced the idea of *number strings*. Number strings were first described in their book *Young Mathematicians at Work: Constructing Number Sense, Addition and Subtraction* (2001), part of a series that would later extend through rational numbers. Later, number string resource books were written for teachers, starting with *Minilessons for Early Addition and Subtraction: A Yearlong Resource* (2007), and also extending through rational numbers. In this book, we deepen this well-known mathematical practice by bringing problem strings to the high school math class, supporting mathematicians to make sense of their algebra and advance algebra courses.

Of these lesson structures, investigations and math discussions typically involve messier, bigger questions that can be more involved and complicated, and therefore take longer to work with and solve. Students tackle these bigger, non-routine problems in context in small groups, working together to make sense of the math, and then deciding how to explain their findings to the class for review and comment. Problem strings, in comparison, tend to be shorter and much more targeted.

Discovering Advanced Algebra contains some of the best rich investigations to help students construct complicated concepts and skills. Yet often, when students begin solving these challenging problems in context, they stumble, struggle, get frustrated with themselves, or us, or give up. In these moments, it can be tempting to do one of three things:

1. Pre-teach the anticipated skills and ideas before introducing the investigation.

2. Use the tasks solely to engage students, but then simply show them how to solve them.

3. Or, abandon the use of investigations altogether.

Each of these pedagogical choices is an attempt to minimize the messiness of learning and the evidence of struggle. But without some struggle there is rarely deep learning. To learn is to continually re-organize one's ideas and encounter the unfamiliar until it becomes connected to our ideas, sensible, understood, and eventually familiar.

We believe that the regular use of the powerful routine called *problem strings* helps both students and teachers before and after investigations. Increasingly, problem strings are being used by teachers to:

- preview big ideas that will arise in an investigation,

- solidify the ideas and skills that came up in the investigation,

- create puzzlement, disequilibrium, and curiosity,

- invite students to prove or justify their ideas,

- describe and solidify strategies, and move towards efficient strategies,
- build students' efficacy at choosing strategies, and
- generalize an idea beyond the task at hand.

Problem strings allow students to struggle in a contained, guided, purposeful set of tasks. They compliment and support the work of investigations and math discussions, working together to foster conversations and form conclusions about relationships, structures, and repeated reasoning. This is the work worthy of teachers, to help students develop and grow into mathematicians.

What Is a Problem String

A problem string is a purposeful sequence of related problems, designed to help students mentally construct mathematical relationships. It is a powerful lesson structure during which teachers and students interact to construct important mathematical strategies, models, and concepts. The power of a problem string lies in the carefully crafted conversation as students solve problems, one at a time, and the teacher models student thinking and draws out important connections and relationships.

	Problem strings are...	Problem strings are not...
The problems	a purposefully chosen sequence of related problems.	random nor entirely predictable ($x + 1$, $x + 2$, $x + 3$...).
Lesson format and timing	a mini-lesson that typically precedes or follows an investigation.	a substitute for other forms of problem solving, including inquiry, rich investigations, and math discussions.
Role of the teacher	guided by the teacher to bring out the students' mathematics; the teacher systematically nudges students toward more efficient and sophisticated strategies.	student-led—students do not hijack the conversation.
Type of instruction	facilitative—there is some explicit teaching only when related to social knowledge (mathematical terminology, notation, etc.).	teacher-centered with direct instruction; not an "I do, we do, you do" approach.
Strategies	developed by students. Over time and with lots of experiences, we nudge students to • develop a variety of strategies to draw upon and • choose the most efficient one for the problem at hand.	spaces for teachers to demonstrate their own strategies or thinking (e.g., "This is how I might solve this one..."); places to "practice" using the same strategy for every problem over and over again.
Concepts	based on the ideas noticed by students, though the teacher must be looking and listening for "glimmers" of these ideas during the routine so that they can be made public and then explored together as a class.	opportunities for teachers to explain the big ideas (and why they work) to students.

Algebra Problem Strings
©2017 Kendall Hunt Publishing

	Problem strings are…	Problem strings are not…
Pacing	short routines. Be purposeful, deliberate, respectful of the time it takes to think, but with some energy—don't put the kids to sleep! Teacher celebrates ideas, risk taking, and deep thinking—not speed.	a race or anything resembling a timed test.
Modeling	the teacher modeling students' thinking while the students articulate their ideas to the class; the teacher makes students' thinking visible so that it can be compared and discussed.	intended for students to model their own thinking for each problem—often students do not initially know how to model their thinking.
Engagement	mini-lessons where every student in the class is engaged in thinking about the mathematics.	sparse conversations due to the teacher calling only on students who have a right answer or an "effective" strategy.
Assessment	wonderful experiences to learn about how your students think, what they understand, and are still constructing.	"gotcha" moments where students are categorized as either "right" or "wrong."
Focus	focused on strategies, sense-making, ability to generalize, and convincing others of their ideas.	routines to reward students who get the "right" answers quickly or students who say they "just know it."
Student participation	all students solving every problem. The teacher chooses which strategies are to be shared, developing the students as mathematicians. Sometimes this can be sharing less sophisticated strategies or misconceptions so that students can discuss, compare, and ferret out common pitfalls.	every student sharing their thinking during the whole class discussion, especially if a strategy has already been shared or if it is not helpful to moving the mathematics forward.
Student experience	times for students to try new ideas, to be uncertain, and to question. Learning is happening and that does not always feel like you are on solid ground.	times for students to stick to one strategy just because it works.

Addressing 10 Common Misconceptions:

- Teachers should not present the entire problem string all at once. Teachers may be inclined to hand out the problems all together, list them on the board as students come in the room, or show them on a slide all at once. This approach is unlikely to support many of the goals of problem strings and limits the kind of deliberate and purposeful conversation about mathematical relationships.

- Problem strings are not a collection of random or unrelated practice problems. There may be legitimate space in your teaching practice for this activity, but that is not a problem string.

- Problem strings are not opportunities for a teacher to demonstrate a strategy for students to then "practice." Said differently, a problem string is not the place to introduce a traditional algorithm or a procedure. This is a time for teachers to listen and watch carefully, picking up on ideas and helping to bring them forward by noticing, questioning, and wondering. This noticing, questioning, and wondering is about what the students are saying, doing, and thinking about. If a teacher puts forth "the right" way, this often prevents students from trying out strategies and taking risks in the conversation. Students will pick up on this and wait a teacher out, knowing that "the teacher's way" will come. Mathematizing is not about "the" way; it's about using relationships and connections to solve problems.

- Problem strings are not spaces for direct teaching—with the exception of social knowledge (including mathematical terminology) that students may not have access to. For example, notation or technical language is a social construct that should be mentioned to students (e.g., "Yes, we call that cube root.") and recorded publicly so that students can see and hear the new ideas. Don't make students guess about convention! However, the logical mathematical knowledge that must be constructed cannot be passed down by simply telling, but must be experienced so that connections and relationships are constructed in the learner's mind.

- Teachers should not expect students to all use the same strategy. If you find that this occurs, you may be too prescriptive, leading, or rigid in your facilitation. Problem strings are designed with multiple entry and exit points, meaning that students with a variety of skills and understanding can access the problems. Likewise, the conversations in problem strings are designed to meet students where they are and nudge them along their mathematical journey. Hence, we expect multiple exit points, meaning students will not all construct the same relationships at the same time. The goal is to help grow judicious problem solvers who choose strategies based on the numbers or structure of the problem, or on what they infer and understand—based on relationships they own. When students take up this type of noticing-then-choosing-a-strategy approach, we find they are better able to tackle unfamiliar problems and tinker with new ideas. If students use the same strategy (regardless of the numbers or the structure of the problems), they are likely mimicking, not mathematizing. Our goal is thoughtful, flexible, creative problem solvers who possess a bank of strategies, not just one. Problem strings often have "sister strings" that allow students to revisit the big ideas on successive occasions. Students are enabled to construct the ideas when they are ready instead of on any one day.

- During the sharing time, students should not model their own thinking. Often the teacher models a student's thinking using a different representation than the student because the teacher understands that certain models can help bring relationships to light, help students make comparisons, and help models become a realized tool that students can begin to use. Many students do not know how to model their own thinking. Thus, this can be an opportunity for the teacher to help make the reasoning and relationships visible for the rest of the class.

- The teacher should not take time to figure out a student's unknown strategy. Be prepared so you will have thought deeply about possibilities, but if something takes you by surprise and you cannot make sense of it in a reasonable amount of time, don't leave the rest of the students hanging too long. Taking too long to decipher an unknown strategy at the moment (problem strings are short) can derail the discussion. Don't discount student thinking, but if it's taking you too long in the moment, take note, tell the student that you need to think about it (how cool is that—you have to think about what they've come up with!) and study it later. Bring it to the class later if it is generalizable or if it is an example of a common misconception for students to parse out.

- The sharing time is not an opportunity to share every student's strategy. We definitely want students engaged, solving, and sharing, but it is not a free for all. Allowing every strategy to be shared—regardless of its usefulness or novelty—incentivizes lazy repetition rather than constructive mathematizing. Conversely, we do not want to share only the 'right' strategy. By striking a balance between showing everything and showing one thing, you set the pattern you want your students to follow. You are demonstrating that students should be flexible enough in their problem solving to consider multiple approaches, but discriminating enough not to be satisfied with a strategy just because it comes to mind.

- It is not necessary to share every student's strategy to honor every student's thinking. By encouraging students to solve problems using their own thinking rather than requiring them to regurgitate yours, you are already honoring them far more than you would be otherwise. In addition, sharing every student's strategy regardless of its usefulness cheapens your regard. Not always, but often the question, "Did anyone do it another way?" (just trying to get lots of strategies on the board) indicates a teacher who has not thought deeply about the big ideas involved, the mathematical terrain ahead, or how to purposefully guide the conversation. This is different than a teacher who asks, "What does everyone think about the strategies we have on the board?"

- The student asked to share his or her strategy is not always, or even often, the student who "got it right" or had the clearest explanation. Sometimes a student who only got started can share an important beginning or can highlight the complexity of finding a starting point. Sometimes students who are a little muddy but on a fruitful track can bring the ideas to the class in such a way that the rest of the students get the opportunity to

Algebra Problem Strings
©2017 Kendall Hunt Publishing

weigh the ideas and make sense of them as the class works out the ideas together. Sometimes a student can be the "canary in the mine shaft" and bring a strategy for the class to weigh, compare, and bring to light misconceptions. Such instances must be handled with the utmost respect and gratitude. While the student may have gotten something wrong, his or her blunder has created the opportunity for the entire class to better understand the topic. Likewise, questions from students can provide similar opportunities. This can be done with strategic turn and talks and by creating a community where students are encouraged to ask questions, challenge ideas, and try new things.

Modeling and Models

Within mathematics, models and modeling have always played a prominent role. Yet, the terrain of models and modeling can be complex and hard to understand. The terms can be used within (and beyond) mathematics in several different ways, making it hard to distinguish what is meant by each in its context. The idea of "model" itself can be used both as a noun and a verb, sometimes describing the actions of the teacher and other times describing students' mathematical activity.

Yes, a model is a person walking down a runway in fashionable clothing. And yes, to model good behavior, we copy or mimic what others are doing. But these are not the meanings we intend here. When leading problem strings, we use the words model and modeling very specifically in *two different, but related ways*:

Model students' thinking—an action, performed by the teacher

Here we are describing what a teacher is doing during a problem string. To begin, we are listening and trying to make sense of students' strategies. Then, using what we understand, we make that thinking visible for the community. Sometimes we say, "I'm going to make a picture of your idea so that we can all see it and hear it at the same time." This is a deliberate and essential part of what it means to lead problem strings—and part of what makes them powerful. As students speak, we are capturing their ideas in a visual image that allows more students to make sense of the ideas, and may even help the "authoring" student to better understand his or her own idea. To see one's idea made public can be a powerful boost of status to any member of a community.

Use a model—a noun or object, introduced by the teacher and eventually taken up by students

When teachers are modeling students' thinking (e.g., making it visible) we are not just drawing whatever we choose, nor are we writing down a symbolic transcript of what the student says. Instead, we are typically using well-known mathematical models. These are generalizable mathematical representations that help us think about and solve problems in more than one way and one context, and communicate our ideas to other mathematicians. Examples of the models we draw from include, but are not limited to:

- open number lines
- open double number lines
- open arrays
- ratio tables
- rectangular diagrams
- graphs
- tables
- expressions
- equations
- functions

By drawing upon a menu of mathematical models we are helping students to see their thinking in new ways.

How do models move from teachers to the minds of students?

We share the belief with our colleagues Fosnot & Dolk (see *Young Mathematicians at Work* series, 2001) and the research traditions of the Freudenthal Institute that mathematical models typically begin as *models of thinking*. When a teacher "makes a picture of my thinking" he or she is gently suggesting that what I am saying and this particular model are related. When students see their teachers using models to represent their thinking over and over again—and when they are allowed to investigate rich contexts where these models arise naturally—they begin to transition from *model of thinking* to *model for thinking*. That is, students begin to embody or envision the model on their own, without our prompting, and use that mental model to solve new problems. When students begin to say things like, "I thought about it on a number line," or "I imagined the graph of $y = x$ and shifted it up two places," we know that students are moving towards *models for thinking*. This, of course, is our eventual goal, that these mathematical models are constructed in the minds of students and used strategically by them whenever they encounter new and unfamiliar problems.

Facilitating Problem Strings

The first time you try a problem string in class might feel like sailing into the uncharted ocean of student thinking and reasoning, and to some extent it is. But it is important to realize, that whether we know it or not, students have always been trying things, thinking of alternatives, or wishing they could see the big picture so it could all make sense.

Facilitating a problem string requires careful attention to the mathematics as well the ability to really listen to students and model their thinking for all to see. Thinking is often in development—not fully polished or formed and sometimes idiosyncratic or just tricky to understand. Because we believe in a mathematical community that includes all learners, our role is to bring before the class what is helpful for development—whether it is clear, clean, and polished, or messy, incomplete, and developing, or even incorrect. The goal is always to give students the chance to articulate their ideas and to see each others' thinking—to give the class the chance to respond to, challenge, and make sense of someone else's strategy or idea. This means that as teachers we are withholding our authoritative stamp of approval and giving the mathematics back to kids to reflect on and sort out. This requires both restraint and the artful use of questions.

As teachers of algebra, there is often institutionalized pressure to teach kids "the steps"—to give students access to algebra by offering generic, one-size-fits-all strategies that "work," particularly in testing situations. We may find ourselves working in isolation from our peers, focused on "getting our kids to…." [say the right answer, do the right strategy] and not willing to explore a variety of strategies because we believe we do not have the time. Teaching under these conditions is not easy, but we invite you to resist many of these tendencies, which are informed by a culture of testing more so than a culture of learning. Too often we encounter students who mimic the teacher's mathematics, sometimes using procedures when they are not relevant, and often having no opportunity to explore why they do (or do not) make sense. We want to offer a different mathematical experience and we believe that the regular use of problem strings provides:

- a chance for kids to make sense of algebra,

- opportunities to build algebraic relationships that will extend past our particular course, and

- a way to build a classroom community where knowledge is explored, validated, and constructed together.

Our belief in problem strings stems partially from the idea that telling or showing students mathematics does not produce learning—and in fact, it never has. Disrupting this pattern of "delivering" or "showing" students mathematics is a bold undertaking that will require planning, restraint, and real trust in students' ability to think for themselves. We must believe that our students are full of interesting mathematical ideas, insights, and questions—and that what they offer is enough to begin the work of formal algebra together. Allowing students to solve problems any way they want, asking students to share their thinking, and pushing students to justify can be foreign, new, and unsettling. And, downright fun. In the next section, we offer some pedagogical "moves" that will support you on your journey.

Algebra Problem Strings
©2017 Kendall Hunt Publishing

Sample Dialogs

The sample dialogs were written to help you get a "vivid image" of what the string might look like and sound like in a classroom. We tried to highlight the important parts of the mathematics, the questions to have ready to ask, and the ways you might model student thinking. These transcripts are based on our own experiences leading problem strings with students, as well as observation of our colleagues leading strings with their students.

We varied the sample dialogs because we know different teachers may need different types of support. There are three formats of sample dialogs:

	What It Is	Why We Included It	What You Might Find
Full dialog	A sample exchange (resembling a transcript) between the teacher and students, with accompanying modeling of strategies.	We wanted to provide a clear window into the classroom interactions. Sometimes there were new or atypical strategies present. Or we felt that it would be helpful to study a possible way of steering the conversation. Or, we wanted to highlight how the timing of certain questions and modeling might be important to the success of the string. All of the full dialogs offer an important window into the possibility of a vibrant community of learners.	You will find purposeful examples of students expressing ideas in vague, non-mathematical, student vocabulary where the teacher responds by restating, agreeing, or questioning the ideas using precise mathematical language and helping students name important concepts.
Partial dialog	Many of the questions a teacher might ask alongside the modeling of student strategies.	We provide these examples when the modeling was more important than every interaction between the teacher and students. Perhaps the models were new, or we are suggesting juxtaposing two models or strategies.	Sometimes partial dialogs accompany the second string in a series of strings because the interaction has already been established in the first string, but the problem-by-problem modeling remains important.
Important questions	A list of central, critical questions that can help guide the facilitation of the problem string.	We wrote these for problem strings where teachers tended to need less guidance, especially on the problem-by-problem modeling.	These often accompany strings in a series, where related strings have already been discussed through full or partial dialogs, and many of the pedagogical moves have already been explored.

Preparing

Leading your first few problem strings requires equal parts courage, curiosity, spontaneity, and preparation. Despite your careful planning, anticipate that you will hear ideas and strategies that you do not expect, may not fully understand, or may struggle to model. In these moments, you want to just listen and then you can pivot to the class—"Who understands what Henry is saying and can put his strategy in their own words?" Work as a class, share the task of understanding kids' thinking with your class, and do not feel that just because you understand it that they will.

As you gain experience, you will develop greater confidence. You will begin to know which ideas from students will lead you away from the mathematical goals of the problem string, and which ideas are central to building new mathematical ideas. Even if you do not choose to pursue an idea you can always respectfully honor students' contribution to the conversation. It's acceptable to tell a student, "That's interesting. Let me think about that and get back to you." And then choose to model the strategies you were planning on.

It's ideal to prepare a problem string with a colleague or other adult who can "play" the student as you try to model their thinking, or who can hear your thinking as you think aloud with them. If you have never led the string before, it is essential that you take time to model students' thinking in advance—yes, take time to draw what different strategies will look like. This will allow you to feel more comfortable and confident as you are leading the problem string with students. As you prepare, ask yourself the following questions:

- How is each problem related to the previous problem(s)? Describe the progression of the string. What's going on here?

- Why these numbers? These problems?

- What big ideas could emerge during this string? How might I encourage kids to articulate these big ideas as they are solving problems?

- What strategies do I expect students to use to solve the problems? How will I model, or represent, this thinking?

- What kinds of questions or prompts might I use to encourage students to consider and explore the big ideas?

- Are there problems I want to insert or add? Why?

- Would a context support my students to reason about the mathematics? If so, what context? Why?

- What is the "rhythm" of the string? Where is the energy? Where will I slow down, speed up, or really try to engender good conversation?

Questioning

You will notice that often the dialogs have examples of questions that seem open but are actually purposefully focused. A mistake that teachers sometimes make is to ask questions that are too open. This can end up with students confused about what you are asking. It can also result in students just trying to guess what you're after. This is not about playing Twenty Questions. The questions should be specific enough where the answers can be based on reasoning, not guessing. Following are some things to think about when you plan the questions you will ask.

If You Want to…	You Might Prompt With or Consider…
Nudge towards more efficient strategies	After putting the strategies on the board, you might ask, "What about the problem or the numbers make this an efficient or not very efficient strategy?" "Which one was more helpful here and why?" This is very different than asking a vague, "Explain your thinking." Or "How did you do it?" Or even, "Which is the best strategy?" **Note:** Best is a subjective term. We want to get out of personal preference and into the space of justifying a strategy choice based on some insight about the problem. Using words like clever, efficient, and elegant can help convey a sense of searching for sophistication in thought, not a mimicking of procedure nor an absolute "best" strategy.
Move quickly through some problems to get to some more interesting mathematics	Do not belabor the conversation in the beginning of a problem string. Acknowledge that the questions are not so tricky, and that more challenging problems are on their way. You might say something like, "Okay, it seems like you solved this one quickly, so let's keep going." or "Tell us what you're thinking when you see this one. Yep. Onto the next one." If you wait too long or ask for reasoning or multiple strategies on the simpler problems, you may lose students' interest in the string. Make sure you have planned where the "energy" of the string should be—meaning which problems to spend more time discussing because there is something more interesting to discuss.

Algebra Problem Strings
©2017 Kendall Hunt Publishing

If You Want to...	You Might Prompt With or Consider...
Juxtapose two strategies	"Did anyone think of the problem this way?" If no students volunteer, you might ask, "*Could* you think about the problem this way?" "These are two really different ways to think about this problem. What is the same about them and what is different?" Alternatively, sometimes you can follow with, "Last year, I had a student who used this relationship. What do you think about that?" You can also insert a problem that might prompt that strategy or plan to write a "sister" string for the next day that would bring that strategy to the forefront.
Build a community of learners	Remove yourself as the mathematical authority in the room by changing your pronouns. Go from "tell me, show me" to "show <u>us</u>, tell <u>us</u>." Along with your body language, this deliberate use of a collective pronoun can dramatically change the dynamics of the conversation. One measure of the strength of a classroom community is the extent to which kids are talking to each other, unfiltered by the teacher, about each other's ideas. Whenever you are puzzled by a student's ideas, you can begin by asking, "What do <u>we</u> think of this idea?"
Move from "algebraic rules" to sense-making	Algebra has been commonly associated with an unknown mathematical authority (usually the teacher, sometimes a more abstract figure) who provides a set of often confusing laws or rules to follow. This has prevented students from making sense because they are more worried about whether they can or cannot "do" something or whether "it is allowed." We want to move away from the paradigm of "Is this allowed?," and into the space of "What makes sense to us?" So we invite students to paraphrase—"Who understood that idea/strategy and could say it again for us?"—as a way to get students to really listen to each other as sense-makers. Additionally, we include questions such as: • Would this make sense? Say why it makes sense to you… • What would this mean? • Are we okay with this? • Is this helping us to (solve, graph, envision, simplify)? • Do you think this would help us? • How do we know that this is an equivalent form? • Will this always work? • What allows us to do this? • Are these still equivalent? How do we know?
Bring the string to a close	At the end of the problem string, we suggest that you question students to help them reflect and pull together what they learned during the string. We provide suggested answers that can help you focus the problem string as you facilitate it.

Sample Display

Each problem string includes a sample display section. This is not meant to be prescriptive, but as a possible end display when planning the string. It might not seem important to plan how your display will end up, but the very nature of a problem string is enhanced when things are displayed in such a way as to help suggest connections, patterns, and relationships. You can encourage sense making as you model strategies using deliberate color, organization, labeling, and placement. You will notice that some of the strings suggest a rather large area. Try to arrange for sufficient space for each string. If you must write on paper under a document camera, write in pencil so you can erase if needed. Plan ahead so that you write in such a way that students can see as much as possible. Keeping the record of the work in front of students encourages students to use what they've learned in prior problems, find patterns between problems or answers or strategies, and connect multiple representations.

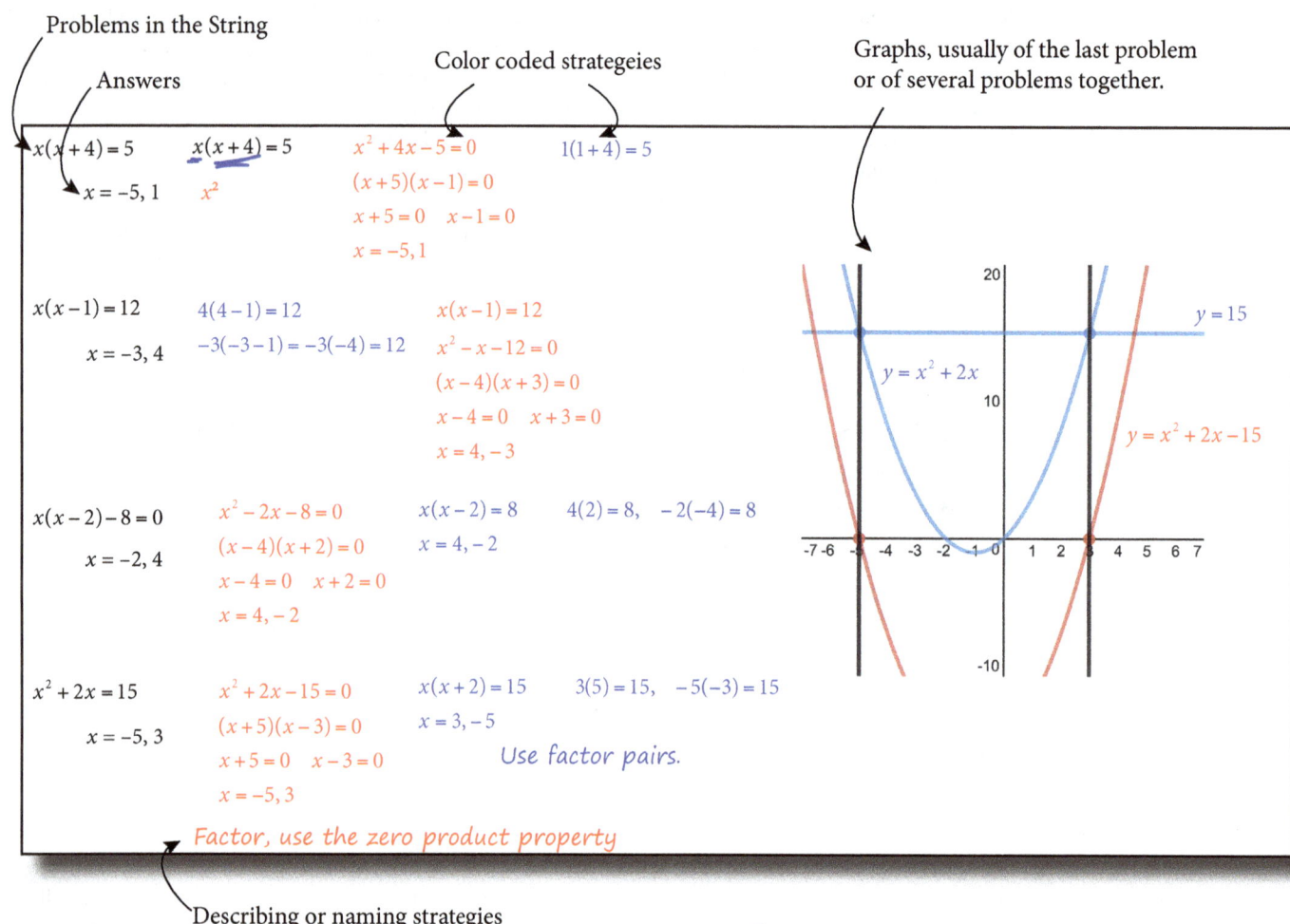

Problems in the String

Answers

Color coded strategeies

Graphs, usually of the last problem or of several problems together.

$x(x+4)=5$ $x(x+4)=5$ $x^2+4x-5=0$ $1(1+4)=5$

$x=-5,1$ x^2 $(x+5)(x-1)=0$

$x+5=0 \quad x-1=0$

$x=-5,1$

$x(x-1)=12$ $4(4-1)=12$ $x(x-1)=12$

$x=-3,4$ $-3(-3-1)=-3(-4)=12$ $x^2-x-12=0$

$(x-4)(x+3)=0$

$x-4=0 \quad x+3=0$

$x=4,-3$

$x(x-2)-8=0$ $x^2-2x-8=0$ $x(x-2)=8$ $4(2)=8, \quad -2(-4)=8$

$x=-2,4$ $(x-4)(x+2)=0$ $x=4,-2$

$x-4=0 \quad x+2=0$

$x=4,-2$

$x^2+2x=15$ $x^2+2x-15=0$ $x(x+2)=15$ $3(5)=15, \quad -5(-3)=15$

$x=-5,3$ $(x+5)(x-3)=0$ $x=3,-5$

$x+5=0 \quad x-3=0$ Use factor pairs.

$x=-5,3$

Factor, use the zero product property

Describing or naming strategies

$y=15$

$y=x^2+2x$

$y=x^2+2x-15$

Sample final display from Lesson 8.6 Factors and Factoring.

Sample Facilitation Notes

The facilitation notes at the end of each string are meant to be an abbreviated version of notes for you to use at the time of doing the string. When you prepare to lead, you might study the entire problem string, first learning the math and noting the relationships, and then go back over the string a second time attempting to capture the modeling, the flow, and the possible questions. Finally, you might review the entire plan, noting the important changes and questions from problem to problem, and how you will model student strategies using the sample display. Compare your notes with our facilitation notes and add any that you need.

Problems in the String

Things to ask

Things to do

$x(x+4)=5$	First, what do you think the solutions might be? Solve for x. Find factor pairs, factor w zero product property. Graph given and general form. Make connections.
$x(x-1)=12$	Now what's x? Start describing strategies. Might notice that numbers are 1 apart. What do the strategies have in common?
$x(x-2)-8=0$	Different structure. What about the structure influences your strategy choice? Notice that numbers are 2 apart. Which is more efficient?
$x^2+2x=15$	Different structure. How do solutions relate to each other? Factor pairs? Tag the strategy name. Add to anchor chart. Notice patterns when using factor pairs. Graph.

Reminders about the structure of the string.

Sample Facilitation Notes from Lesson 6.8 Factors and Factoring.

Sample Anchor Charts

Some of the problem strings have sample anchor charts. An anchor chart is a semi-permanent fixture in the classroom designed to codify some learning that the class has developed together. While they are typically "teacher-made" they are full of students' ideas and often give citations and credit to the student or students who first articulated the ideas (e.g., "George's conjecture", "Amaia's strategy"). Anchor charts are useful when students have a number of strategies and are beginning to determine which strategy to use when. In this way, having these strategies named and clearly modeled with sample problems allows students to refer to them during any act of problem solving.

An anchor chart might also be about big ideas or conjectures. When a student has an important mathematical idea that the group is wrestling with, or is convinced of, a nice way to honor this thinking and to reinforce this practice is to make a poster of this big idea or conjecture for all students to see.

Unlike images on an electronic board that tend to disappear quickly, anchor charts are designed to be displayed in such a way that every student can see and refer to them throughout the unit or year. Teachers can refer to them, "We are working on solving quadratics right now, and we all know where there is a wall of strategies that we can draw upon." These anchor charts evolve over time as the class brings new strategies on board.

$8x+2y=60$ $-8x-5y=72$	Elimination • When you have "opposite terms" like 9x and -9x • When you can add or subtract the equations to get one of the variables to be zero
$-4x+3y=-20$ $y=5x+8$	Substitution • When one equation is in slope-intercept form and the other is not • When a variable is isolated with a coefficient of one • When substituting one variable for the other will be friendly
$2x+4y=72$ $3x+5y=80$	Combinations • When you can tell the story of the combinations • When both of the variables are on one side and there's a value or quantity on the other side • If you can make new equations by combining and those new equations make it easy to isolate a variable.
$y=2x-12$ $y=3x-12$	Envisioning the graph • Same slope—lines are parallel • Same y-intercept—it's the solution

Sample Anchor Chart from Lesson 5.4 Choose a Strategy.

Problem String Structures

Our colleague Rachel Lambert (Chapman University) noted that strings have various structures. If you have used problem strings you know that there isn't a single "formula" for designing strings. However, there are some general structures that we use. In this section we highlight a few of those structures—giving some insight into how and why certain sequences of problems were chosen. The examples below are provided to give you some ideas, if or when you begin to write your own problem strings.

Name of Structure	What Is It	An Example
Helper-Challenge OR **Helper-Challenge-Clunker**	Providing access to all students at the beginning of the routine is not only a smart pedagogical move towards building a classroom community, but it also speaks to issues of equity in mathematics. Typically this problem is followed by one whose structure is similar, but with a small new twist, the *challenge problem*. This *helper-challenge* structure can be repeated throughout the problem string. We have found that when students first begin to notice this pairing, they later make use of it—drawing upon something in the helper problem to make sense of the challenge problem. Sometimes this structure will conclude with what we playfully call *the clunker*—chosen to be challenging and not to immediately resemble previous problems. This sudden unfamiliarity forces students out of the habit of looking back to a previous problem, and nudges them to think about a helpful strategy to use.	**String 8.1—At a Glance** $x^2 = 64$ — Helper problem $x^2 - 17 = 64$ — Challenge problem $x^2 + 15 = 64$ — Related challenge problem $(x + 3)^2 = 49$ — New, but related problem—note the 64 is replaced with a different, but friendly perfect square, 49. $(x - 7)^2 = 25$ — New, but related problem $(x + 2)^2 + 1 = 82$ — Clunker—The slightly new structure of the equation may make this problem feel challenging for some students, which is why we save it for the end of the problem string.
Equivalence	Problems in this structure may result in the same solution. Equivalent problem strings—such as doubling and halving in multiplication, constant difference in subtraction, keeping the ratio constant in division, or yielding the same solution to equations—are designed to give students opportunities to construct big ideas. "Why are we continuing to get the same answer here?," a teacher might ask, nudging students into the space of wondering and puzzlement. When students encounter problems that have the same answer, they tend to pay attention and wonder. This shifts the mathematical goal away from answer-seeking towards investigating the relationships that would explain this phenomenon. Later, they may use strategies related to equivalence—such as solving the equivalent reciprocal equation—to solve new messy problems.	**String 6.9—At a Glance** $\dfrac{8}{4}$ — The first problem sets the stage for division. $\dfrac{32}{4}$, $\dfrac{2 \cdot 2 \cdot 2 \cdot 2 \cdot 2}{2 \cdot 2}$ — The next two problems are equivalent by design. This supports students to ponder why each is equivalent to 8 or 2^3 and how the format of the equations relate to each other. $\dfrac{3^4}{3}$, $\dfrac{2^m}{2^n}$ — The next problem is similar, but with a different base, 3. This gives students an opportunity to take up using repeated multiplication and division. $\dfrac{x^m}{x^n}$, $\dfrac{2^5 \cdot 3}{2^2}$ — The next two problems set the stage to generalize, first using the familiar base of 2 from above and then with the more general base of x. $\dfrac{2^4 3^3}{12}$ — The last two problems give students an opportunity to use the patterns and relationships to find equivalent expressions.

Algebra Problem Strings
©2017 Kendall Hunt Publishing

Name of Structure	What Is It	An Example
Building a Model	For some problem strings, the purpose is to help students build a model as a mental construct. These are important models that begin as models of student thinking that can then transition as tools with which to reason and use as a mechanism for thinking. In other words, these strings build a model to represent thinking so it can later become a model for thinking. These problem strings consist of what may seem to be random problems, but the problems are based on specific relationships. As students solve the problems, a model begins to emerge. The order of the problems matter—they are usually not in order (too predictable, not intriguing), but are purposefully sequenced so that as answers are plotted, graphed, or even listed, a visual representation emerges. Students begin to realize that as their brains make and use certain relationships, those connections can be represented outside of their brain. These representations built and synthesized from mental relationships begin to solidify into tools that help with other problems. These *building a model* problem strings are intended as introductions. We should not expect mastery of all aspects or even facility with the model after the first string. Rather, continue to build the model with more rich experiences.	**String 7.1—At a Glance** $(2,7)$ $f(-1)=1$ $f(-4)=-5$ $(\frac{1}{2},4)$ $f(1)=$ ____ $f(\underline{\hspace{0.5cm}})=0$ $f(x)=$ ____ In this problem string, students start with an ordered pair, representing it in a table and on a graph and the teacher introduces function notation. As students continue alternating between representations, a familiar line emerges. By the fifth problem, the teacher confirms that f represents a line, which gives students the opertunity to make sense of finding $f(1)$ using the line and then $f(\underline{\hspace{0.5cm}}) = 0$, helping students solidify the meaning of function notation. The last problem of the string ask student to generalize the y-value for any x in f. Students are now primed to begin to use function notation.

Challenges

As we've mentioned, problem strings take some preparation and practice to engender the kinds of vibrant mathematical conversations we envision. If you study the problem strings in advance, you will typically have a good sense of the mathematics that are likely to emerge. The conversation should never feel like a complete surprise. However, students' ideas can often be idiosyncratic, unformed, and even confusing. Knowing what to do when the problem string you've planned begins to feel different than the one that is unfolding takes experience and insight. In the table below we frame some of the most common challenges you may experience and offer a few suggestions about what you might say or do in these moments.

The Challenge	What You Might Consider	What You Might Say or Do	Try to Avoid
Students are not talking much, if at all.	• Have you given students something interesting to think about? • Is the math accessible to the students? • Have students done this math before and are not being asked to think about it in a new way? • Are students with partners that are "just right" for them? • Are you asking them about a problem that is too easy? • Have you created a class climate of respect and a low cost of failure, while rewarding risk taking?	• Ask students to turn and talk to their partner and listen to their pair talk. Then invite partners to share together, or ask a pair of students if you can tell the class what they discussed. • Change the conversation from "What's the answer?," to "Let's not talk about the answer yet. Who can get us started?" This may make some students feel safer and more equipped to share their ideas. • Get students paying attention to the logic of the string: "What problem do you think I'm going to put up next and why?" • Circulate and see what math students have recorded, if any. Ask individuals about insights.	• Saying for students what you hoped they would say. • Telling or showing students how you would solve the problem. • Randomly calling on students who may not be ready to participate in the routine.
You do not fully understand what a student is saying when describing a strategy.	• Do other students seem to be making sense of it? • Would hearing the strategy again help you to make sense of it? • Is what you are understanding going to help the whole class?	• "Who can put Dominic's strategy in their own words so that we can all try to make sense of it?" • "Who understands this idea? Will you say it in your own way?" • "Let's slow down for a moment. What parts of Dominic's ideas do you understand?" Then later, "What questions do you have for Dominic? What doesn't make sense to us yet?" • "That seems like a really different strategy. Will you hold onto it and see if it helps you with the next few problems?"	• Devaluing the idea just because it is not straight-forward. • Thinking that you are the only one who needs to understand the strategy—remember you are working as a community.

Algebra Problem Strings
©2017 Kendall Hunt Publishing

The Challenge	What You Might Consider	What You Might Say or Do	Try to Avoid
You understand a student's idea but are not sure how to model it.	• What about the strategy makes it challenging to model? • If you had a moment to think, would you be able to model it? • Is it worth taking the time now to model it—will you lose the interest and attention of the class as a result?	• Acknowledge the complexity of making a picture of the thinking: "Okay, I'm going to try to make a model of Monica's idea. It's a little new for me so let's see if I can capture it. Be thinking about whether this model shows what Monica just said to us." • "Turn and tell your partner what parts of Monica's strategy make sense to you. While you are doing that I'll make a model of it." • Decide that even when modeled, the strategy or idea won't help most of the students. Encourage the student to hold the idea or strategy to test out with other problems, or huddle with this student right after the string.	• Passing the pen to a student to model his or her own thinking. • Asking the class how to represent it (they are likely still making sense of the model for themselves). • Dismissing the idea entirely.
The energy of the conversation is low.	• Are you clear about what students understand and are still constructing? • Are you and one student engaged in a conversation that excludes everyone else? • Are students just solving but perhaps not being challenged to think about the relationships in new ways? • Are students waiting for you to solve the problem for them? Or for you to choose a student with the "right" strategy?	• Use partner talk strategically. • Try to intrigue students or get them wondering about whether something is true in other cases, or why something is happening. Your energy can be contagious. • Play a skeptic yourself, "You all seem pretty convinced that this strategy will always work, but what about with fractional or negative coefficients? Will it work then? I'm just not sure. Who thinks it might and wants to try to convince us?" • Take a very quick "stand and stretch" break and perhaps ask students (with their partners) in the back of the room to take a seat at the front of the room and vice versa.	• Taking over the conversation yourself. • Calling on the same students all the time. • Staying at the board the whole time instead of circulating. • Emphasizing students' note taking over thinking and reasoning.
The string is taking too long.	• What part of the string took longer than you expected and why? • Are you letting every student have the chance to share their strategy (not the goal)? Or are you purposely choosing a few strategies to model, based on what you are seeing and hearing? • Are you trying to have all students master everything during one problem string? • Are you clear what the goal of the string is?	If it's appropriate you can: • jump ahead in the problem string to a more challenging problem. • decide to end the string wherever you are and return to it another day, or not at all. • move into generalizing or summarizing with students instead of getting to the end of the string. • pass out index cards and ask students to record a question they have about the string and a strategy that is making more sense to them. • post the next problem in the string and ask students to think about it but not solve it (come back to it tomorrow).	• Feeling the need to record every single idea and strategy that is in the room. • Taking too long on easy problems that are only meant to set the stage, not be belabored.

The Challenge	What You Might Consider	What You Might Say or Do	Try to Avoid
Some students are done working the problem quickly while other students need more think time.	• Have all students gotten a good start? • Would inserting a helper problem help or be too pointed? • Are quicker finishers focused only on getting an answer? • Would it benefit slow starters to clarify the question?	• Ask early finishers to consider efficiency, connections to other problems, or to begin to generalize. • "Now that you've solved it, look back to see if you can identify any relationships you could use to solve it more efficiently." • "How does this problem connect with the previous problems?" • "Will your strategy work all of the time? How do you know? When is it a great strategy and when might it not be very efficient?"	• Moving on before the majority of students have had time to get a good start on the problem or get far enough to make the conversation fruitful. • Telling slower students how to work the problem.
Students do not make the connections or offer up the strategies you are intending.	• Are you modeling their work in a way that connects to the big idea? • Are your students still working with other ideas and not ready to construct your target idea? • Are the students moving forward even if it does not match your anticipated goal?	• "Did anyone use the previous problem to help them? Could you? Explain how." • Use a follow up "sister" string later to come back to it.	• Telling students what you wanted to hear from them. • Skipping the big idea of the string because no one offered it up.

Algebra Problem Strings
©2017 Kendall Hunt Publishing

1.1 Subtraction as Difference

At a Glance

92 – 60

91 – 59

90 – 58

190 – 158

1090 – 1058

80 – 16

380 – 316

170 – 119

104 – 99.5

600 – 489

Follow up problem string

52 – 4

61 – 57

202 – 18

581 – 47

2081 – 47

71 – 37

73 – 39

121 – 87

Objectives

The goal of this string is to support students as they calculate the range when provided a set of data by associating it with subtraction on an open number line and to help develop or support efficient strategies for calculating difference.

Placement

Use this problem string as students work with dot plots and bar graphs to offer students quick calculation strategies for finding the range of a data set.

You could use this string during your work in textbook Chapter 1 Data Exploration and particularly in Lesson 1.1 Bar Graphs and Dot Plots.

Guiding the Problem String

As you are leading the string listen for, highlight, and name the variety of strategies to make sense of subtraction in flexible and clever ways. Encourage students to do these calculations mentally—without paper and pencil.

The first problem is a set up for the next four problems. Model it on an open number line so that students can potentially use it to reason about the next four equivalent problems. The next two problems, 80 – 16 and 380 – 16, are related. The last three problems can be presented simultaneously and are designed to elicit a variety of strategies.

If the traditional algorithm comes up, you might acknowledge it but not spend time on making sense of it. Instead, you want to focus on the other strategies that students might use to quickly and accurately find the difference.

About the Mathematics

In their work with data, students will encounter several statistical measures (mean, median, mode, minimum and maximum values, and range). Some of these measures will require students to do some calculations, while others will be a matter of finding the value within the given data set. Where calculations are involved, we want to support students to do those with ease and flexibility—neither overly dependent on algorithms nor calculators.

Finding the range in a set of data means finding the distance between two values, which suggests subtraction. There are several strategies for subtraction and considering the values first before choosing a strategy.

These include:

- removing a friendly (landmark) number that's too big or small and then adjusting,
- removing to a friendly (landmark) number and then adjusting,
- finding the difference by building up from the subtrahend or down from the minuend (in a subtraction problem, minuend – subtrahend),
- creating an equivalent problem by shifting the numbers so that the subtrahend is friendly (constant difference), 91 – 59 = 92 – 60, and
- creating an equivalent problem by shifting the whole problem to smaller values on the number line, 190 – 158 = 90 – 58.

(continued)

Sample Interactions

Use the following as you plan how to elicit and model student strategies. This is not meant as a script, but as a view into the relationships involved and the intent of the problem string.

Teacher: *Okay, everyone. Let's get our brains warmed up with a problem string. You won't need paper and pencil, but be sure you are sitting near your math partner so you can talk. I'll put a problem on the board and ask that you think about the answer and be able to explain how you know. Here we go! What is 92 subtract 60? Or another way to ask that is what is the difference between 92 and 60?*	$92 - 60$
Student: *I would say 32.* **Teacher:** *How do you know?* **Student:** *Because 92 minus 60 is 32.* **Teacher:** *I'm going to model that on an open number line. You said that 92 minus 60 is 32, so I'm going to start on 92 and subtract or remove 60 and you said that place you land is 32. Could this represent what you did with the number?* **Student:** *Yeah.*	$92 - 60 = 32$ 60 ⌒ ├─────────────┤ 32 92
Teacher: *Did anyone envision it a different way? Or solve it differently?* **Student:** *Yeah, I thought about the space between 60 and 92. That would be 32.* **Teacher:** *Let me model what you said. You thought about the distance between 60 and 92 so I'll put them both on the number line and you said that they are 32 apart. Does that represent what you said?* **Student:** *I guess.*	32 ⌒ ├────┤ 60 92
Teacher: *Does that make sense to everyone, how you could think of this problem as the difference between the numbers?* **Student:** *It's kind of like saying 60 plus what is 92.*	$60 + \_\_\_\_ = 92$
Teacher: *Okay, now we have two representations for the same subtraction problem. Turn and talk with your partner about whether both could represent the problem and what makes you think this.* Students turn and talk, while the teacher listens in. **Teacher:** *Come on back everyone. Let's hear your thinking.* **Student:** *My partner and I thought both represent the problem.*	60 32 ⌒ ⌒ ├──────────┤ ├────┤ 32 92 60 92 $60 + \_\_\_\_ = 92$

Algebra Problem Strings
©2017 Kendall Hunt Publishing

Teacher: *What about the one on the left? How would you describe the thinking?* **Student:** *It's like you're taking away the 60. Removing or minus.* **Teacher:** *What about the strategy on the right?* **Student:** *They found the distance between the numbers.* **Teacher:** *Sometimes we can call that the difference between the numbers. Do all of these words make sense?*	removal, subtract, minus distance, difference 60 32 ⌢ ⌢ 32 92 60 92 60 + ____ = 92
Teacher: *What about the answer? Where does it show up?* **Student:** *On the right, the 32 is the space between and in the other 32 is the place where you land.* **Teacher:** *That sounds important. Can some reiterate that for us.* **Student:** *When you remove, the answer is where you land. When you find the distance, that's the answer, the distance!* **Teacher:** *Does this idea of the landing spot versus the space between make sense to you? Yeah?*	removal, subtract, minus distance, difference 60 ⟨32⟩ ⌢ ⌢ ⟨32⟩ 92 60 92 60 + ____ = 92
Teacher: *Here's another problem. Be thinking about what changes. Visualize a number line.*	$91 - 59$
Student: *So at first I didn't like this problem, but then I thought about the difference strategy and it feels like the whole problem just scooted down one—do you know what I mean?* **Teacher:** *Let's find out. Who understands what she is saying here and can put the idea in their own words?* **Student:** *So she is saying that the space between 92 and 60 is the same space or distance as the space between 91 and 59. Everything just moved to the left.* **Teacher:** *It looks like the first problem was helpful in solving the second one if you think about the distance or difference between the numbers.*	$91 - 59 = 32$ 32 32 ⌢⌢ 59 60 91 92 $92 - 60 = 91 - 59 = 32$
Teacher: *All right, so keeping this idea of a constant difference in mind, how would you envision and solve this problem?* After brief think time, the teacher elicits conversation around the relationship between the three problems focusing on the distance between the numbers of each problem. Since the distance remains constant, the problems are equivalent and thus have the same solution. We also call the answer to a subtraction problem "the difference."	$90 - 58$ 32 32 32 ⌢⌢⌢ 58 59 60 90 91 92 $92 - 60 = 91 - 59 = 32 = 90 - 58$

(continued)

Teacher: *What about now? Take 30 seconds with your partner to think about and solve this problem. Go!* After brief think time, the teacher elicits conversation about the equivalence between the problems and models the equivalence by shifting the problem left 100.	$190 - 158$ $190 - 158 = 90 - 58 = 32$ *number line: 32 over 58—90, shift 100, 32 over 158—190, Shift 100*
Teacher: *So what would you do with this one?* **Student:** *That's the same problem, just shifted into the thousands. It's still 90 – 58 so we know the answer is 32.* **Teacher:** *Anyone want to challenge this idea? Give us another way to think about it? No? That's pretty nice, isn't it?*	$1090 - 1058$ $1090 - 1058 = 32$
Teacher: *Okay, so what about this one? Again, how do you envision it, and how does that help you to solve it?* **Teacher:** *Amalia, will you share your strategy with all of us?* **Student:** *For this one, I was going to find the space between 16 and 80, but then I thought about the numbers and I decided to start at 80, jump back 10 and then jump back another 6, landing at 64.* **Teacher:** *Am I capturing your thinking here?* **Student:** *Yes.*	$80 - 16$ $80 - 16 = 64$ *number line: 6, 10 jumps, 64 70 80*
Teacher: *Let's talk about why this jumping back 16 strategy made sense to Amalia here. Did her decision make sense to you?* **Student:** *Yeah, it made sense to me since 16 and 80 are pretty far apart.* **Teacher:** *Is it more a removing strategy or a distance, difference strategy?* **Student:** *Removing because she was taking away 10 and then 6.*	*number line: ? arc over 16 ——— 80*
Teacher: *So could we have solved it using a distance or difference model? Could someone walk us through that strategy?* **Student:** *You would mark 16 and 80 on a number line and then you have some choices—jumping up from 16, or jumping back from 80.* **Teacher:** *Okay. And what if you didn't like either of those options and wanted to make an easier, equivalent problem for yourself. Is there one here? Turn and talk with your partner about this—how could we make this problem easier?* The teacher solicits students' ideas about an easier, equivalent problem, based on the big idea of constant difference. One example might be adding 4 to the minuend and subtrahend turning 80 – 16 into 84 – 20.	*number line: ? arc over 16 ——— 80* $80 - 16 = 84 - 20 = 64$ *number line: 64 64 16 20 ——— 80 84*

Algebra Problem Strings
©2017 Kendall Hunt Publishing

Teacher: *So this one doesn't seem too bad, right?* The teacher allows students time to think, then elicits the relationship to the previous problem.	$380 - 316$
Teacher: *The last three I'll put up at all once, which I normally do not do, and I'm going to ask you to talk with your partner about how your solved it and maybe how you made them easier to think about. Ready?* The teacher listens in as students solve, compare, and discuss. Look for students who are using the constant difference. The teacher continues the class discussion, modeling student thinking.	$170 - 119$ $104 - 99.5$ $600 - 489$

$$170 - 119 = \underline{171 - 120} = 51$$

$$104 - 99.5 = \underline{104.5 - 100} = 4.5$$

$$600 - 489 = 111$$

$$600 - 489 = 100 + 11 = 111$$

$$600 - 489 = \underline{611 - 500} = 111$$

Teacher: *How would you summarize some of the things that came up in this string today?*

Elicit the following:

- *Subtraction can be thought of as removal and difference.*

- *When the numbers are far apart, it can be easy to remove.*

- *When the numbers are close together, it can be easy to find the difference.*

- *You can turn the subtraction problem into an equivalent problem that is easier to solve by shifting the problem to numbers that are easier to subtract. We call this strategy "constant difference."*

Note: Use the follow up problem string if needed to give students more experience thinking and reasoning about subtraction.

(continued)

Sample Final Display

Your display could look like this at the end of the problem string:

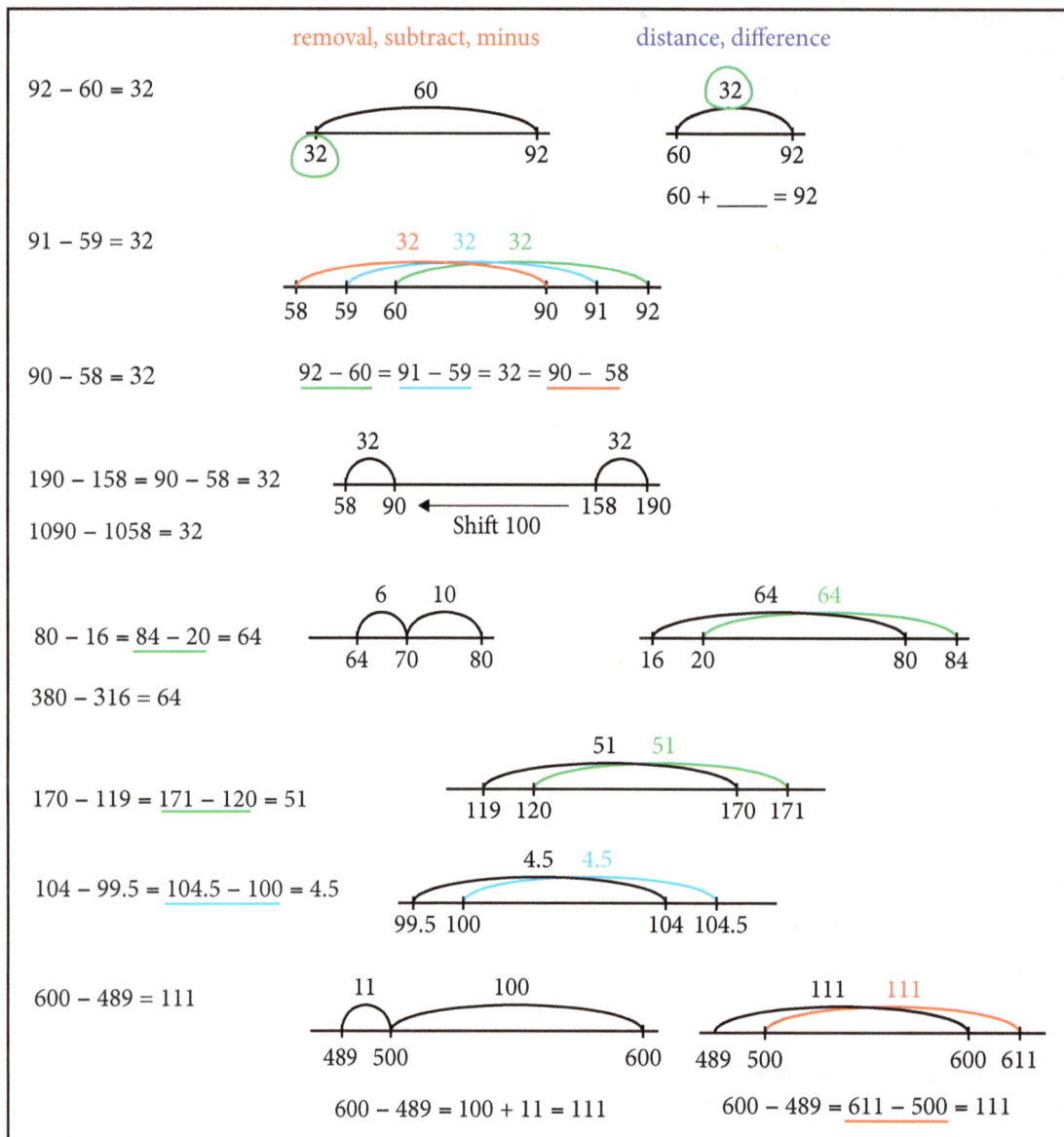

removal, subtract, minus   distance, difference

$92 - 60 = 32$

$60 + \underline{\hspace{1cm}} = 92$

$91 - 59 = 32$

$90 - 58 = 32$

$92 - 60 = 91 - 59 = 32 = 90 - 58$

$190 - 158 = 90 - 58 = 32$

$1090 - 1058 = 32$

Shift 100

$80 - 16 = 84 - 20 = 64$

$380 - 316 = 64$

$170 - 119 = 171 - 120 = 51$

$104 - 99.5 = 104.5 - 100 = 4.5$

$600 - 489 = 111$

$600 - 489 = 100 + 11 = 111$ $600 - 489 = 611 - 500 = 111$

Algebra Problem Strings
©2017 Kendall Hunt Publishing

Facilitation Notes

This version of the problem string lists short notes for important teacher moves during the string. After you've done the string yourself and studied the relationships involved, you might make similar notes for the things you want a reminder of or deem important.

$92 - 60$	What is 92 subtract 60? Another way to say that is, what is the difference between 92 and 60? Models both subtraction strategies (removal and difference/distance). label both strategies. Identify where the answer is.
$91 - 59$	Shift by one. Constant difference
$90 - 58$	Shift by one. Constant difference
$190 - 158$	Shift by 100. Still constant difference
$1090 - 1058$	Shift by 1,000. Still constant difference
$80 - 16$	New problem. How do you envision it? How does that help you to solve? See what strategies make sense to kids. Model both.
$380 - 316$	Shift by 300. Constant difference
$170 - 119$ $104 - 99.5$ $600 - 489$	All three together. Move around the room, listen, document smart strategies. Emphasize the idea of making the problem easier. Encourage partner talk.

1.2 | Finding the Mean

At a Glance

Find the mean:

10, 10, 10, 10, 10

9, 10, 10, 10, 11

5, 5, 15, 15

6, 7, 9, 10

16, 14, 10, 8

12, 8, 10, 12, 8

24, 24, 20, 16, 16

12, 12, 12, 10, 10

12, 10, 12, 10, 12

Objectives

The goal of this string is to build students' number sense around finding the mean of a set of numbers by paying attention to the values before choosing a strategy.

Placement

Use this string as you are about to work with students reviewing and using measures of center, specifically the mean. The purpose of this string is to help students be more fluent and flexible when students find the mean from a set of data and for students to be able to reason about what finding the mean entails.

You can use this problem string before or during textbook Lesson 1.2 Summarizing Data with Measures of Center.

Guiding the Problem String

In this string, de-emphasize the common strategy of finding the mean—adding all values together, then dividing by the number of values. You can give this string a context by saying the values represent denominations of money spent on meals over several days and you are looking for the average money spent.

The first problem sets the stage and should be very quick. The next problem has the same mean and almost identical values and can help students begin to look to the numbers before reflexively adding and dividing. The next three problems have four values with wider ranging differences and are not listed in order, but still easy to distribute out the extras evenly. The next problem goes back to five values where the amounts are not listed in order, but are easy to level out. The next problem lists doubles of the previous, nudging students to wonder if doubling the data results in a doubled mean. The last two problems have rational number means. Encourage students to think about splitting up the leftover dollar into five chunks. Spend time on this problem. The last problem is just a re-arrangement of the previous and should go quickly.

About the Mathematics

When working with data, students will encounter several statistical measures (mean, median, mode, minimum and maximum values, and range). Where calculations are involved, we want to support students to do those with ease and flexibility—neither overly dependent on algorithms nor calculators. While the traditional algorithm of dividing the sum of the values by the number of values to find the mean always works, it can be inefficient and it doesn't encourage students to pay attention to the values or think about creative, flexible, and smart strategies for finding the mean. The evening out strategy works because the associative property of addition allows us to redistribute the values of the numerator until they are all equivalent, thus making the division unnecessary.

$$\text{The mean of } a_1, a_2, ..., a_n = \frac{a_1 + a_2 + ... + a_n}{n}$$

So, the mean of 10, 10, 10, 11, 9 is $\dfrac{10+10+10+(10+1)+9}{5} = \dfrac{10+10+10+10+(1+9)}{5}$

Algebra Problem Strings
©2017 Kendall Hunt Publishing

Sample Interactions

Use the following as you plan how to elicit and model student strategies. This is not meant as a script, but as a view into the relationships involved and the intent of the problem string.

Teacher: *Good morning! Check in with your partner and get ready for a problem string. We ready? Okay. So here's a situation I want us to think about—and like always I'm going to encourage you to think deeply and cleverly about how to make sense of these situations.*

Over the [break/summer/vacation] my family and I spent some time away and had to eat most of our meals out. I kept track of what I spent on meals and our job today is to think about the average or mean cost of my meals. Someone remind us what we mean by mean?

Student: *If you spent $10 one day and $20 the next, your mean or average would be $15 which is right in the middle.*

Teacher: *So one way of thinking about mean might be to think about those measures of center or the middle. Other ways?*

Student: *Yeah, I thought you add everything up and divide by the number of things. So here $10 + $20 is $30, spent over two days which is $30 divided by 2 or $15.*

Teacher: *Okay, so using either of these ways or another strategy, be thinking about how much my average meal was on vacation and how you know. For simplicity I'm going to list the five meals using commas and no dollar sign but we'll remember what they represent, yes?*

Let's start with a friendly one. What's the mean meal cost here?

Student: *It's 10. You maybe went to the same place over and over again and got the special.*

Teacher: *Probably, huh? I'm going to model this by quickly coloring in a 10 for each of the 5 days.*

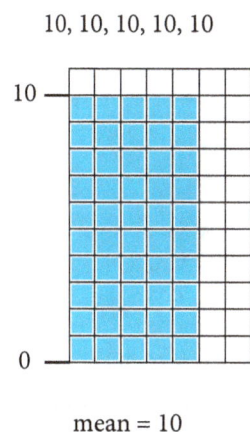

10, 10, 10, 10, 10

mean = 10

Teacher: *What about now? Here are five different meal prices.*

Student: *Still 10 because the 9 and the 11 kind of balance each other out.*

Teacher: *Who knows what Daniel is saying here and can explain it more.*

Student: *Daniel means that if you take $1 from one meal and add it to another meal, the sum stays the same, and therefore the mean does, too.*

Teacher: *If we start out with coloring in the five meal costs, then could we model that balancing, that shifting, like this? Just move this $1 over here? Do they look even again?*

Students: *Yes.*

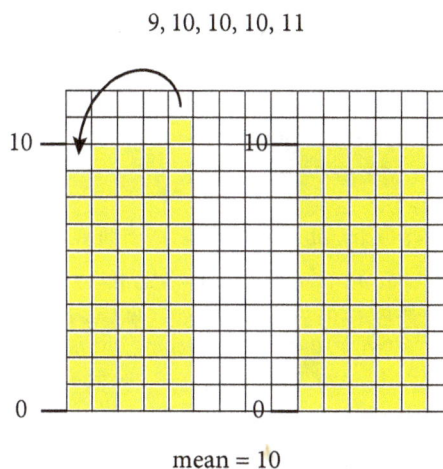

9, 10, 10, 10, 11

mean = 10

(continued)

Teacher: *In this situation I had four meals, two fast food meals and two fancier meals. What's the mean of these, and how do you know? Turn and tell your partner how to do this with very little calculation.* The teacher encourages efficiency and reasoning and models a leveling out strategy. **Student:** *You could add them all up. An easy way to do that is that 5 and 15 are 20, double that to 40. Then 40 divided by 4 is 10.* **Student:** *Yeah, I did the same except added the 5s to get 10 and then the 15s to 30. That's how I got 40.* **Teacher:** *Did any of you think about this the way we did the last one, evening out?* **Student:** *Yeah, I thought if each 15 gave 5 to a 5, then they would all be 10. And you're done.* **Teacher:** *That seems really efficient.*	5, 5, 15, 15 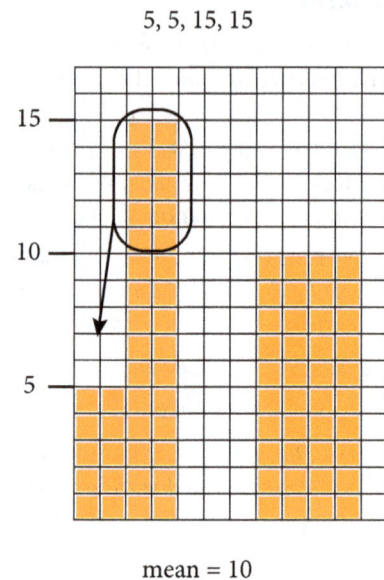 mean = 10
Teacher: *Let's keep going with this idea. What do you think about this situation?* Think time. **Student:** *So I think the average is 8 for two reasons. One is that I made two 16s, 6 plus 10 and 7 plus 9, which gives you 32 divided by 4 is 8. But the other way I thought about it is that you made a list with 8 left out and everything else is balanced—I'm not sure if I'm explaining that right.* **Student:** *I thought the same thing. It's like if the 10 gave 2 to the 6 they would both be 8. And then the 9 could give 1 to the 7 and they would both be 8, so then everything is an 8. It makes sense that your meals averaged out to $8.* **Teacher:** *I'm going to try to draw a model of this trading and evening out that you both are talking about. Turn and tell your partner what is going on in this model and what it has to do with finding the mean.*	6, 7, 9, 10 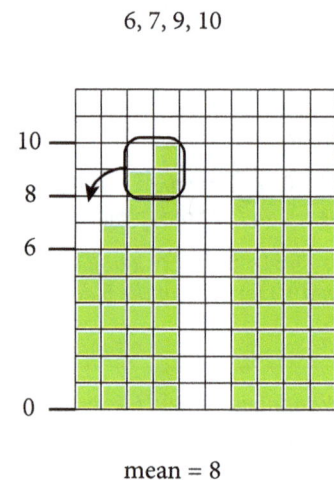 mean = 8
Teacher: *Okay, but what if these were my meals? What now?* **Student:** *Same idea as before—it's like you are skip counting down by twos and you forgot the 12, which is right in the middle. Or, if you use this evening out idea, then the 8 and the 16 do some trading to each become 12 and same with the 10 and the 14. So the mean is 12.*	16, 14, 10, 8 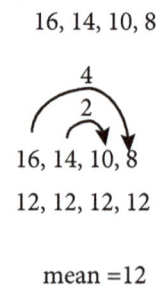 16, 14, 10, 8 12, 12, 12, 12 mean =12

Teacher: *Here's the next one. Let's give folks a few seconds to think about it.*

Brief think time.

Teacher: *What changed in the scenario?*

Students: *You ate more! It is one more meal.*

Teacher: *What is the mean amount I spent for these five meals? Turn and tell your partner how you are finding the mean here.*

The teacher listens in, strategically identifying a set of students who evened out the values.

Teacher: *Tell us about your thinking.*

Student: *We moved two from the 12 to an 8 and then did that again and then they were all 10, so 10.*

12, 8, 10, 12, 8

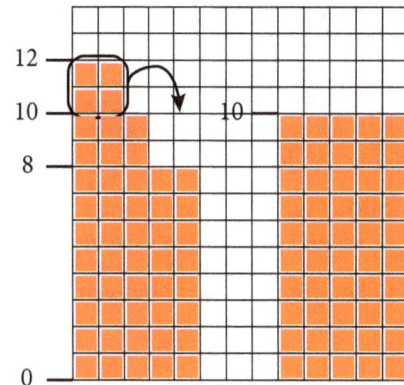

mean = 10

Teacher: *Let's do a few more to just study this situation. Imagine I spent these amounts on meals, over five days. What's the mean amount I spent?*

Think time.

Student: *If you rearrange the last one, all of your meals doubled. So my guess is that the mean doubled, too.*

Teacher: *Interesting idea. If you double the price of all of the meals, does the average price double, too? What do we think?*

24, 24, 20, 16, 16

mean = 20

Student: *It makes sense to me, but I'd want to test a few more examples.*

Teacher: *Good instinct. Why don't you hang onto that idea today as you are working and we'll come back to it.*

Teacher: *Here's one more to think about.*

Student: *Ughh, I don't like this one.*

The teacher encourages sense-making and estimating strategies to start, before soliciting the answer in order to focus on reasoning.

Teacher: *Without solving this one can you tell us about how big the mean is and how you know?*

Student: *I think it's going to be in between 12 and 10.*

Student: *Yeah, and it's going to be bigger than 11, but not 12.*

Student: *Right, so it has to be in between 11 and 12.*

Teacher: *Talk with your partner about whether this one is friendly or not and what you might do to solve it.*

Think time.

Student: *I started to add them, so I could divide, and I got to 56 ÷ 5 and we decide to try evening out.*

Student: *Once we moved $2 to the 10s then we have all 11s except one more whole dollar. That whole dollar split up is like 20 cents or ⅕. So the mean is 11⅕ or $11.20.*

12, 12, 12, 10, 10

$1 ÷ 5 = ⅕ = $0.20 $0.20 = ⅕

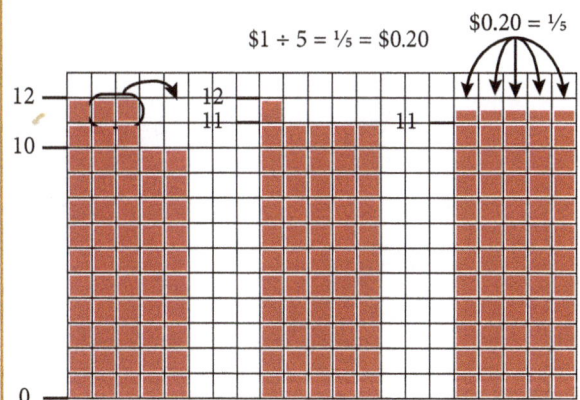

mean = 11.20 or 11⅕

(continued)

Teacher: *Last one. Any way to make this easier to find the mean?*	12, 10, 12, 10, 12
Student: *Yeah, it's the same as the last one, but the order of the meals changed. Since that doesn't matter, the mean is still 11⅕ or $11.20.*	mean =11.20 or 11⅕

Teacher: *How would you summarize some of the things that came up in this string today?*

Elicit the following:

- *The mean is one of the ways to talk about average. It is one of the measures of center.*

- *You can find the mean by evening the amounts out or spreading the amounts out evenly.*

- *You can think of evening the amounts by trading the extra bits around until everything is even.*

- *We are wondering if you double all of the data, does the mean always double also?*

- *Adding the values and dividing by the number of values is not the only way to find the average.*

Algebra Problem Strings
©2017 Kendall Hunt Publishing

Sample Final Display

Your display could look like this at the end of the problem string:

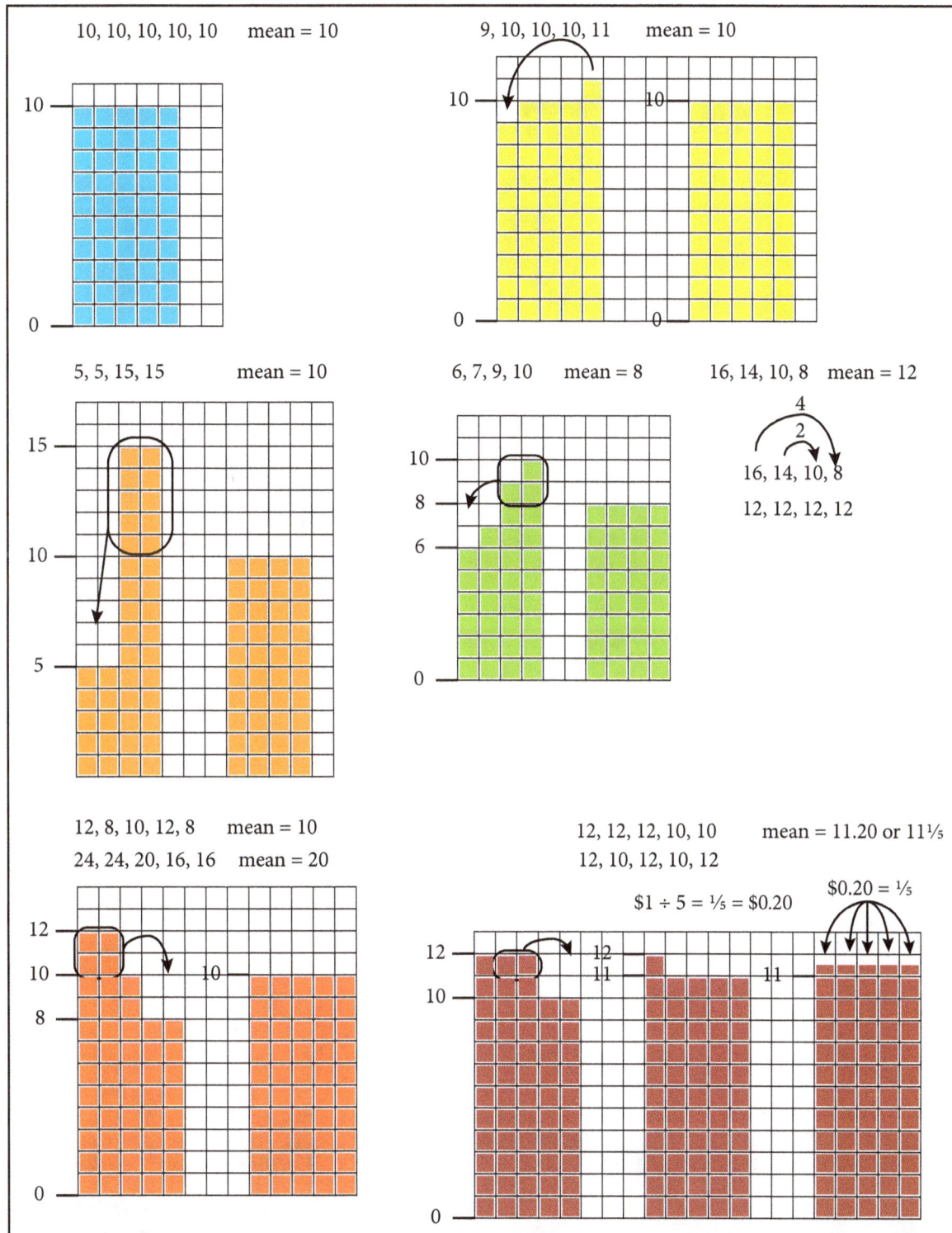

10, 10, 10, 10, 10   mean = 10

9, 10, 10, 10, 11   mean = 10

5, 5, 15, 15   mean = 10

6, 7, 9, 10   mean = 8

16, 14, 10, 8   mean = 12

16, 14, 10, 8
12, 12, 12, 12

12, 8, 10, 12, 8   mean = 10
24, 24, 20, 16, 16   mean = 20

12, 12, 12, 10, 10   mean = 11.20 or 11⅕
12, 10, 12, 10, 12

$1 ÷ 5 = ⅕ = $0.20 $0.20 = ⅕

(continued)

Facilitation Notes

This version of the problem string lists short notes for important teacher moves during the string. After you've done the string yourself and studied the relationships involved, you might make similar notes for the things you want a reminder of or deem important.

10, 10, 10, 10, 10	Friendly—getting into the context. What is the mean?
9, 10, 10, 10, 11	Leveling out the 9 with the 11.
5, 5, 15, 15	Making 20s, total 40; leveling out.
6, 7, 9, 10	Making 16s, total 32; leveling out.
14, 12, 8, 6	Making 20s, total 40; leveling out.
12, 8, 10, 12, 8	What changed? More meals, more values.
24, 24, 20, 16, 16	What happens when we double everything?
12, 12, 12, 10, 10	Tricky: 56÷5 = 11.2 or 11 1/5. Estimate and reason. Just get to 11 and then 1÷5.
12, 10, 12, 10, 12	Same—just rearranged.

Algebra Problem Strings
©2017 Kendall Hunt Publishing

1.3 | Mean, Median, Mode

At a Glance	Objectives

At a Glance

Find the mean, median and mode:

20, 20, 20, 20, 20

18, 20, 20, 20, 22

18, 18 20, 22, 22

18, 18, 18, 22, 22

18, 18, 22, 22, 22

18, 18, 18, 18, 22

Objectives

This string follows String 1.2 with the goal of expanding students' ability to reason about mean, median, and mode. The purpose of this string is for students to pay attention to the values within a data set and to begin to develop relationships between the measures of center.

Placement

Use this problem string after students have some experience finding the measures of center of mean, median, and mode to help increase students' facility in finding the mean using a leveling out strategy, especially when the mean is a rational number.

You can use this string during your data work in textbook Chapter 1 following the facilitation of Problem String 1.2.

Guiding the Problem String

In the previous string students had some experiences finding the mean within a set of data. This string gives students another chance to build on those strategies, while beginning to think about the relationship between median and mean within a set of data.

The first problem sets the stage. The rest of the problems build from that first problem, changing values. Encourage students to consider which measures change as the data changes in different ways.

About the Mathematics

The string is designed so that the measures of center are relatively easy to find and so big ideas can emerge through questioning and discussion. The numbers chosen are not meant to vary widely. Instead, we are simply tinkering with the values from one problem to the next, hoping that by keeping many of the values the same and only changing a few of them each time, students will begin to make some observations and conjectures about what is happening. It is these conjectures that will serve them well in future explorations of mean, median, and mode, and not simply the ability to calculate or find these measures when required.

(continued)

Sample Interactions

Use the following as you plan how to elicit and model student strategies. This is not meant as a script, but as a view into the relationships involved and the intent of the problem string.

Teacher: *Let's get our brains warmed up today with another problem string. This will build on the work we did thinking about mean yesterday and get us ready to find other measures of center. To start, tell your partner what each of these ideas means to you and how you would find them in a data set.*	Mean Median Mode

The teacher gives students a brief time to discuss and draw upon their own understandings of the terms. Then the teacher asks students to share until the class has a working definition to use during this problem string.	

Teacher: *Now that we remember what all of these mean, I'm going to put up a data set and ask you to jot down the mean, median, and mode for each set. Put your thumb up when you have some ideas and are prepared to share your thinking with the group. These shouldn't require lots of calculations so if you find yourself doing lots of computation on your paper, I encourage you to look for some shortcuts.*	20, 20, 20, 20, 20

Student: *In this set 20 is everything. It's the mean, the median and the mode.* **Teacher:** *Thoughts about this. Anyone want to challenge this idea?*	Mean 20	Median 20	Mode 20

Teacher: *What about now? Change the numbers, change everything, right? Give everyone a few seconds to think and jot down first.* Brief think/work time. **Student:** *Actually the measures didn't change. In this set, the mode is still 20 because it occurs the most, the mean is also 20. I thought back to what we did yesterday and brought 2 over from the 22 and gave it to the 18, making them all "towers" of 20. And the median, I didn't finish that one.*	18, 20, 20, 20, 22

Teacher: *No problem. Who can keep going?* **Student:** *So the median here is also 20. It's the one in the middle if they are in order from smallest to biggest.*	Mean 20	Median 20	Mode 20

Teacher: *How about if I change a few of the 20s to another 18 and another 22? What happens then? Turn and tell your partner whether the mean, median, or mode will change and why.*	18, 18, 20, 22, 22

The teacher listens for and draws out students' reasoning about how the mode of 20 has changed. Now there are two modes (18 and 22) but the median and mean stay the same as 20.	Mean 20	Median 20	Mode 18, 22

Algebra Problem Strings
©2017 Kendall Hunt Publishing

Teacher: *Here's another one. What happened? How will this change the mean, median, and mode? Let's let everyone really think about this one for a moment.*	18, 18, 18, 22, 22

Student: *So the only thing that changed is that the 20 became an 18. This means a few things. The mode is now just 18 because there are three of them and only two 22s. Then I'm kind of picturing the mean as "towers" and I can see that we could level out the 22s with the 18s making four towers of 20. But we still have that one tower of 18 to deal with. Ughh, I'm stuck.*

Teacher: *I tried to capture what you just said because I'm hoping that if we hear it and see it in a model we might be able to reason about the mean. Does anyone else have an idea?*

Student: *Well, I'm feeling like the mean is around 19—bigger than 19 but less than 20. Why don't we move one from the 20 to the 18, then each tower could have 19 and then we can divide the rest over the 5 towers.*

Teacher: *Like this?*

Student: *Yes, exactly.*

Student: *Oh, that helped me. It's 19 and ⅗. Every tower has 19 and we have 3 leftover, so divide those over the 5 towers—3 divided by 5 is ⅗.*

Student: *Or like yesterday, when we were talking about money, it's like 300 cents divided among 5 meals, so 60 cents, that's a total of $19.60.*

Teacher: *Turn and tell your partner what they just said and whether you agree with this idea or not.*

Teacher: *What about the median?* **Student:** *It's just 18. The middle number is 18.*	Mean $19\frac{3}{5} = 19.6$	Median 18	Mode 18

Teacher: *Okay, two left. What if I do this? What are the mean, median, and mode now?* Students work while the teacher circulates, listening in and questioning where needed.	18, 18, 22, 22, 22

(continued)

Teacher: *What did you find?*	
Student: *So the mean is going to be a little bit more than 20. Hold on, it's like 20 and ⅖.*	
Teacher: *Who knows where Evan came up with 20 ⅖ and could explain that to all of us?*	
Student: *I'm picturing the towers, and if you even out the towers, you have enough for each one to have 20. Then there is one tower with 22. Those extra two, I'm thinking, is what Evan spread out over the five towers, giving each of them 20 and ⅖.*	
Student: *Yeah, like 200 cents split up into five is 40 cents, or 0.4. So that 20.4 for the mean.*	

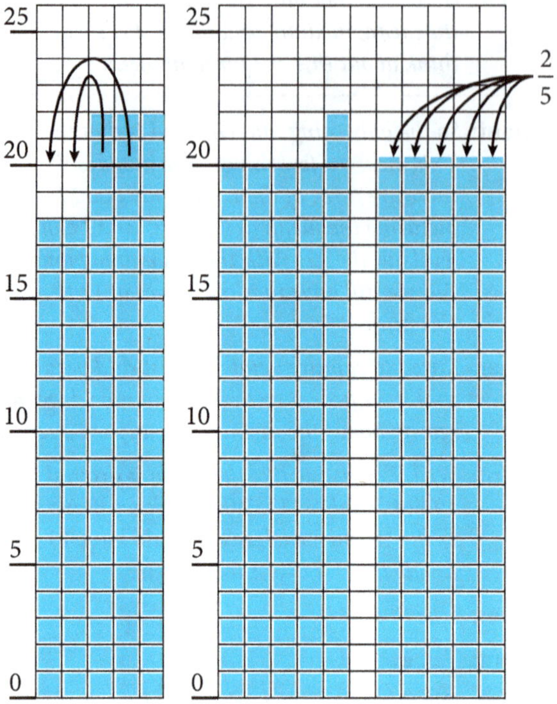

	Mean	Median	Mode
Teacher: *What about the median and mode?*	20⅖ =20.4	22	22
Student: *The mode is 22 because you have the most of them. The median changed to 22 as well because it's the value right in the middle.*			

Teacher: *All right, last one. Think about what changed, what stayed the same, and how all of that affects the mean, median, and mode. Turn and tell your partner what you found. We want to hear from any partners who do not agree after you've had a moment to talk. Go!*	18, 18, 18, 18, 22

Teacher: *What did you find?*	
Student: *The mean is easy if you just take the extra 4 from the 22 and spread it out over the other four numbers. Now they are all 19.*	19, 19, 19, 19, 18
Student: *Wait, if you gave 1 to the 18s, now they are 19 and the last number is 18.*	
Student: *Oh yeah. True. Fine, but then you can take a little bit from each of the 19s to give to the 18.*	
Student: *That would work! Take ⅕ from each of the 19s and that leaves them as 18⅘ and now the 18 also has four ⅕, so they are all 18⅘.*	18 ⅘, 18 ⅘, 18 ⅘, 18 ⅘, 18 ⅘,

	Mean	Median	Mode
Student: *I see what you are doing. I just took off the extra 4 from the 22 and split up the 4 across the 5 numbers, that's 4 divided by 5, ⅘.*	18 ⅘	18	18

The teacher ends the string by building a conversation around ideas such as:

- What are some situations when the mean and the median will be the same? What explains this?
- When is it likely that the mean will be greater than the median? What's going on there?
- When it is likely that the mean will be less than the median? Why would this be true?
- What about the mode? Does it affect the other measures in a significant way?

Teacher: *How would you summarize some of the things that came up in this string today?*

Elicit the following:

- *When the data changes, the mean, median, and mode may or may not change.*
- *When you are spreading everything out, making the data "even," that is finding the mean.*
- *When you look for the middle number you are finding the median.*
- *When you look for the number that shows up the most you are finding the mode.*

(continued)

Sample Final Display

Your display could look like this at the end of the problem string:

	Mean	Median	Mode
20, 20, 20, 20, 20	20	20	20
18, 20, 20, 20, 22	20	20	20
18, 18, 20, 22, 22	20	20	18, 22
18, 18, 18, 22, 22	$19\frac{3}{5} = 19.6$	18	18

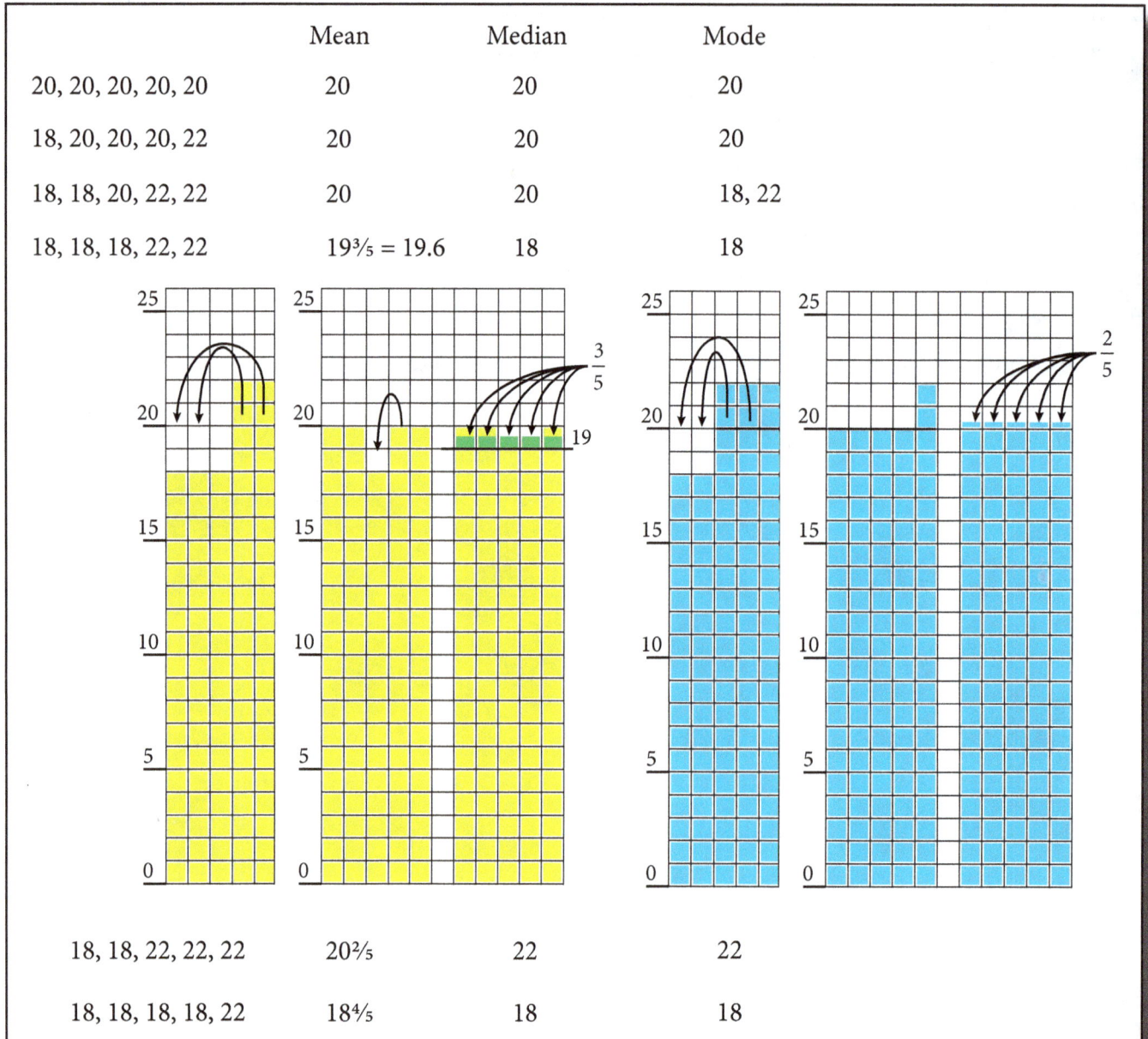

18, 18, 22, 22, 22	$20\frac{2}{5}$	22	22
18, 18, 18, 18, 22	$18\frac{4}{5}$	18	18

Algebra Problem Strings
©2017 Kendall Hunt Publishing

Facilitation Notes

This version of the problem string lists short notes for important teacher moves during the string. After you've done the string yourself and studied the relationships involved, you might make similar notes for the things you want a reminder of or deem important.

20, 20, 20, 20, 20	Super friendly—building the base so we can tinker with it.
18, 20, 20, 20, 22	Tinkering with the "ends" of the set.
18, 18, 20, 22, 22	Continuing to tinker.
18, 18, 18, 22, 22	More tinkering—mode changes, median changes, mean decreases. How to even out when not a whole number? Model.
18, 18, 22, 22, 22	More tinkering—mode changes, median changes, mean increases. Not a whole number again? Model.
18, 18, 18, 18, 22	And now what?

1.4 | Making Sense of Histograms

At a Glance

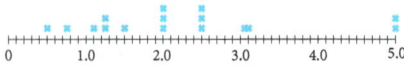

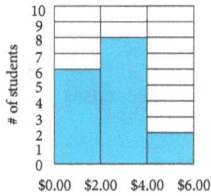

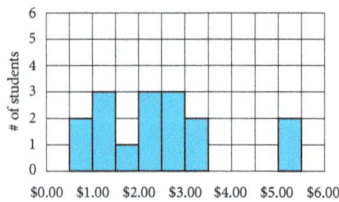

Objectives

The goal of this string is encourage students to strengthen their sense of a possibly new representation—the histogram. Students will be invited to make several true statements about histograms that are displayed to the class, as well as generate questions that they are wondering about.

Placement

While students will likely have significant experience analyzing and constructing dot plots, bar graphs, and line plots, the histogram may be a newer form of data representation. Therefore, this problem string is designed to support them in reasoning about various data that can be displayed on a histogram. Use the string when introducing histograms.

You can use this string as you work with textbook Lesson 1.4 Histograms and Stem-and-Leaf Plots.

Guiding the Problem String

Students have been encouraged to consider graphs and other forms of data representations as telling stories about situations. The same approach is used here. We ask students to study two histograms, one at a time, trying to make sense of what is being represented while also trying to craft a narrative about what is being displayed. This will feel more open-ended than other problem strings that follow a neat, sequential approach. Here you are:

- learning what your students know about histograms,

- gently suggesting some language they can use to describe what they see (e.g. bins, frequency),

- comparing histograms to other ways to represent data that students have seen (e.g. dot plots), and

- setting them up to reason with more confidence about histograms they will encounter in their studies of algebra and data.

As you talk with students about what they notice, draw out their conjectures about the shape of the graph, as well as what clusters and gaps reveal about the data.

About the Mathematics

The string is designed to highlight the unique features of a histogram. Unlike bar graphs, which use categorical data along the horizontal axis, the histogram uses numerical data and orders this data along the x-axis in a sequence of bins of equal width. Though related to the dot plot (which displays each unique data point) and a box plot (which shows the spread of the data but obscures the number of data points), the histogram captures both features of spread and number of data points.

Sample Interactions

Use the following as you plan how to elicit and model student strategies. This is not meant as a script, but as a view into the relationships involved and the intent of the problem string.

Teacher: *I'm going to start today's problem string by showing you a dot plot called "Pocket Money." Before we go further, what do you think pocket money means?*

Draw out students' ideas about money carried on a person, spare change, money in pockets or bags, etc.

Teacher: *Your job is to study the dot plot and try to tell the story that is being displayed. How many people were in this data set? What do we know about them? What trends are emerging? What questions do you have about them? Maybe you have a theory about who they are, based on the data.*

We are going to just look and notice for about 30 seconds, then I will ask you to talk with your partner about what you are seeing, and what you think it means.

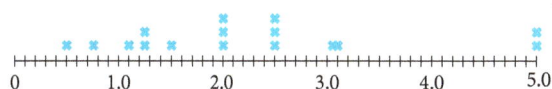

When the teacher brings students back together, the teacher crafts a conversation about the data and what the representation reveals and does not reveal using questions such as:

• *Does this seem reasonable?*

• *Can we tell how many people were surveyed?*

• *What could we tell about pocket money if this data were typical of all of us?*

• *Does this data seem reasonable?*

• *Do you think this data is typical of everyone in our class? Everyone at our school? Why or why not?*

Teacher: *Okay, now I'm going to show you a histogram that is also called "Pocket Money" and I'm going to ask you to do the same thing—first study and notice and talk with your partner, then we will talk as a class. Be thinking about how this data is similar to or different from the dot plot we just saw. Ready?*

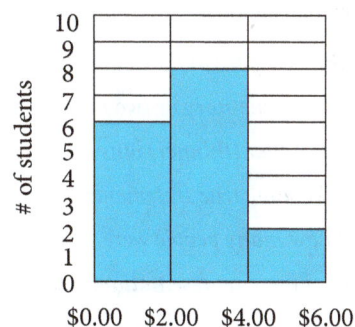

(continued)

When the teacher brings students back together, the teacher crafts a conversation about the representation and how it relates to the previous graph using questions such as:

- *What's different about a histogram representation?*
- *How does it compare to the dot plot?*
- *Do they tell the same story or two different stories?*
- *What information do we lose that we could see in the dot plot?*
- *Can you tell if anyone has no money?*
- *Where do we see the frequency of the $2–$4 bin?*
- *Can you make some true statements based on the data?*
- *Can we tell how many people were surveyed?*
- *What could we tell about pocket money if this data were typical of all of us?*

Teacher: *Finally, I'm going to show you a second histogram, called "Pocket Money." Be thinking about what's going on with this one. I'm going to leave the image of the last histogram up so that you can compare them more closely.*

Again, we will just look and study on our own for about 30 seconds, then talk with our partners. I'll record anything that you are thinking about so that we can think about it together.

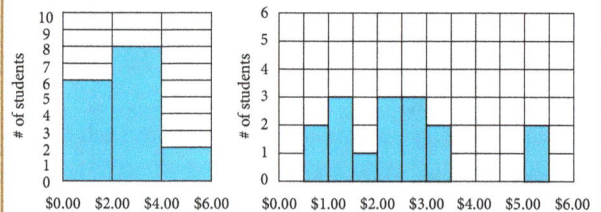

When the teacher brings students back together, the teacher crafts a conversation about the representation and how it relates to the previous graphs using questions such as:

- *What about this graph?*
- *What's different about these two histograms? (Bins: width, height.)*
- *How does this compare to the dot plot?*
- *Do they tell the same story or two or three different stories?*
- *How are the smaller (thinner) bins related to the larger (thicker) bins?*
- *Can you make some true statements based on the data?*
- *Can we tell how many people were surveyed?*
- *What do all of these representations make you think more about?*

Teacher: *How would you summarize some of the things that came up in this string today?*

Elicit the following:
- *In a dot plot you can see individual responses. The responses don't fall in a bin, they are at a location of a specific value. You can count the total number of responses by ones. A dot plot represents the raw data, organized, but not grouped. Because of this you could use it to find the mean, median, and mode.*

- *In a histogram, you can see broader groups and trends because you're not focusing on single data points, but rather bins of data. The range can be obscured if you have large bins where the maximum values are not near the edges of the bins. This could hide outliers better, which may or may not be a good thing.*

- *In a histogram with smaller bins, the bins will probably be shorter than in a histogram with larger bins. You can be a little more precise with a histogram with smaller bins, but you can see broader trends with larger bins.*

Algebra Problem Strings
©2017 Kendall Hunt Publishing

Sample Final Display

Your display could look like this at the end of the problem string:

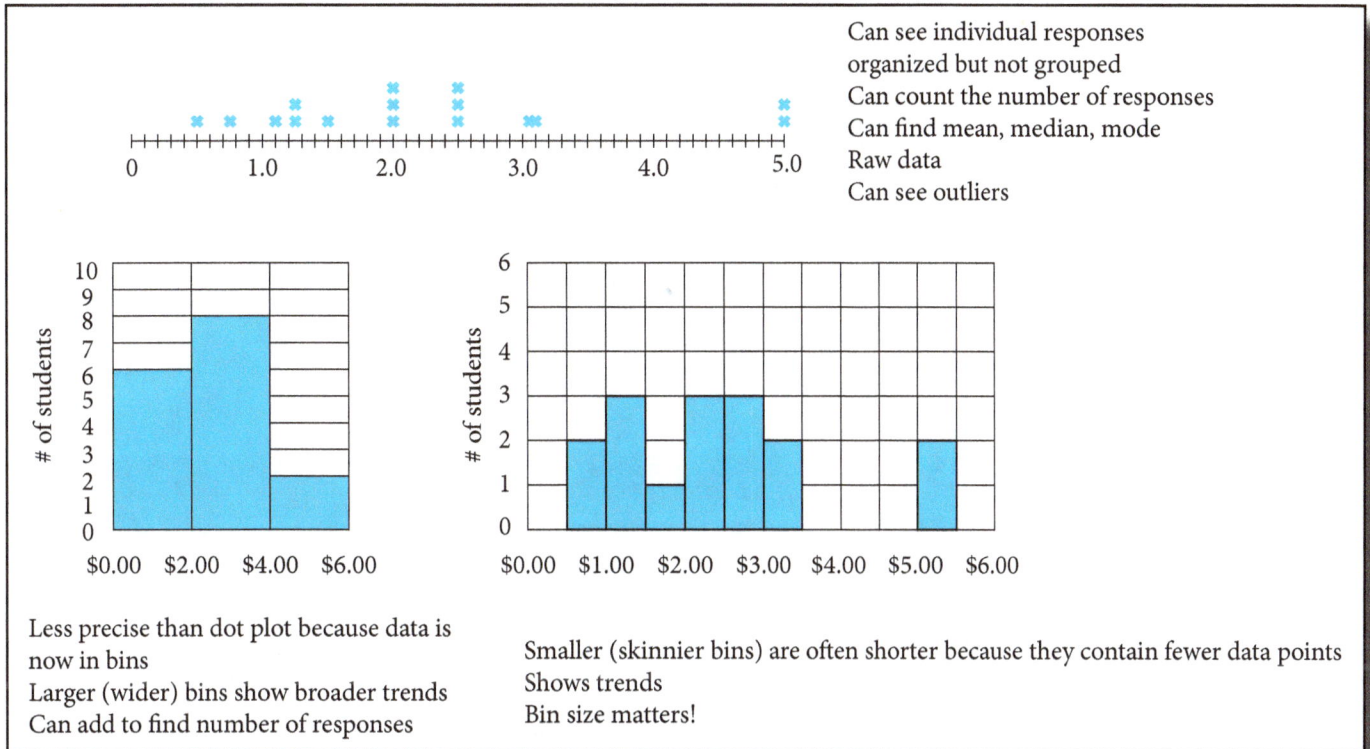

Can see individual responses organized but not grouped
Can count the number of responses
Can find mean, median, mode
Raw data
Can see outliers

Less precise than dot plot because data is now in bins
Larger (wider) bins show broader trends
Can add to find number of responses

Smaller (skinnier bins) are often shorter because they contain fewer data points
Shows trends
Bin size matters!

Facilitation Notes

This version of the problem string lists short notes for important teacher moves during the string. After you've done the string yourself and studied the relationships involved, you might make similar notes for the things you want a reminder of or deem important.

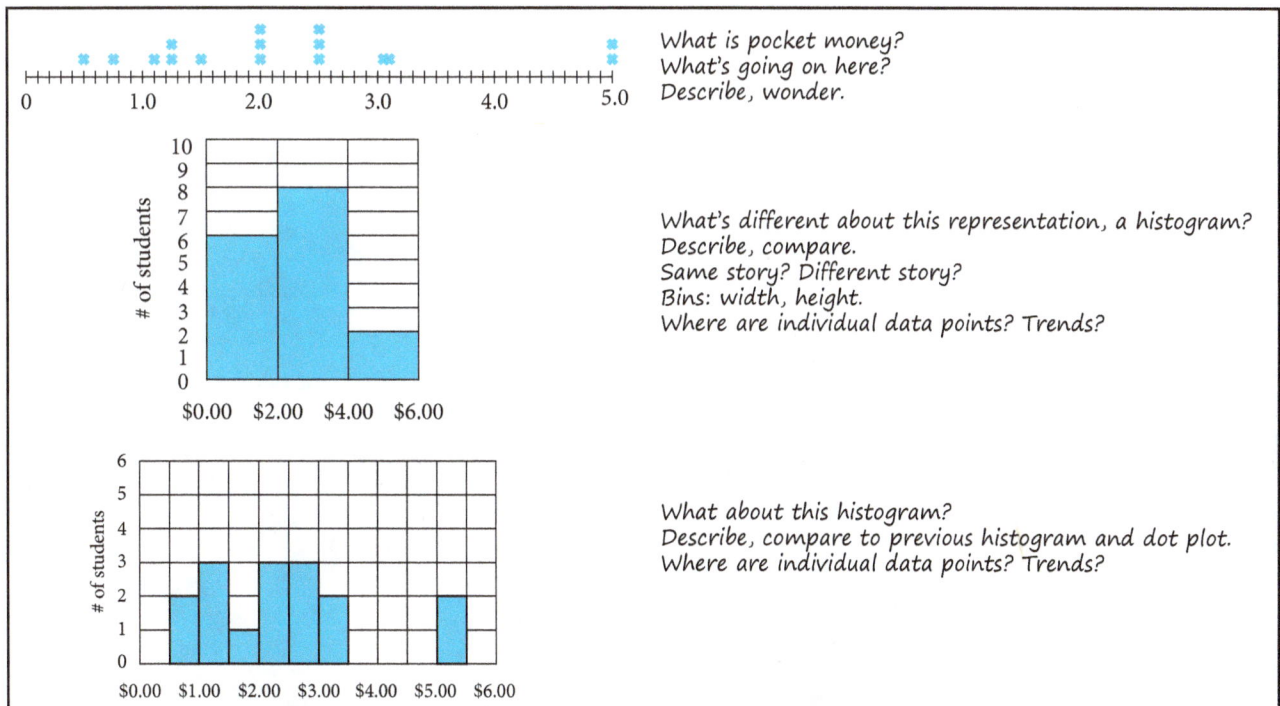

What is pocket money?
What's going on here?
Describe, wonder.

What's different about this representation, a histogram?
Describe, compare.
Same story? Different story?
Bins: width, height.
Where are individual data points? Trends?

What about this histogram?
Describe, compare to previous histogram and dot plot.
Where are individual data points? Trends?

For Display

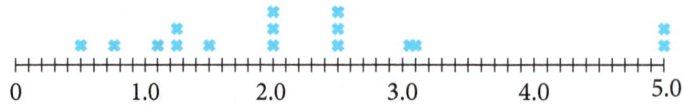

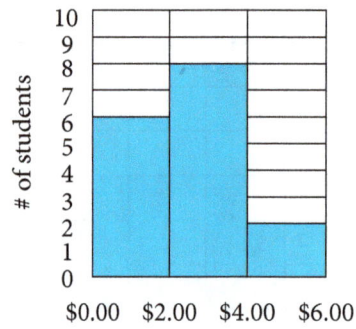

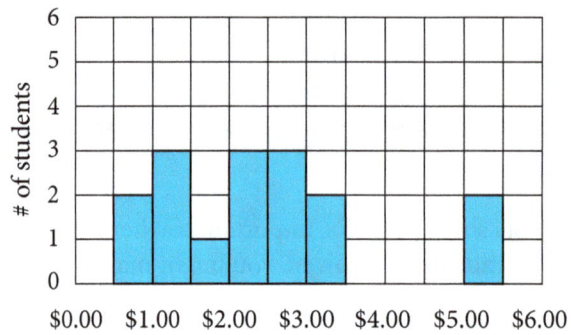

1.5 | Percents of 360

At a Glance	Objectives

At a Glance

100% of 360°

50% of 360°

25% of 360°

75% of 360°

10% of 360°

5% of 360°

15% of 360°

35% of 360°

144° is what percent of 360°

Objectives

The goal of this string is to support students as they reason about percents of whole numbers and to encourage flexible strategies for finding percents of 360°.

Placement

Use this string as students are working with circle graphs (pie charts), which are relative frequency graphs. This problem string can help students construct circle graphs using what they know about angle measures within a circle.

You can use this problem string as you work with textbook Lesson 1.5 Relative Frequency Graphs.

Guiding the Problem String

This string is designed so that students can use friendly landmark percents to construct new percents of a whole. It is designed to support students' work with circle graphs, so you should model student thinking on a circle showing relative portions.

The first problem establishes the model of a circle. The next three problems should go quickly, but throughout the string, label all of the numbers, keeping the units clear, 180 *degrees* is 50 *percent*, not 180 is 50. The next four problems build off of each other with multiples of 5%. The last problem turns things around by giving 144 degrees and asking for the percentage of 360. Keep students reasoning by asking how they can use what they know.

About the Mathematics

One strategy students may have learned (or memorized) is to convert all percents to their decimal equivalents (e.g., 50% becomes .50) and multiply by the whole. While this strategy works every time, it can be inefficient with many numbers and is not the focus of this problem string. Instead we suggest students reason about the quantities involved and find smart, efficient ways to calculate percents without algorithms and calculators whenever possible.

In later courses, students will look at angle measures on a unit circle. Traditionally the unit circle is oriented with 0° at the right most point of the circle and the angles opening counter-clockwise. Even though this orientation is not necessary for circle graphs in general, for consistency and to give students a preview of relationships to come, you could orient your class model the same way. Do not force students to use the same orientation, but keep yours consistent.

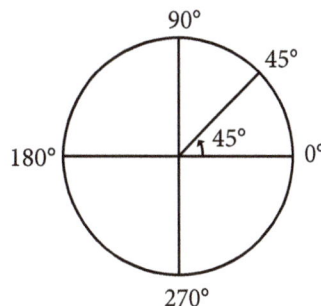

(continued)

Sample Interactions

Use the following as you plan how to elicit and model student strategies. This is not meant as a script, but as a view into the relationships involved and the intent of the problem string.

Teacher: *All right, find your partner so we can do a problem string together. You won't need paper or pencil for this one. Just think along with us. Today we are going to play with percents and try to connect them to a circle graph.* *I'd like you to do two things: solve these and try to envision them as well. Remember that there are 360 degrees in a circle, so if I asked you to think about 100% of 360°, what do you imagine?* **Student:** *Like, the whole thing.* **Teacher:** *Okay, a quick model of the whole circle.*	100% of 360°= 360°

Teacher: *And what do you picture when I ask for 50% of 360°?*

Student: *Half of the circle or 180°*

Teacher: *Does it matter where the half goes?*

Student: *Not really, as long as you draw a straight line and it goes through the center, you have a half.*

Teacher: *So, I'm going to start here and label it 0% and put that half over here. This is a way that you'll see circles in later math. If you think about starting here at 0% and opening up to cover half of the circle, you would cover from this 0° to this 180°.*

Teacher: *How about 25% of 360°? How many degrees now and what does this look like?*

Students may split either half of the circle. You could model the split as shown if you want to stay consistent with a trigonometry view of circles and angles.

50% of 360°= 180°

25% of 360°= 90°

90°

180°   0°

Teacher: *And what about this one? Can you do it easily without much computation? Tell your partner how you solved it and what you are picturing.*

Students turn and talk while the teacher listens for students who are using the relationships already established.

Teacher: *What are you thinking?*

Student: *We know that a quarter of the circle is 90° so three-quarters is three times 90° or 270°.*

Student: *We thought about the half being 180° and added a quarter, 90°, to get 270°.*

Student: *And we looked at the whole. The whole circle minus a quarter leaves three-quarters, so 360° minus 90° is 270°.*

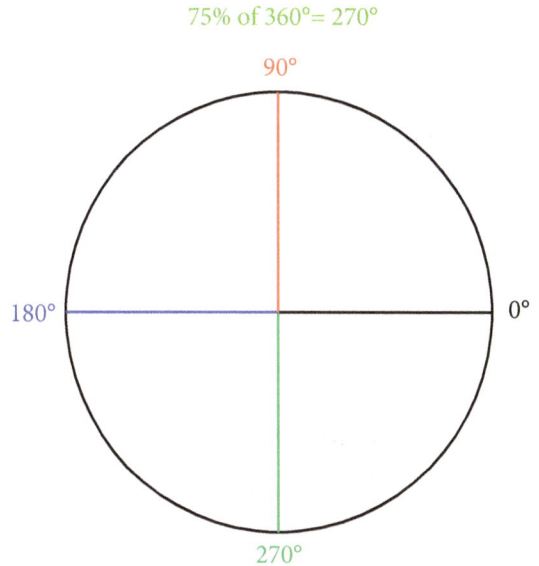

75% of 360°= 270°

90°

180°   0°

270°

Teacher: *Now I'm wondering about 10% of 360. What would that look like on a circle graph and how many degrees would that be?*

Student: *We know that the 10% is less than 90° because that's 25%. So, it's got to be smaller than the quarter circle.*

Student: *It's like the circle was a pizza cut into exactly 10 slices and you have one of them. I think it is 36°.*

Student: *10% is 100% divided by 10 so 360° divided by 10 is 36°.*

Teacher: *Those slices you're talking about, we call them sectors. Where shall I draw it? How does 90° compare with 36°? How does 25% compare with 10%? About here?*

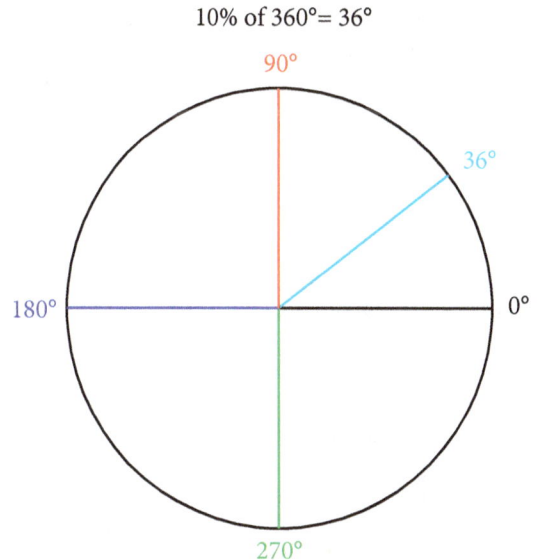

10% of 360°= 36°

90°

36°

180°   0°

270°

(continued)

Teacher: *So, what do you picture when I say 5% of 360°.*

Student: *Well, 36° is 10%, right? So isn't 5% just half of 10%?*

Student: *Yeah, so you just need half of that pie piece, half of the 36°, so 18°.*

Teacher: *So about half of the 36° is right here and we'll label it 18°.*

Teacher: *So we know several parts of the circle, 10%, 5% ... What would 15% of this circle look like—and how many degrees would that be?*

Go ahead and tell your partner how you are picturing this one.

The teacher looks for students who used the 5% sectors three times, or added the 5% sector to the 10% sector. The teacher wants to encourage this kind of flexible thinking, so ask for other strategies if only one appears. It may be helpful to sketch in the 45° line that corresponds to 12.5% to help estimate where to draw the 54°.

5% of 360°= 18°
15% of 360°= 54°

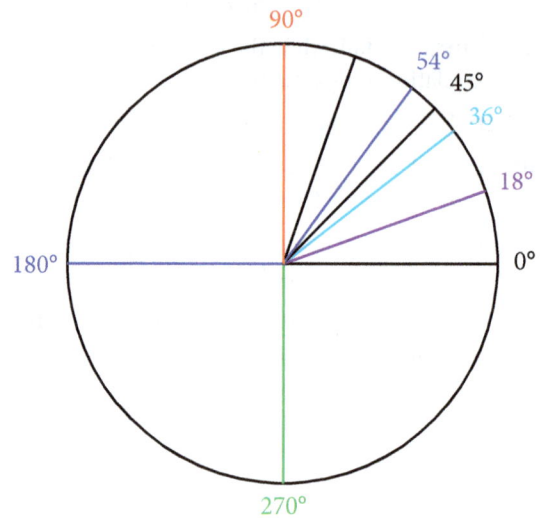

Teacher: *So, now I'm thinking about this one, 35% of the circle or 35% of 360°. What's your strategy for finding this one? Let's think for a few seconds first and then we can talk with partners.*

Think and partner talk time, listening for students who are using the 25% and 10% and/or 5% and noting other strategies.

Teacher: *Where is 35% of this circle?*

Student: *We know that the 90° is 25%, so 35% is bigger.*

Student: *Yeah, it's 10% bigger and we know 10% is 36°. So 90° and 36° is 126°.*

Student: *And you would draw it 36° past the 90°.*

35% of 360°= 126°

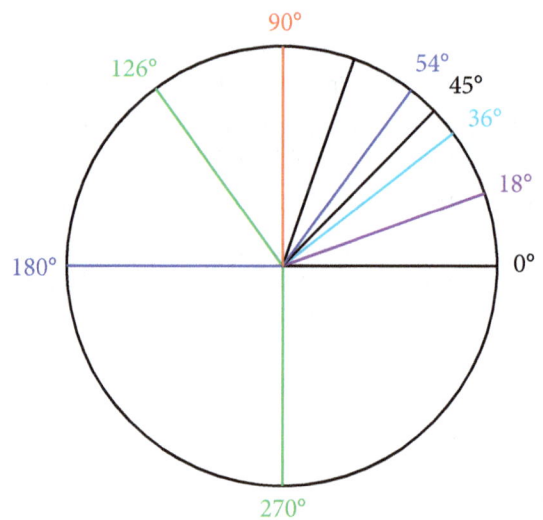

Algebra Problem Strings
©2017 Kendall Hunt Publishing

Teacher: *Now I'm going to flip the routine a bit and ask you to think about 144° within a circle. What would that look like? And what percent of the circle? Use anything here we have talked about to help you. After you have thought about it, turn and talk to your partner.*

Students talk to partners while the teacher listens in, looking for students who are reasoning using the relationships they already know.

Teacher: *What are you thinking?*

Student: *We knew it was past the 126° and we thought about 144° as 90° and 54° since we had those already. So 90° is 25% and 54° is 15% so it must be 40%*

Student: *Oh yeah, that's nice. We knew that 144° is less than 180° and we realized that it's just 36° less! So we went back from the 180° about 36° to draw it. And since 180° is 50%, then 50% minus 10% is 40%.*

Student: *We looked at 144° and know that 144 is 12 times 12, so it's also 36 times 4. We looked at four 36° chunks to help us know where to draw it and 4 times 10% is 40%.*

Teacher: *Really nice work, nice connections. Now that you've seen what your classmates have done, what do you wish your brain would be inclined to do the next time you run into numbers like these? Which of these strategies seems particularly clever and why?*

144 is ____% of 360°

144 is 40% of 360°

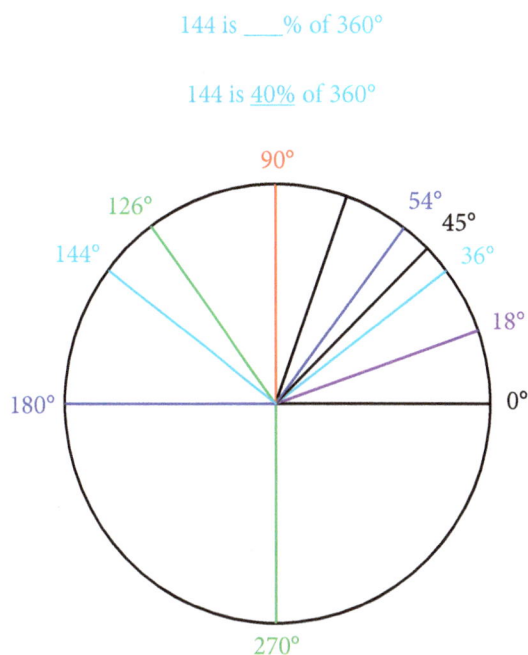

Teacher: *How would you summarize some of the things that came up in this string today?*

Elicit the following:

- *We can use chunks or sectors we know to figure out ones we don't know.*

- *A whole circle is 360°.*

- *Parts of a circle, called sectors, can be found by thinking about percentages of the circle.*

- *10% of any number is handy! We can find 10% by dividing the number by 10, a place-value shift.*

(continued)

Sample Final Display

Your display could look like this at the end of the problem string:

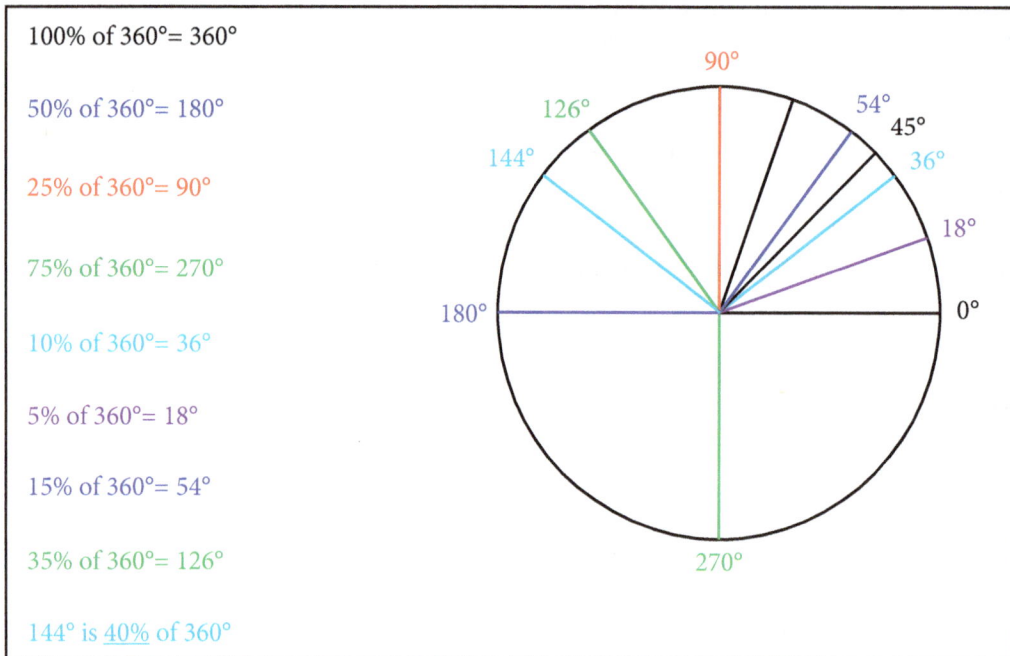

100% of 360°= 360°

50% of 360°= 180°

25% of 360°= 90°

75% of 360°= 270°

10% of 360°= 36°

5% of 360°= 18°

15% of 360°= 54°

35% of 360°= 126°

144° is 40% of 360°

Facilitation Notes

This version of the problem string lists short notes for important teacher moves during the string. After you've done the string yourself and studied the relationships involved, you might make similar notes for the things you want a reminder of or deem important.

100% of 360°	Quick. Sketch a large circle.
50% of 360°	Quick. Could be either half. Label 0° and 180°.
25% of 360°	Quick. What's a quarter of 360°? Half of 180°?
75% of 360°	Three 1/4s, Half + 1/4, Whole – 1/4.
10% of 360°	One-tenth of the circle. How can we find 1/10 of anything? About where? How many degrees?
5% of 360°	Half of the tenth. Quarter divided by 5.
15% of 360°	Elicit a few strategies.
35% of 360°	Elicit a few strategies.
144° is what percent of 360°	Challenge. Where? How can you use what you know?

Graphs: What's the Story?

At a Glance

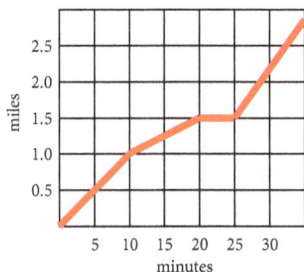

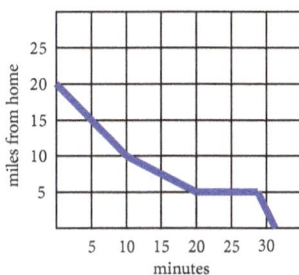

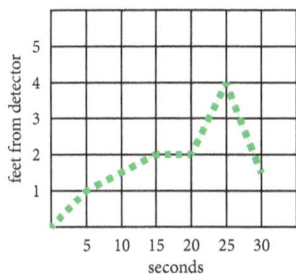

Objectives

In this problem string we offer students three different time-distance graphs and invite them to reason about what they see and develop a narrative or story about the possible meanings of the graphs. This ability to interpret two-variable graphs, based on evidence from the graph, will support their work in algebra in this lesson and beyond.

Placement

Use this problem string as you begin to work with two-variable data, (x, y), and graph reading to help student learn to interpret graphs and the meaning of ordered pairs. Ideally students will have prior experience with motion detectors and the resulting graphs.

You can use this problem string with textbook Lesson 1.7 Two-Variable Data.

Guiding the Problem String

If students have had experiences using a motion detector to construct and analyze time-distance or time-speed graphs, they may draw upon those experiences here. If not, it will require some work on your part to help students "tell the story" of each graph. Start with open-ended questions. What's going on here? What feels important? Let students take risks and put their ideas out to the class community. When possible, do not evaluate these ideas but give that work back to the class with questions such as: What do we think of this idea? Are you convinced? What questions do you wonder about?

Additionally, you might think of this problem string as a pre-assessment string, where you are listening and recording the various ideas (including misconceptions) that your students have about two-variable data such as these graphs. Students will have many opportunities to work out those misconceptions so it isn't vital that you correct each of them during this string. This may also not be a place to correct students' language related to graphs. Instead take note of the language that students do use for describing this complex data and build on it in future problem strings.

About the Mathematics

Beginning graph readers or students with less experience reading time-distance graphs will often interpret graphs as pictures of the situation rather than as the relationships between the distances as time changes. They might interpret the first graph as "going up a hill." Students need experience analyzing the meaning of ordered pairs and sections of the graph to help them learn to look relationally at the graph, instead of pictorially.

(continued)

Sample Interactions

Use the following as you plan how to elicit and model student strategies. This is not meant as a script, but as a view into the relationships involved and the intent of the problem string.

Teacher: *Today I'm going to show you a series of graphs and in each case, I'm going to ask you to think about what it going on in the graph—what the story is and what makes us say what we say. You won't need paper and pencil, but you do have your partner and all of us, and I'll record any of the things we think are true or are wondering about. Here's the first graph. What do you think is going on here? What makes you say this?*

Allow students to study the graph—perhaps make one copy per pair so that students can study it up close together before you begin any partner talk or whole class conversation. Or this might mean asking students in the back of the room to move up to the front half of the room where you have projected or displayed the graph. It's critical that all students can see the graph and all of its details in order to contribute to the class conversation, so plan accordingly.

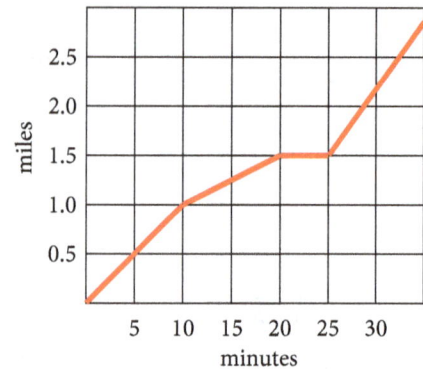

Teacher: *Take 45 seconds with your partner to talk about what is going on here. What could be true and be thinking about what you want to share with all of us. Go!*

Students talk with a partner while the teacher listens for student ideas about rates.

Teacher: *I was listening in to this group's conversation and was wondering if they would start us off.*

Student: *Sure, we thought it was someone training for a race. Because they start out doing 1 mile in 10 minutes but at the end they got faster.*

Teacher: *Tell us where this 10-minute mile is and where is the section of the graph that you say they were running faster.*

Student: *In the first 10 minutes, they went 1 mile, but at the end there, in the last 10 minutes they went further, so that means they went faster.*

Teacher: *Class, what do we think of their idea? Are you convinced?*

Student: *If you go more distance in the same amount of time, you are going faster, so yes, I agree.*

Teacher: *What else could you say for sure about this graph? Let's get some more voices in the conversation.*

Student: *It looks like they took at break between the 20-minute and 25-minute mark.*

Teacher: *Who understands this idea? Okay, and who agrees with it? Anyone want to ask a question about this or challenge it?*

Teacher: *What's going on in the middle of the graph, from minutes 10 to 20? Turn and talk to your partner about what's happening there.*

After students talk to partners briefly, the teachers builds a conversation about the 10- to 20-minute section, noting important ideas on the graph.

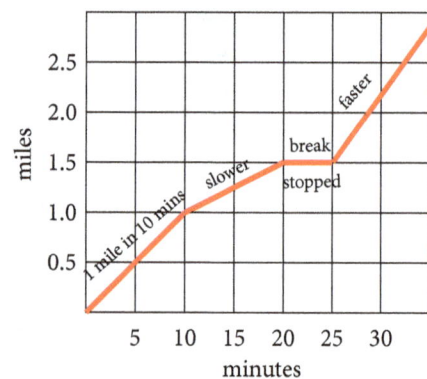

- Training for a race
- 1 mile in 10 minutes = 10-minute mile?
- Started fast, then went slower
- Took a five minute break from minutes 20–25
- Sped up at the end—maybe sprinted
- Probably running or run/walking

Teacher: *I'll leave this graph up along with our claims as we continue.*

Teacher: *Okay, I'll project a second graph. Be thinking about the same questions as before: What do you think is going on here? What makes you say that?*

Let's really study it on our own for a full minute, then we can check in with a partner. Sound okay?

Private think time, then students talk with a partner while the teacher listens for student ideas about rates.

Teacher: *What's going on here, with this second graph?*

Student: *My partner and I talked about how the y-axis was different— the numbers are way bigger and the label says miles from home. So we are thinking someone is driving back home from being away and it looks like they stopped on the way home.*

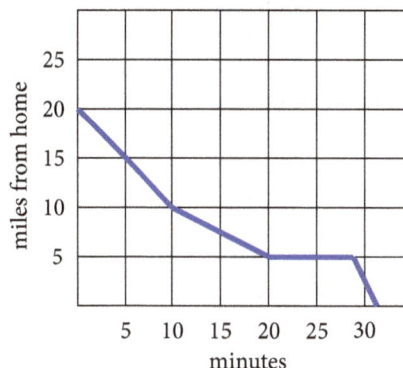

Teacher: *Who understands these ideas and could put them in their own words?*

Student: *She is saying that the scale is different and the y-axis changes the story totally. This is no longer a story of running or walking, this has to be driving and it looks like the person is driving home. They were 20 miles away and now they are finally home, with a short stop near the end.*

Teacher: *Okay, so who agrees with these claims so far? Anyone want to play a skeptic—you are just not convinced?*

Student: *Yeah, I'm not sure that we know if this is driving or not. I mean, couldn't you be biking really fast.*

Teacher: *What do we think of this challenge? Could this be a story about biking—turn and talk to your partner about this?*

Teacher: *I've added these rates to the graph and now I'm going to keep this graph and our notes so that we can still see it all.*

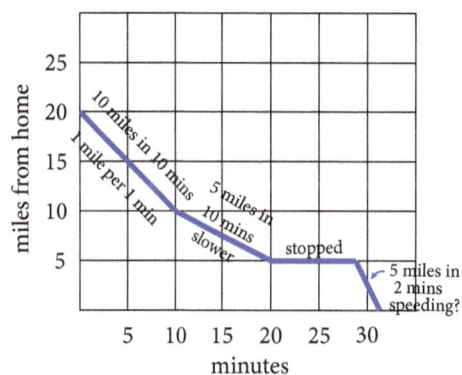

- Not training for a race—likely driving home
- x-axis stayed the same
- y-axis changed a lot
- 10 miles in 10 minutes is fast— probably on the highway
- Then maybe got off the highway
- Speed/rate changes to 5 miles in 10 minutes (slower)
- Stopped for 6 minutes
- Maybe stopped to get gas on the way home
- Really fast at the end. Is this reasonable?
- Skeptic—how do we know this is a car? Could it be a fast biker?

(continued)

Teacher: *And here's the last graph. Some of you may have used a motion detector in middle school or in another high school class. If not, a motion detector sends out a signal that bounces off a person walking in front of the detector. It sends that information to a grapher and produces a graph of the distance of the person from the motion detector over time. Let's study this first on our own and be thinking about: What do you think is going on here? What makes you say that?*

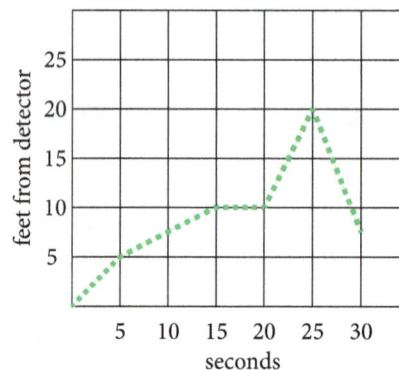

The teacher builds a conversation about the motion detector graph, pulling in diverse voices and encouraging students to support or challenge ideas.

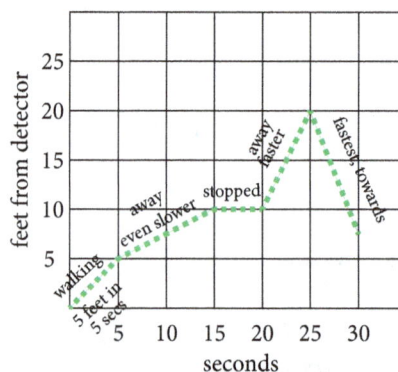

- A motion detector graph
- Both *x*- and *y*-axes changed
- Scale on *x*-axis didn't change. Units change from mins to secs
- Walked away from and towards the motion detector
- Units changes (seconds and feet)
- Also took a break but for seconds, not minutes
- Fastest at the end, but didn't make it back to the motion detector when time stopped at 30 seconds

Algebra Problem Strings
©2017 Kendall Hunt Publishing

Teacher: *How would you summarize some of the things that came up in this string today?*

Elicit the following:

- *It's important to look at the labels, scale, and units.*

- *We can make sense of going faster or going slower or even standing still by looking at the slope of the lines.*

- *If you cover more distance in the same amount of time, you're traveling faster.*

- *If the lines are decreasing, you are moving toward home or toward the motion detector. If the lines are increasing, you are gaining distance or moving away from the motion detector.*

- *If the line is horizontal, the distance is not changing. As time keeps going, you're not moving away or towards.*

Sample Final Display

Your display could look like this at the end of the problem string:

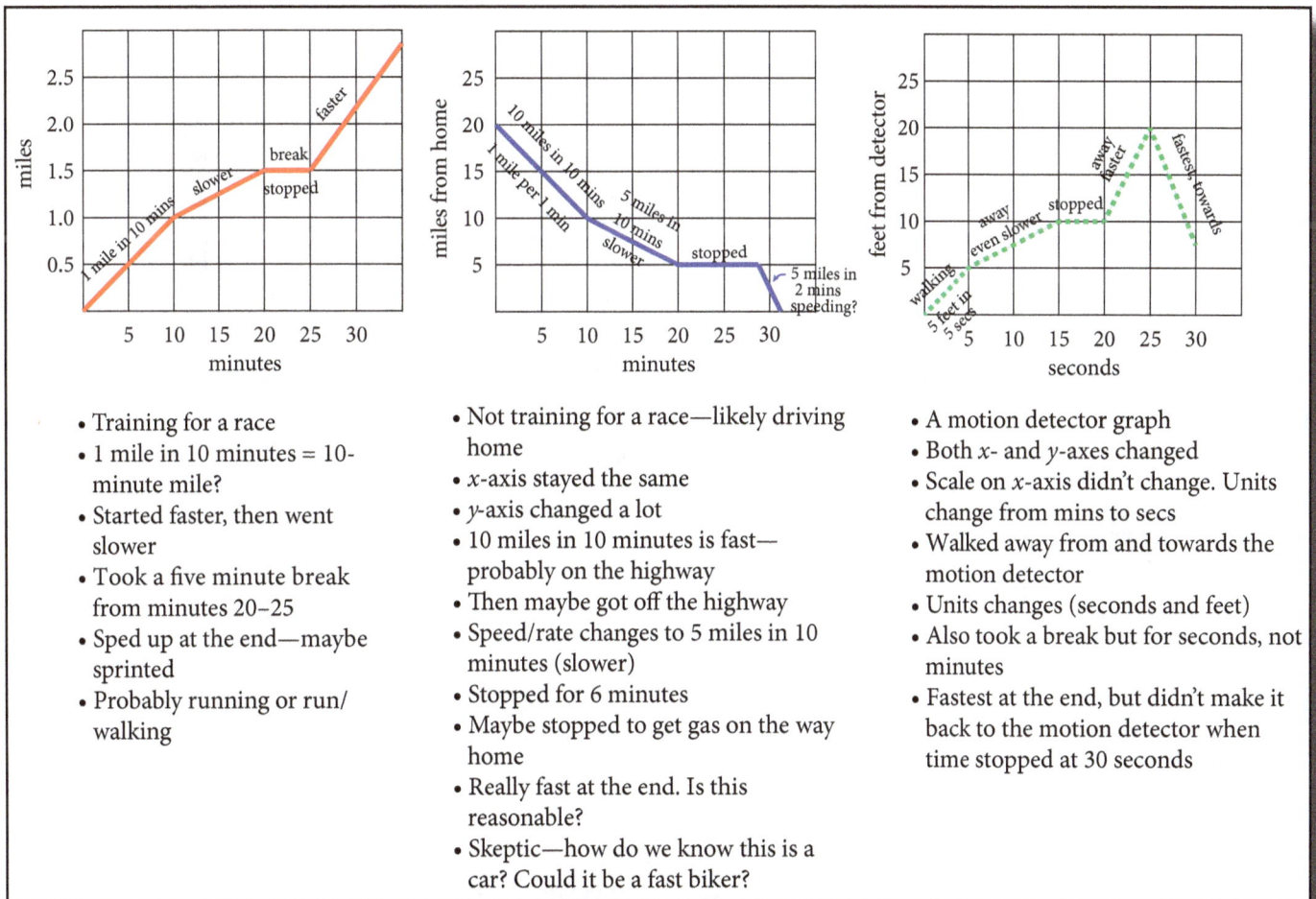

- Training for a race
- 1 mile in 10 minutes = 10-minute mile?
- Started faster, then went slower
- Took a five minute break from minutes 20–25
- Sped up at the end—maybe sprinted
- Probably running or run/walking

- Not training for a race—likely driving home
- x-axis stayed the same
- y-axis changed a lot
- 10 miles in 10 minutes is fast—probably on the highway
- Then maybe got off the highway
- Speed/rate changes to 5 miles in 10 minutes (slower)
- Stopped for 6 minutes
- Maybe stopped to get gas on the way home
- Really fast at the end. Is this reasonable?
- Skeptic—how do we know this is a car? Could it be a fast biker?

- A motion detector graph
- Both x- and y-axes changed
- Scale on x-axis didn't change. Units change from mins to secs
- Walked away from and towards the motion detector
- Units changes (seconds and feet)
- Also took a break but for seconds, not minutes
- Fastest at the end, but didn't make it back to the motion detector when time stopped at 30 seconds

(continued)

Facilitation Notes

This version of the problem string lists short notes for important teacher moves during the string. After you've done the string yourself and studied the relationships involved, you might make similar notes for the things you want a reminder of or deem important.

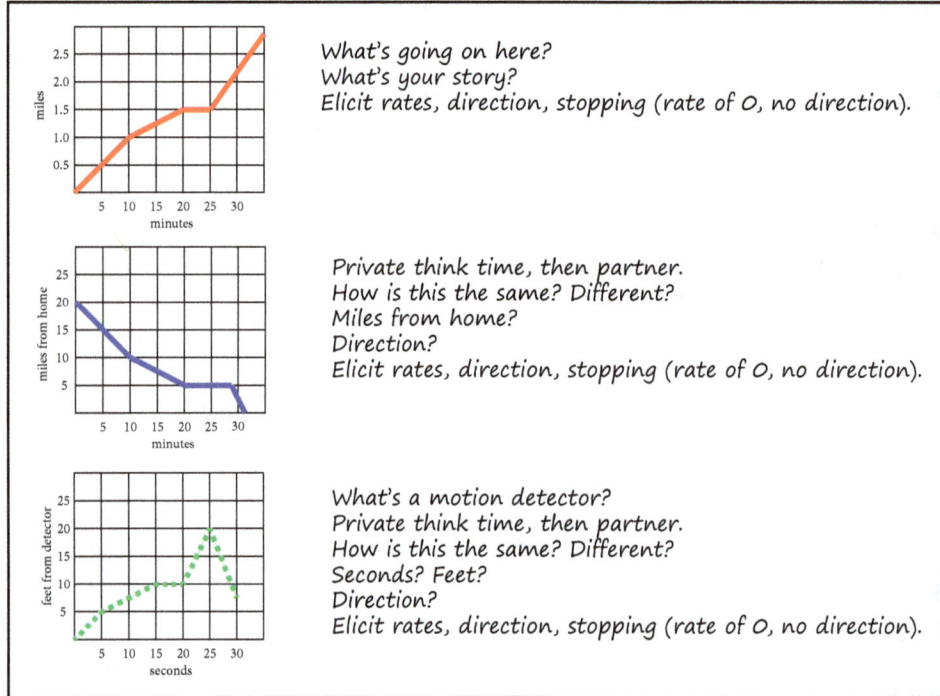

What's going on here?
What's your story?
Elicit rates, direction, stopping (rate of 0, no direction).

Private think time, then partner.
How is this the same? Different?
Miles from home?
Direction?
Elicit rates, direction, stopping (rate of 0, no direction).

What's a motion detector?
Private think time, then partner.
How is this the same? Different?
Seconds? Feet?
Direction?
Elicit rates, direction, stopping (rate of 0, no direction).

Algebra Problem Strings
©2017 Kendall Hunt Publishing

For Display

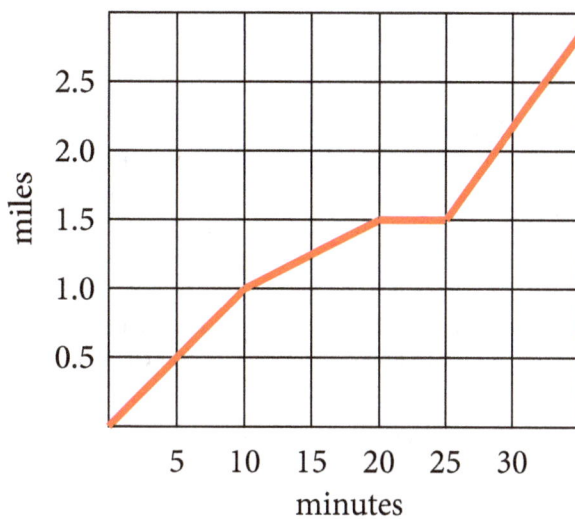

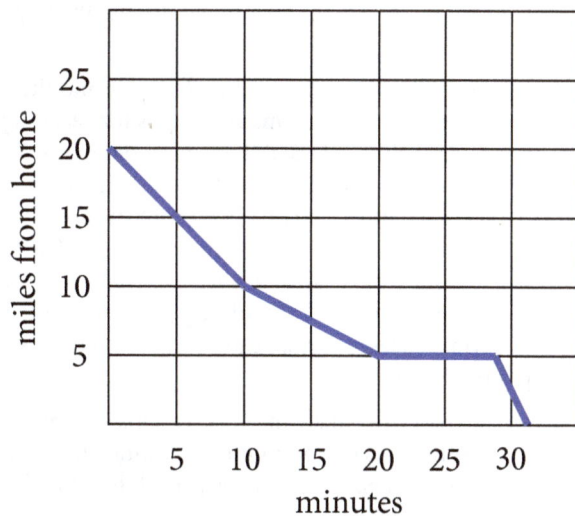

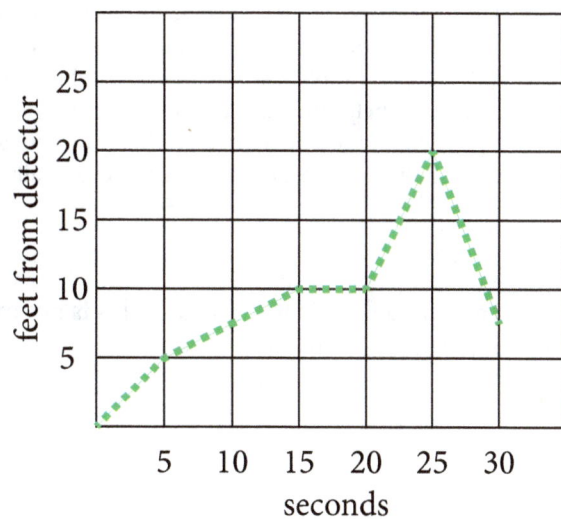

Playing the Game "Guess It!"

At a Glance

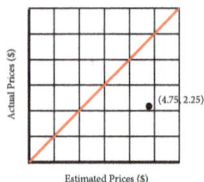

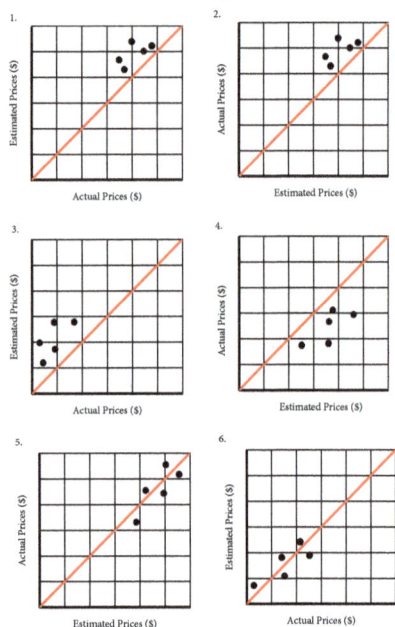

Objectives
The goal of this problem string is to build students' facility with ordered pairs and the parent function $y = x$ by comparing estimates and actual prices of items in a game show game called "Guess It!"

Placement
Use this problem string just as students are learning about scatter plots. It is designed to preview that work and help students make claims about scatter-plot data based on how the data relates to the line $y = x$.

You can use this problem string before or during your work with textbook Lesson 1.8 Lines and Data.

Guiding the Problem String
In this problem string students analyze scatter plots of estimated and actual prices of various household items and discuss the important features of the scatter plot and data.

The first problem establishes the game and the meanings of an ordered pair. The rest of the problems are graphs that each represent one contestant's game of estimating the prices of five household items. Throughout the string, the axes will switch, encouraging students to carefully consider the meaning of the ordered pairs each time. The first two problems are identical except the axes have been switched, making the meanings different. The third and fourth problem switch the axes again and change the prices of the items. The next two graphs represent pretty good estimates and provide a good time to ask about winning the game. If time permits, a last problem can be introduced for students to invent a scenario that is not pictured.

Have students make the observations, claims, and wonderings, and justify these ideas to the class community. Continue to ask questions like, "What do we think of this idea?" and "Are you convinced that this claim is true? What convinced you."

About the Mathematics
The linear parent function is the line $y = x$. This very important function could be argued to be the parent of all functions, with each function being a composition of $f(x) = x$. This problem string uses ordered pairs (estimates, actual) and (actual, estimates) to help students develop a sense of the relationship when the estimates are equivalent to the actual prices, estimates = actual, $y = x$. Upon this basis, students can solidly build linear functions and eventually all other functions.

Sample Interactions

Use the following as you plan how to elicit and model student strategies. This is not meant as a script, but as a view into the relationships involved and the intent of the problem string.

Teacher: *Today, I'm going to tell you the story of a game on a game show we'll call Guess It! In this game, the contestant is presented with five common household items and is asked to guess the price of each. Once their guesses, or estimates, are locked in the host of the show reveals the actual prices. If the difference between the total of the estimates and the actual prices is $2.00 or less the contestant wins the game.*

Check out this graph. This point represents one item's actual price and the contestant's estimate. Tell me about it.

Does this point represent an overestimate or an underestimate? But it's below the line... shouldn't it be an under estimate?

So, the axes might matter. We had better keep an eye on that. Since this graph was just to explain the game, I'm not leaving it up on display.

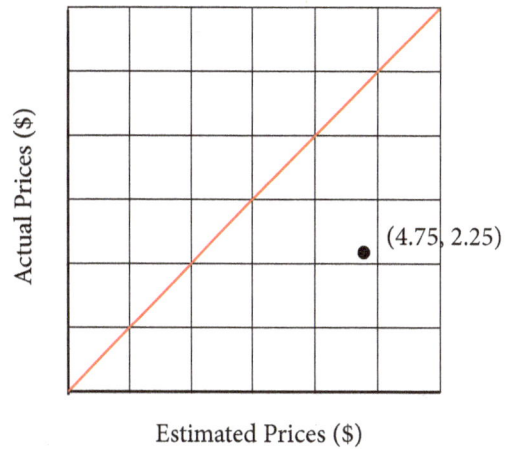

(4.75, 2.25)

Actual Prices ($)

Estimated Prices ($)

Teacher: *Here are the results for Contestant #1. Tell me about this contestant's estimates and the prices of the items. What can you tell by this graph?*

What are the labels on the axes? Does that matter?

Where is a perfect estimate?

What were the actual prices, high, medium, or low?

Were contestant #1's estimates, high, medium, or low?

Let's choose one point. What are its approximate coordinates? What does it mean?

Where would it be if Contestant #1 would have guessed the price exactly? How far from the actual price is it?

So, we have determined that Contestant #1 had high priced items but guessed too high. An over-estimator. Do we have that right? Does that make sense to everyone?

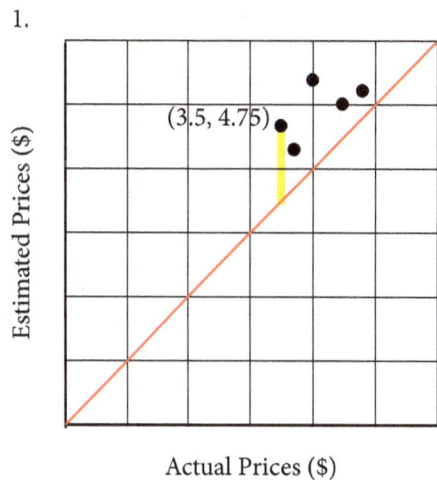

1.

(3.5, 4.75)

Estimated Prices ($)

Actual Prices ($)

Teacher: *Okay, tell me about Contestant #2. Here are his results. What does this graph mean in the story of the game? Turn and tell your partner.*

I'm hearing some of you say that there must be a typo or else Contestant #2 and #1 got exactly the same results. Is that what everyone thinks? There are no other differences?

Some of you are shaking your heads. What do you think?

Does it matter that the axes have switched? How?

Now what does our point mean? Now the point (3.5, 4.75) is a low guess. How do you make sense of that?

So, we have determined that Contestant #2 had high priced items but guessed too low every time. An under-estimator. Axes matter!

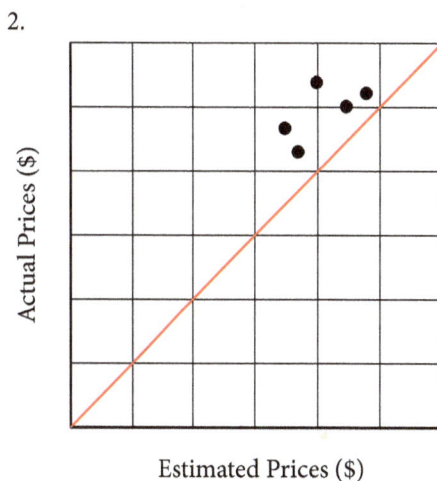

2.

Actual Prices ($)

Estimated Prices ($)

(continued)

Teacher: *Here are the results for Contestant #3. Think for a bit about this one. Then we'll come together and describe the estimates and the actual prices.*

Okay, who can get us started? Tell us about the items and Contestant #3.

What are the axes this time? How does that affect everything?

But wait, the points are above the line. I thought that with Contestant #2, we decided that the points were not where we thought. If they were over the line, the contestant was an under-guesser? Oh, only when the axes are switched? Ahhh, that seems important. So, maybe instead of trying to remember which is which, actually reasoning about the axes might work well.

The teacher crafts a conversation that includes the axes, the relative prices of the items, and the relative guess accuracy of the contestant. This graph describes a contestant who consistently overestimated the prices of relatively low priced items, making Contestant #3 an over-estimator.

3.

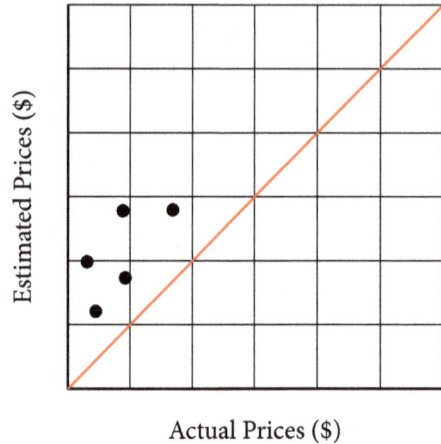

Teacher: *And next up is Contestant #4! Go!*

The teacher crafts a conversation that includes the axes, the relative prices of the items, and the relative guess accuracy of the contestant. This graph describes a contestant who consistently overestimated the prices of relatively middle-range priced items, making Contestant #4 an over-estimator. Because the points represent (estimates, actual), points under the line represent overestimates.

4.

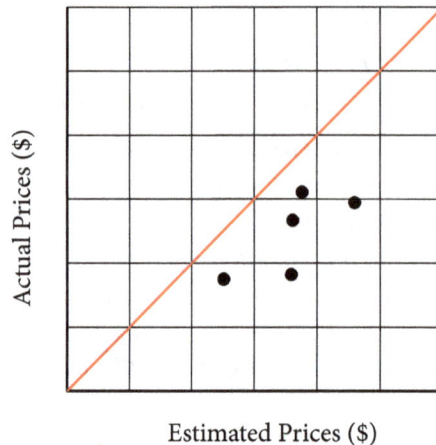

Algebra Problem Strings
©2017 Kendall Hunt Publishing

Teacher: *What can you tell about Contestant #5! Go!*

The teacher crafts a conversation that includes the axes, the relative prices of the items, and the relative guess accuracy of the contestant. This graph describes a contestant who estimated pretty closely, but not exactly, the prices of relatively high priced items, making Contestant #5 a pretty good estimator of the high priced items.

If time permits, you could ask students to think about whether the contestant wins or not. The tick marks represent dollars.

> *Since this contestant was a pretty good estimator, I wonder if Contestant #5 won? Remember that the sum of the estimates need to be within $2.00 of the sum of the actual prices. What do you think?*

> *You could look at the difference between the estimates and the actual prices on the graph. Where would those be?*

5.

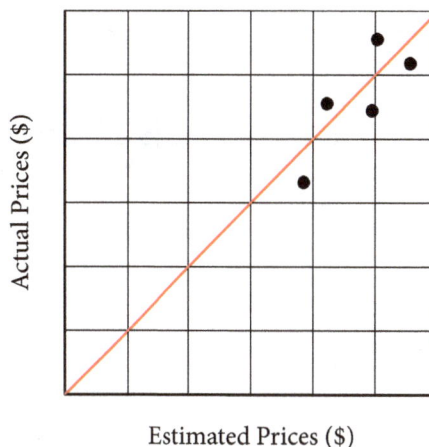

Teacher: *Last contestant for today, Contestant #6. Dive in!*

The teacher crafts a conversation that includes the axes, the relative prices of the items, and the relative guess accuracy of the contestant. This graph describes a contestant who estimated pretty closely, but not exactly, the prices of relatively low priced items, making Contestant #6 a pretty good estimator of the low priced items. The main difference between Contestant #5's and #6's games is that #5 had high priced items and #6 had low priced items.

If time permits, you could ask students to think about whether the contestant wins or not. The tick marks represent dollars.

> *Some of you are saying that we don't need to do any work to decide if she won or not? What's going on? Ah, since this contestant's scenario is just like Contestant #5 except low prices versus high prices, you think the differences will be about the same? Nice thinking.*

6.

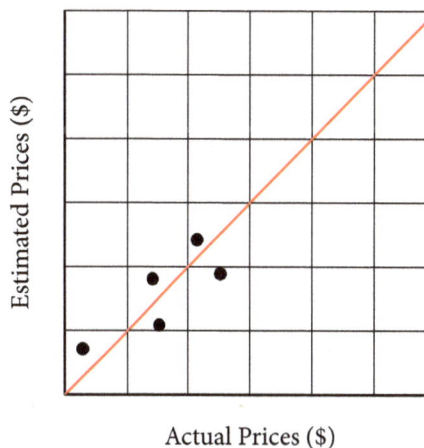

Teacher: *Finally, since we have time, let's all make up a story where Contestant #7 has a different game than all of the previous contestants.*

> *What would your situations look like on a scatter plot. You can work with your partner and I'll walk around to answer any questions and I'll be looking for a few that we can talk about together.*

Listen to students' conversations about what these situations look like as scatter plots. As you walk around the room, identify which pieces of student work would be good to discuss together. You might juxtapose two different pieces and have students try to determine what happened to the contestant in each game, or have the authors of the scatter plot make a claim and try to convince the rest of the class.

Teacher: *How would you summarize some of the things that came up in this string today?*

Elicit the following:

- *The axes matter! The meaning of the ordered pairs depends on the axes.*
- *The line y = x represents perfect estimates.*
- *Points close to the line y = x are good estimates.*
- *It does not matter if the items were low, medium, or high prices. It only matters how close the estimate is to the actual price, how close to the line y = x the point is.*

(continued)

Sample Final Display

Your display could look like this at the end of the problem string:

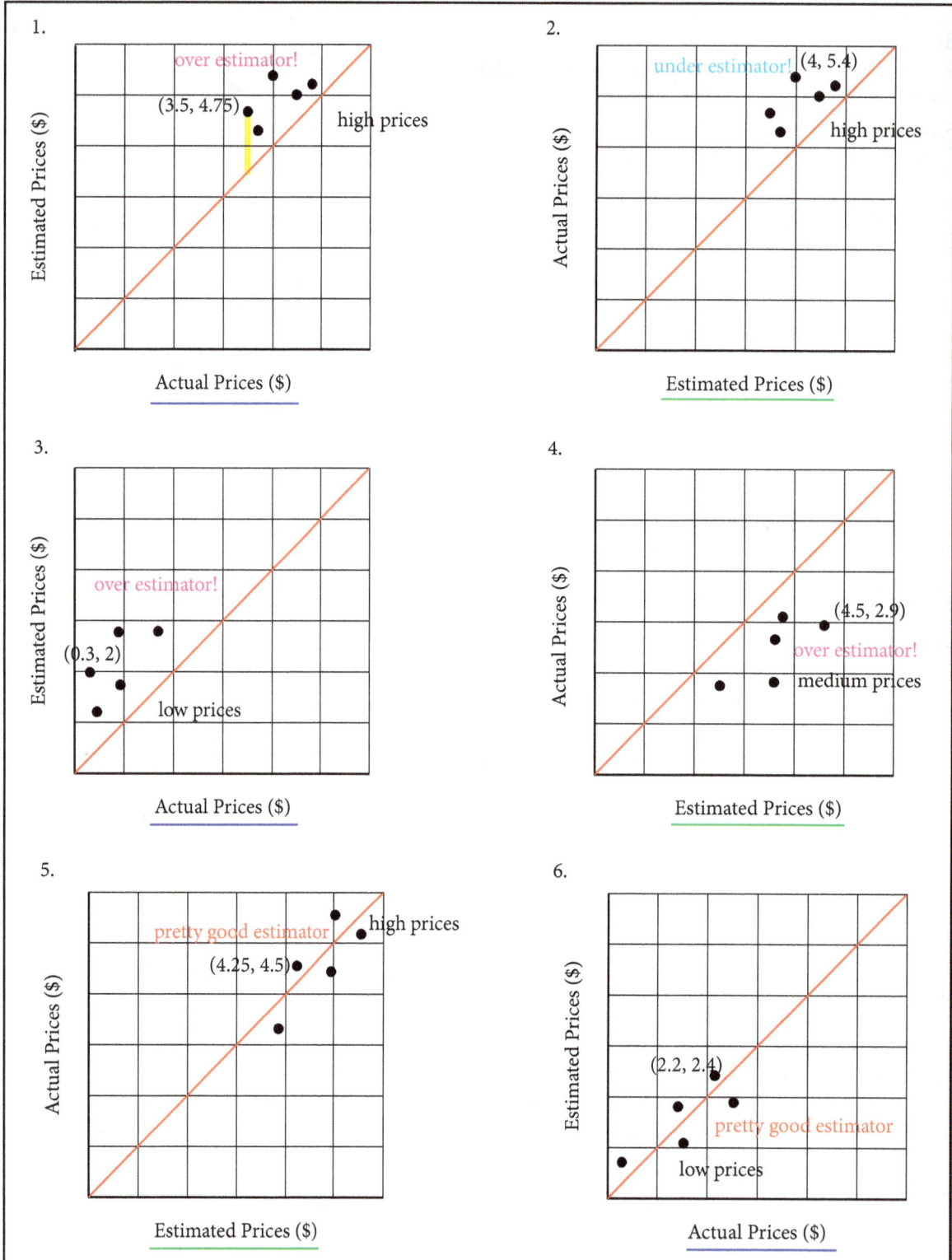

1. over estimator!
(3.5, 4.75) high prices
Estimated Prices ($) / Actual Prices ($)

2. under estimator! (4, 5.4)
high prices
Actual Prices ($) / Estimated Prices ($)

3. over estimator!
(0.3, 2)
low prices
Estimated Prices ($) / Actual Prices ($)

4. (4.5, 2.9)
over estimator!
medium prices
Actual Prices ($) / Estimated Prices ($)

5. pretty good estimator high prices
(4.25, 4.5)
Actual Prices ($) / Estimated Prices ($)

6. (2.2, 2.4)
pretty good estimator
low prices
Estimated Prices ($) / Actual Prices ($)

Algebra Problem Strings
©2017 Kendall Hunt Publishing

Facilitation Notes

This version of the problem string lists short notes for important teacher moves during the string. After you've done the string yourself and studied the relationships involved, you might make similar notes for the things you want a reminder of or deem important.

Introduction graph	Game: Guess it! Rules. Show graph w/ one point. What is this point? (4.75, 2.25)
Contestant #1	Show Contestant #1. What can you tell about this game/player? Axes switch! Look at an ordered pair. Prices? Estimates?
Contestant #2	Show Contestant #2. What can you tell about this game? Same as #1? Axes switch again! Look at an ordered pair, (prices, estimates) or (estimates, prices)?
Contestant #3	Repeat. Don't guess—figure out what an ordered pair means. Compare to previous.
Contestant #4	Repeat. Compare to previous.
Contestant #5	Repeat. Did this contestant win?
Contestant #6	Repeat. Did this contestant win? Compare to previous.
New Contestant	Invent a contestant who has not been shown yet. Describe. Defend.

(continued)

For Display

Estimated Prices ($)

1.

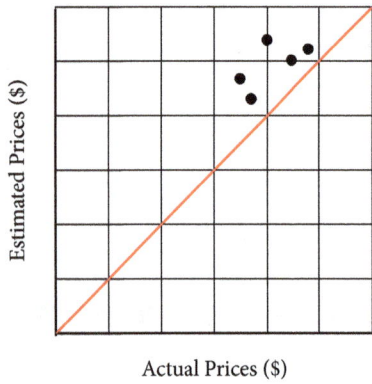

Actual Prices ($)

2.

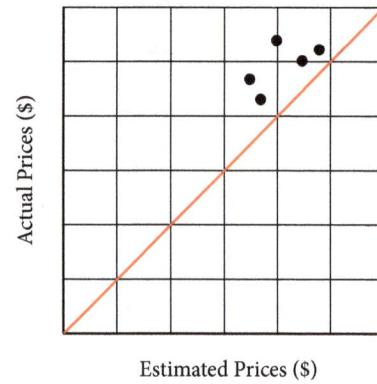

Estimated Prices ($)

3.

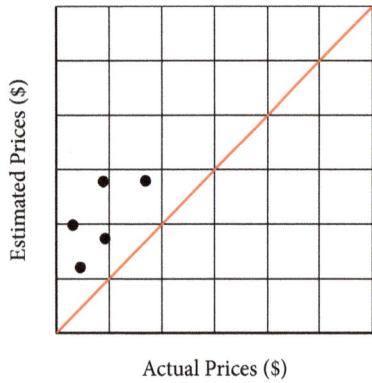

Actual Prices ($)

4.

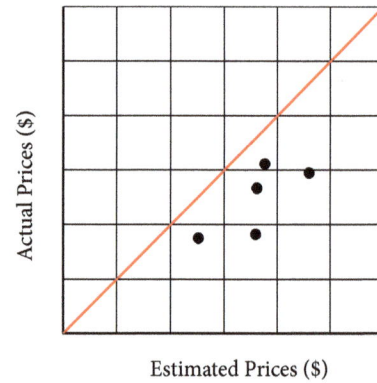

Estimated Prices ($)

5.

Estimated Prices ($)

6.

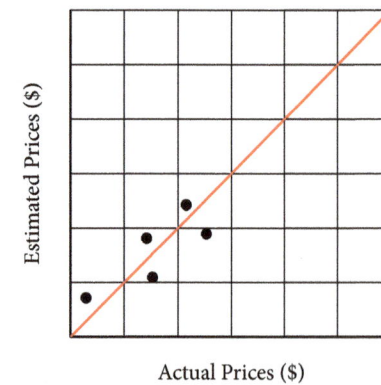

Actual Prices ($)

Algebra Problem Strings
©2017 Kendall Hunt Publishing

Proportional Reasoning with Ratio Tables 1

At a Glance

# of Packs	# of Sticks
1	17
2	
4	
20	
5	
15	
	153

Objectives

The goal of this string is to help students construct the ratio table as a generalizable model for reasoning about equivalent ratios.

Placement

This string is the first of four ratio table problem strings. Use this problem string to remind students of or to build upon proportional reasoning and relationships.

You can use this problem string to introduce textbook Lesson 2.1 Proportions. If your students have experience using a ratio table to solve problems, you may want to skip to Proportional Reasoning with Ratio Tables 2, which starts with a non-unit rate.

Guiding the Problem String

This string should proceed quickly. It is intended to link a context (packs of gum) to a model (the ratio table) while reasoning proportionally. Encourage flexibility, reasoning about new combinations of packs and sticks of gum. Encourage students who are over-reliant on the unit rate to take up other interesting relationships that the string makes available to them.

If we are not purposeful, ratio table strings can become quite cluttered with students' strategies. It is important to record strategies, but you want the class to be able to follow the progression of ideas in the string and keep the table as a useful tool. Erase before the table gets too smothered with arrows and calculations.

It is vital to use ratio tables flexibly, pointing toward efficiency—modeling how to scale up and scale down ratios, and to reason about rates even when the unit rate is not given. Step back to ask which strategy we wish our brains would gravitate toward next time or without the other problems first. Too often students see ratio tables that begin with the unit rate and proceed only by "chunking" these quantities to make new equivalent ratios. This form has been codified for some students as the signifier of a ratio table (e.g., it must begin with the unit rate), although this isn't true, as will be seen in the next string.

About the Mathematics

A ratio table is a special paired number table where all of the pairs are equivalent ratios. In this scenario, the unit rate is given, 1 pack to 17 sticks, but a unit rate is not always given.

$$\frac{1 \text{ pack}}{17 \text{ sticks}} = \frac{2 \text{ packs}}{34 \text{ sticks}} = \frac{4 \text{ packs}}{68 \text{ sticks}} = \frac{20 \text{ packs}}{340 \text{ sticks}} = \frac{5 \text{ packs}}{85 \text{ sticks}} = \frac{15 \text{ packs}}{255 \text{ sticks}} = \frac{9 \text{ packs}}{153 \text{ sticks}}$$

If you are not familiar with using ratio tables as a tool for computing, study the dialog to find examples of using the distributive and associative properties with the table to find equivalent ratios.

(continued)

Sample Interactions

Use the following as you plan how to elicit and model student strategies. This is not meant as a script, but as a view into the relationships involved and the intent of the problem string.

Teacher: *Alright, let's warm up with a short problem string today. I'm thinking about packs of gum. You know how they come in different sizes—some have 5 sticks, some have 10 and then some are like those jumbo packs?* [Optional: Quickly display an image of various popular forms of gum to invite students into the context. Find out who chews gum and what kinds/flavors of gum they enjoy.] *Great. So let's imagine a pack with 17 sticks of gum inside. And I'm going to make a table to keep track of our ideas here.* *So far, so good?*	**# of Packs** | **# of Sticks** 1 | 17
Teacher: *So, how would you find the number of sticks in 2 packs?* Brief think time. [Here you see the teacher write the "2" as the teacher asks about 2 packs. During the rest of the sample interaction, the teacher continues to write the value as the teacher asks the question, but the sample model will show that and what happens next.]	**# of Packs** | **# of Sticks** 1 | 17 2 |
Teacher: *Did anyone just double?* **Student:** *Yeah, that's 34. I doubled 10 and got 20 and doubled 7 to get 14, so 34.* **Teacher:** *So, if you double the number of sticks, you double the number of packs? Yes? That's an example of the distributive property, where you break up one of the factors into chunks that make sense.*	$\times 2$ (**# of Packs** | **# of Sticks**) $\times 2$ 1 | 17 2 | 34 $2(17) = 2(10+7) = 20+14 = 34$ distributive property
Teacher: *What about the number of sticks in 4 packs? What's a nice way to find that?* Brief think time. **Teacher:** *What is the number of sticks in 4 packs? And how did you find your answer?* **Student:** *68. I doubled the 34. Double 30 plus double 4.* **Teacher:** *So you doubled the packs to get double the sticks again. Nice.*	**# of Packs** | **# of Sticks** $\times 2$ (1 | 17) $\times 2$ $\times 2$ (2 | 34) $\times 2$ 4 | 68 $2(34) = 2(30+4) = 60+8 = 68$

Algebra Problem Strings
 ©2017 Kendall Hunt Publishing

Teacher: *Okay. Let's go big. How about one of those jumbo bags with 20 packs inside? Can you picture it? Enough gum for a year. Maybe.... You tell us. How many sticks is that? How can you use what you know?*

Think time.

> *Turn and tell your partner two ways you might have solved this.*

Students turn and talk while the teacher listens in.

Teacher: *Tell us about your thinking.*

Student: *My partner and I did different things. I tried to take the last problem, with 4 packs and multiply that by 5, but that was taking longer. My partner just took the 2 packs and just multiplied by 10—340 sticks of gum. Cool.*

Teacher: *Nice. We know something about times 10, yeah? That can be handy.*

# of Packs	# of Sticks
1	17
2	34
4	68
20	340

$\times 10$ (4 → 20), $\times 10$ (68 → 340)

Teacher: *Anyone try anything else? Was using the original pack of gum—the unit rate—friendly here?*

Student: *Not really. I guess unless you think of 17 times 20 as 17 times 2 times 10. Then it's a bit easier.*

Teacher: *So you can think of 20 as 2 times 10? So, 17 times 20 is equivalent to 17 times 2 times 10. There's a big idea here. Why we can do this?*

Student: *You kind of doubled the 17 and halved the 20. I remember doubling and halving.*

Teacher: *Yes, and that's called the associative property. Does anyone notice anything connected between the first strategy up here, where you multiplied the 2 packs times 10 and the last strategy, where we've been talking about the 17 times 20?*

Student: *That's cool. They use the same 2 times 10 is 20 relationship. If you already have two 17s, then you can scale that up by 10. If you are thinking about 17 times 20, you can think about 17 times 2 and then times 10. Same thing!*

Student: *And in this case, it's also doubling and halving.*

# of Packs	# of Sticks
1	17
2	34
4	68
20	340

$\times 20$ (1 → 20), $\times 10$ (2 → 20), $\times 10$ (34 → 340), $\times 20$ (17 → 340)

$17(20) = (17)(2 \cdot 10) = (17 \cdot 2)10 = 34(10) = 340$

$17(20) = 34(10)$

associative property

doubling/halving

(continued)

Teacher: *Maybe that could help us again sometime. Next question. How many sticks in 5 packs?*

Brief think time.

Teacher: *How many sticks in 5 packs?*

Student: *85.*

Teacher: *Did anyone use the 4 packs to find 5 packs?*

Student: *Yes, I added one pack to 4 packs, so 17 plus 68 is 85.*

Teacher: *Did anyone use the 20 packs to find 5 packs?*

Student: *You can divide 20 packs by 4 to get 5 packs, so 340 divided by 4 is like 320 divided by 4, which is 80 and the leftover 20 divided by 4 is 5, so 85.*

Teacher: *What do we think? Give me a signal if you follow this thinking.*

# of Packs	# of Sticks
1	17
2	34
4	68
20	340
5	85

$1+4$ $17+68$ $\div 4$ $\div 4$

$$5(17)=(1+4)(17)=17+68=85$$

$$\frac{340}{4}=\frac{320}{4}+\frac{20}{4}=80+5=85$$

Teacher: *Did anyone use 10 packs? Yeah, I thought I heard you talk about that. Tell us about that.*

Student: *I know how to find 10 easy, that's just 170. Then half of that is 85.*

Teacher: *So, you added a row to the ratio table? At least in your mind? That seems handy too!*

# of Packs	# of Sticks
1	17
2	34
4	68
20	340
5	85
10	170

$1+4$ $17+68$ $\div 4$ $\div 4$ $\div 2$ $\div 2$

$$5(17)=\frac{10(17)}{2}$$

Teacher: *Great, the next problem is 15 packs. Look for a clever way to find the number of sticks.*

Think time.

Teacher: *Turn and tell your partner and decide who had the more clever way to do it.*

The teacher asks for and models adding the 10 and 5 packs and multiplying the 5 packs times 3.

# of Packs	# of Sticks
1	17
2	34
4	68
20	340
5	85
10	170
15	255

$\times 3$ $5+10$ $85+170$ $\times 3$

$$15(17)=(10+5)17=170+85=255$$

$$15(17)=(3\cdot5)17=3(5\cdot17)=3(85)=255$$

# of Packs	# of Sticks
1	17
2	34
4	68
20	340
5	85
10	170
15	255
9	153

4 + 5   68 + 85

Teacher: Alright, let's end this gum scenario with a twist. What if we have 153 sticks. How many packs? Not 153 packs, mind you. How could you find the number of packs that hold 153 sticks?

Think time.

Teacher: How many packs have 153 sticks?

Student: 9 packs.

Teacher: Did anyone use the 4 packs to help?

Student: I did. I looked for numbers that could get me to 153 and I saw that the 68 and 85 would add to 153, so 4 packs and 5 packs is 9 packs.

$$\frac{153}{17} = \frac{85}{17} + \frac{68}{17} = 5 + 4 = 9$$

Teacher: Did anyone use the 10 packs?

Student: Yes, the 153 is close to the 170. And so I figured out it's just 17 under, so one pack under 10 is 9 packs.

Teacher: That sounds like a really important question that you both just asked. Once you looked at the 68 sticks, you had to find out how much more you needed. Once you looked at the 170 sticks, you had to find out how much too much you had. Finding the difference between what you know and what you're looking for seems like a really important idea. How could we make a note of that?

Student: Asking, "How far away are we?"

Teacher: Which of these strategies do you like? Is there one that might be easier if we didn't already have some of the packs?

Student: The four and five were pretty easy, but only because we had already figured them out. I know what ten 17s is, 170; I could've used it even if I didn't have anything else.

Teacher: That also sounds like an important idea. Someone who is making sense of that strategy restate it for us, so we can hear it in a new way. Go ahead.

Student: We can look for things to use in the table and that's fine. When we don't have stuff already, we can think about what we know.

# of Packs	# of Sticks
1	17
2	34
4	68
20	340
5	85
10	170
15	255
9	153

10 − 1   4 + 5   68 + 85   170 − 17

$$\frac{153}{17} = \frac{170}{17} - \frac{17}{17} = 10 - 1 = 9$$

"How far away are we?"

"Use what you know."

(continued)

Teacher: *Okay, last question. Earlier I referred to this table of packs and sticks as a ratio table because all of the ratios in the table are equivalent. In each of these rows lives the same relationship of one pack to 17 sticks. We could write these relationships as a statement of equivalence.*

$$\frac{1 \text{ pack}}{17 \text{ sticks}} = \frac{2 \text{ packs}}{34 \text{ sticks}} = \frac{4 \text{ packs}}{68 \text{ sticks}} = \frac{20 \text{ packs}}{340 \text{ sticks}} = \frac{5 \text{ packs}}{85 \text{ sticks}} = \frac{15 \text{ packs}}{255 \text{ sticks}} = \frac{9 \text{ packs}}{153 \text{ sticks}}$$

Teacher: *What kinds of things did we just do with a ratio table? Turn and talk with your partner.*

The teacher leads a brief conversation about the multiplication problems and division problems that the students just solved using the relationships in the table, bringing out the points that what you do to the packs you also do to the sticks, that the table did not need to go in order, and you could insert values that you find helpful.

Teacher: *How would you summarize some of the things that came up in this string today?*

Elicit the following:

- *Every combination on the table makes an equivalent ratio (1/17 = 2/34 = 4/68 = 20/340 = 5/85 = 15/255) and could be written as one statement of equivalence.*

- *The 17 sticks in each pack show a relationship between every value on the left (packs) and every corresponding value on the right (sticks)—multiplying or dividing by 17.*

- *Even though this ratio table had the unit rate given, 1 pack to 17 sticks, that is not necessary for the table to be a ratio table.*

- *We can use ratio tables to multiply and divide.*

- *Useful ideas: Use what you know and how far away are we?*

Sample Final Display

Your display could look like this at the end of the problem string:

# of Packs	# of Sticks
1	17
2	34
4	68
20	340
5	85
10	170
15	255
9	153

Important: "How far away are we?"
"Use what you know."

$2(17) = 2(10 + 7) = 20 + 14 = 34$ distributive property

$2(34) = 2(30 + 4) = 60 + 8 = 68$

$17(20) = (17)(2 \cdot 10) = (17 \cdot 2)10 = 34(10) = 340$

$17(20) = 34(10)$ associative property
doubling/halving

$5(17) = (1 + 4)(17) = 17 + 68 = 85$

$5(17) = \frac{10(17)}{2}$

$\frac{340}{4} = \frac{320}{4} + \frac{20}{4} = 80 + 5 = 85$

$15(17) = (10 + 5)17 = 170 + 85 = 255$

$15(17) = (3 \cdot 5)17 = 3(5 \cdot 17) = 3(85) = 255$

$\frac{153}{17} = \frac{85}{17} + \frac{68}{17} = 5 + 4 = 9$ $\frac{153}{17} = \frac{170}{17} - \frac{17}{17} = 10 - 1 = 9$

Algebra Problem Strings
©2017 Kendall Hunt Publishing

Facilitation Notes

This version of the problem string lists short notes for important teacher moves during the string. After you've done the string yourself and studied the relationships involved, you might make similar notes for the things you want a reminder of or deem important.

1 pack: 17 sticks	As you intro the context, draw a ratio table with the 1 and 17. Super quick.
2 packs	As you ask, "How many sticks in 2 packs?", write the 2 in the ratio table. Ah, so if you double the number of packs, you double the number of sticks? Write equation and write "distributive property."
4 packs	Write in 4 as you ask. You doubled the number of packs to get double the sticks again. Did anyone think about 4 times 17? Which strategy do you like? Which is easier if we didn't have anything already?
20 packs	How can you use what you know? Think, then turn and talk. Anyone use 2 packs? Scale up. Anyone use the unit rate, one pack? Bring out the associative property. Connect to doubling and halving.
5 packs	Anyone use the 20? Anyone find 10 packs to help? How? That seems handy.
15 packs	Seek for a clever strategy. Think, then turn and talk. Elicit 5 + 10 and 3 x 5 packs.
153 sticks	Mixing things up—now you've got 153 sticks. How many packs? Elicit 4 + 5 and 10 - 1 packs. Important question: "How far away are we?" This is a ratio table. What kinds of things did we just do with the ratio table? Order doesn't matter. What you do to the packs, you do to the sticks. You can find and use values that are helpful.

Proportional Reasoning with Ratio Tables 2

At a Glance

Time in Car (hr)	Distance (mi)
0.75	36
1.5	
3	
4.5	
1	
	108

Objectives

The goal of this string is to continue to help students construct the ratio table as a generalizable model for reasoning about equivalent ratios. This scenario begins with a non-unit rate.

Placement

This string is the second of four ratio table problem strings. Use this problem string to remind students about, or to build, proportional reasoning and relationships.

You can use this problem string to introduce textbook Lesson 2.1 Proportions. If your students have no experience using a ratio table to solve problems, you may want to use the problem string Lesson 2.0 Proportional Reasoning with Ratio Tables 1 first.

Guiding the Problem String

This string is designed to be more challenging than the previous string, Proportional Reasoning with Ratio Tables 1, in a few ways. The initial rate is a non-unit rate of 0.75 to 36, the placement of the unit rate at the end (not the beginning) of the problem string, and the use of rational number quantities all make the string more difficult.

Plan how you will record students' strategies in advance so that the table does not become too cluttered with students' thinking. Keep the ratio table useful to students as a thinking tool. You might record students' strategies by changing color for each new strategy and erasing before each new problem.

Throughout the string, be clear that the car is traveling at a constant rate and not stopping. The first two problems are simple doubles. Don't waste time with other strategies, but do have students suggest how they doubled the corresponding miles as appropriate. Spend a little longer on 4.5 hours. Spend the most time on finding the speed, which is the unit rate. Elicit and record many strategies, asking students to restate those that are new, especially multiplying 0.75 by $\frac{4}{3}$ and 1.5 by $\frac{2}{3}$. Use unit fractions to help. Make sure on the last problem that students are clear that the scenario has flipped and they are looking for the time it took to go 108 miles.

Dive into the context and ask questions as they arise, such as: Is this a ratio table? Where's the rate? How could we find the speed? Is this the same as the rate?

About the Mathematics

Avoid writing 45 = 0.75 without the labels, 45 minutes = 0.75 hours.

Later when students study direct variation (Lesson 2.4), you can return to this problem string or the previous one and ask, "What would happen if we thought of these as coordinates and graphed this situation? What would the graph look like?" This is an important connection between proportional reasoning contexts and linear functions. These strings provide a context for exploring the meaning of slope, how the rate shows up in the graph, and other key ideas.

A ratio table is a special paired number table where all of the pairs are equivalent ratios. This scenario starts with a non-unit rate.

Algebra Problem Strings
©2017 Kendall Hunt Publishing

Sample Interactions

Use the following as you plan how to elicit and model student strategies. This is not meant as a script, but as a view into the relationships involved and the intent of the problem string.

Teacher: *Today I'd like us to imagine a long road trip with a whole bunch of family members, pretending that you are in the back seat. Someone in your family is driving to your destination. Anyone been on a trip like this before—a really long road trip in the car? The car is all fueled up and off you go. And go. And go. And now it feels like this trip is taking forever. You finally ask how long your family has been in the car and they tell you.*	
Teacher: *I'm going to write what they say in three ways. Turn and tell your partner what these might be suggesting about the trip. Are these related or totally different? What's going on here?*	"Three quarters" "45" "36 miles"

The teacher facilitates a brief class conversation that 45 minutes is the same time in three-quarter hours and they must have driven 36 miles.

As students are convinced, the teacher records the scenario in a ratio table, using 0.75 hours.

Let students know that you are going to keep adding to the story, but will use the unit of hours since the trip lasts so long.

45 min = 0.75 hrs

Time in Car (hr)	Distance (mi)
0.75 hrs	36

Teacher: *After an hour and a half, you are tempted to ask again, "Are we there yet?" Instead, you wonder how far you've gone. If you have been driving at a constant rate, how far have you gone?*

Did anyone double?

Time in Car (hr)	Distance (mi)
0.75	36
1.5	

Time in Car (hr)	Distance (mi)
0.75	36
1.5	64

×2 ( ) ×2

Teacher: *And what about after 3 hours in the car, driving at the same speed, with no traffic. How many miles now and how do you know?*

Did anyone double?

Time in Car (hr)	Distance (mi)
0.75	36
1.5	72
3.0	

Time in Car (hr)	Distance (mi)
0.75	36
1.5	72
3.0	144

×2 ( ) ×2

(continued)

Teacher: *Good news. You finally reach your destination at four and a half hours. Let's think for a few seconds about how many miles this whole trip was. Now, turn and tell your partner how you are thinking about the miles now.*

Did anyone double again? Ahhh, not the problem. Try again.

Did anyone use the 1.5 hours and 3 hours?

Did anyone use the 1.5 hours another way? Times 3?

Time in Car (hr)	Distance (mi)
0.75	36
1.5	72
3.0	144
4.5	

Time in Car (hr)	Distance (mi)
0.75	36
1.5	72
3.0	144
4.5	216

$1.5 + 3$ ×3 ( ) ×3 $72 + 144$

Teacher: *This makes me wonder how fast we were going on this trip. I wonder how we could find our speed? Let's think for a few seconds about that. Okay, turn and tell your partner about how we could use this table to find our speed.*

The teacher elicits from the class that they are trying to find the distance traveled in one hour and then writes "1" in the hour column.

Time in Car (hr)	Distance (mi)
0.75	36
1.5	72
3.0	144
4.5	216
1	

Teacher: *Did anyone use the 3 hours? How?*

Did anyone find some other time that is helpful, like 0.25 or 0.5 hours? How?

Use the 0.75 or 1.5 hours? Could you? How?

Time in Car (hr)	Distance (mi)
.75	36
1.5	72
3.0	144
4.5	216
1	48
0.25	12

÷3 ÷3 ÷3 ×4 ×4

Time in Car (hr)	Distance (mi)
.75	36
1.5	72
3.0	144
4.5	216
1	48
0.25	12
0.5	24

÷3 ÷3 ×2 ×2

Time in Car (hr)	Distance (mi)
.75	36
1.5	72
3.0	144
4.5	216
1	48
0.25	12
0.5	24

$\times \frac{4}{3}$ $\times \frac{2}{3}$ $\times \frac{2}{3}$ $\times \frac{4}{3}$

$$\frac{4}{3}(36) = 4\frac{1}{3} \cdot 36 = 4(12) = 48$$

$$\frac{2}{3}(72) = 2\frac{1}{3} \cdot 72 = 2(24) = 48$$

This is an important moment in the problem string—the explicit linking of speed to unit rate in this context as well as the realization that one doesn't need the unit rate in order to reason about the previous ratios.

	Time in Car (hr)	Distance (mi)
Teacher: *How long had we been driving when we reached the 108 mile mark? Go!*	.75	36
Did anyone find the 2 hour mark and use that? How?	1.5	72
Did anyone use the 144 miles at 3 hours? How?	3.0	144
Did anyone use the 12 miles at 0.25 hours? How?	4.5	216
	1	48
	0.25	12
	0.5	24
		108

Teacher: *Let's look back over our work and connect it to proportions. We've been using a ratio table, where all of the ratios are equivalent. Brainstorm with me—what are some of the proportions we just solved, using thinking and reasoning.*

$$\frac{0.75}{36} = \frac{1.5}{x}$$

$$\frac{0.75}{36} = \frac{x}{108}$$

$$\frac{0.75}{36} = \frac{1}{x}$$

Teacher: *How would you summarize some of the things that came up in today's string?*

Elicit the following:

- *We didn't need the unit rate in order to reason about the ratios.*

- *We were able to tell the story of the trip and use existing relationships to find the elapsed time and the distance traveled, depending on what we were given.*

- *We could find the unit rate using what we know.*

- *We can solve proportions using ratio tables and intermediate values we can find.*

- *The unit rate with miles and hours is the speed in miles per hour.*

(continued)

Sample Final Display

Your display could look like this at the end of the problem string:

Time in Car (hr)	Distance (mi)
0.75	36
1.5	72
3.0	144
4.5	216
1	48
0.25	12
0.5	24
2.25	108
2	96

$3 - 0.75$ [bracketing 0.75, 1.5, 3.0 rows] $144 - 36$

$\times 2$ $\times 9$ (0.25, 0.5) $\times 9$ $\times 2$

$$\frac{4}{3}(36) = 4\left(\frac{1}{3} \cdot 36\right) = 4(12) = 48$$

$$\frac{2}{3}(72) = 2\left(\frac{1}{3} \cdot 72\right) = 2(24) = 48$$

$$\frac{0.75}{36} = \frac{1.5}{x}$$

$$\frac{0.75}{36} = \frac{1}{x}$$

$$\frac{0.75}{36} = \frac{x}{108}$$

Facilitation Notes

This version of the problem string lists short notes for important teacher moves during the string. After you've done the string yourself and studied the relationships involved, you might make similar notes for the things you want a reminder of or deem important.

Time in Car (hr)	Distance (mi)	
0.75		Three quarters, 45, 36 miles. What? Turn and talk. Have students bring the importance of labeling the units.
1.5		Elicit doubling. Did anyone double?
3.0		Repeat.
4.5		Check to see if some students doubled. Nudge back on track. Did anyone use 1.5 and 3? The 1.5 another way? Times 3?
1		How fast were we going? How do you know? Did anyone use 3? Find something new like 0.25? 0.5? Use 0.75 or 1.5? How? Model fraction multiplication using unit fractions.
	108	How long were we driving if we covered 108 miles? Write a few problems as proportions to link representations.

2.2 Reasoning About Percents

At a Glance

Hours left	% of battery
21	70%
	35%
	10%
	45%
	5%
	56%
	100%

Objectives

The goal of this string is to help students think flexibly about percents without requiring a decimal equivalent or other tedious calculations.

Placement

Use this problem string to help students reason about percents by using benchmarks and friendly increments before they solve proportions involving percents.

You can use this problem string to introduce the work in textbook Lesson 2.2 Proportions with Percents.

Guiding the Problem String

The opening problem in this string is the scenario of knowing that 21 hours left on a battery charge is 70% of that battery charge. The first problem is just half, but may be a surprise if students are not used to noticing and using relationships. The second problem is the all important 10%. You can deliver the next two problems, 45% and 5% together and note students finding them in different orders. Linger on the next problem, 56%, and elicit clever combinations and scaling from students. The last problem is to find the number of hours for a full battery charge. Students may have already figured the 100% to use earlier, but bring it to the class's attention here.

The instinct to pause and notice how 21 and 70 are related is an important one that will support students later when they begin to factor algebraic expressions and equations. That the values are related multiplicatively is a first idea; how to use this relationship is the second and more vital idea that this problem string invites students to consider. The goal throughout is to look for, and make use of, important relationships to avoid over-reliance on algorithmic procedures or calculators.

About the Mathematics

The percent bar model is designed to mimic the status icon of a battery on a digital device. We predict that this image will be familiar to any student who owns or uses a digital device—that is, it is believable or imaginable. It is related to the ratio table. Unlike a ratio table where values can increment to any size, the percent bar is limited to show values in order of magnitude between 0% and 100%.

Three types of questions we can ask students about percent are:

*   What is 50% of 250?

*   125 is 50% of what number?

*   125 is what percent of 250?

In Problem String 1.5 students reason about the first type—finding the part of a whole and treating the percent as an operator. In this string, we turn our attention to the more challenging task of thinking about a missing whole. This encourages students to think about relationships.

(continued)

Sample Interactions

Use the following as you plan how to elicit and model student strategies. This is not meant as a script, but as a view into the relationships involved and the intent of the problem string.

Teacher: *Probably like me, you own a digital device like a cell phone, or tablet, or laptop, yes? Of course. And like me, do you ever try to figure out how long you have until your battery dies? Yes, you too?*

So that's what I want us to think about today—how to make these decisions about whether we can finish that movie or make a few more phone calls before the battery dies.

I'm going to put up an image like from a laptop computer that I saw and we'll reason together about what's going on. Okay, first just take a look at this image. Now, tell your partner what is going on here. Maybe, what else you know or could figure out.

21 hours remaining
70%

Teacher: *Okay, so it sounds like this battery, right now, is 70% charged and we have 21 hours left before we need that charger. Is that your thinking? I'll record that on a ratio table.*

Where do the 21 and 70% go on the battery?

21 hours remaining
70%

21 hours
70%

Hours left	% of battery
21	70%

Teacher: *The first problem is to find the number of hours left on the battery if there is a 35% charge.*

Hours left	% of battery
21	70%
	35%

Teacher: *Tell us what you think.*

Student: *The 35% is friendly with 70%. The 35% is just half of the 70%, so half of 21 hours is 10.5 hours.*

The teacher models this in the ratio table and on the percent bar.

Hours left	% of battery
21	70%
10.5	35%

+2
10.5   21 hours

35%   70%

÷2

Teacher: *Alright, the next problem is how many hours are left for the battery if there is only a 10% charge left?*

Student: *That's just divided by 7.*

Teacher: *Who understands why she is saying just divided by 7. What does she mean?*

Student: *I think she's saying that if we divide the 21 and the 70% both by 7 then it's not too bad, just three hours.*

Teacher: *Okay, I'm going to make a note of this on our percent bar model. Ten percent, that seems like a really important landmark to have. Turn and tell your partner what else would be friendly to find now?*

Students will likely say factors and multiples of 10% will be friendly now, such as 5% and 1% or any multiple of 10% up to 100%. Try to encourage these to come up, without saying them yourself.

Hours left	% of battery
21	70%
10.5	35%
3	10%

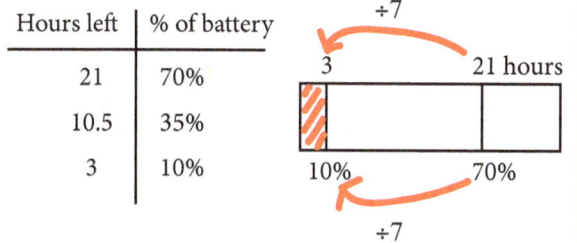

Teacher: *So now we know about this battery at 70%, 35%, and 10%. The next two problems are how many hours of charge are left at 45% and 5%? Do you have a way to find how many hours are left with a 45% charge and a 5% charge. Let's think for a moment about these two.*

The teacher elicits strategies such as using 35% and 10% to find 45% of the battery left, cutting 10% of the battery hours in half to find 5% of the remaining battery hours, and some finding the 5% first, then quadrupling the 10% and adding the 5%, or using the 5% to find 50% and subtracting the 5% (not shown).

Hours left	% of battery
21	70%
10.5	35%
3	10%
13.5	45%
1.5	5%

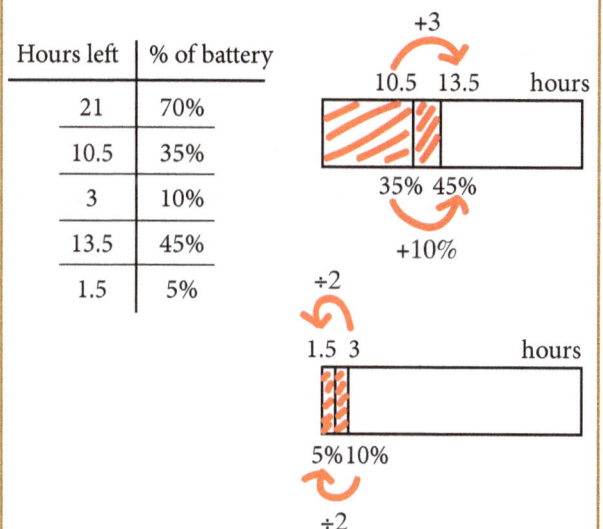

Teacher: *What about 56% battery life? Do we have things up here that might help?*

The teacher elicits strategies such as finding the 50% by scaling up from the 10% or the 5% to 50%, finding 1% and then adding the 5% and 1% or scaling up from the 1% to 6% or adding 45% to 10% and 1% (not shown).

Hours left	% of battery
21	70%
10.5	35%
3	10%
13.5	45%
1.5	5%
15	50%
16.8	56%

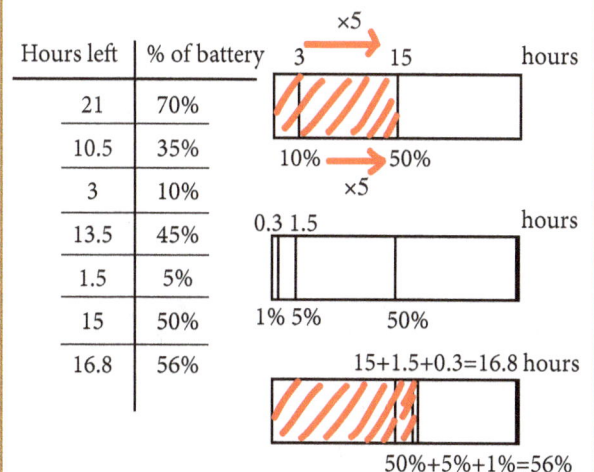

(continued)

Teacher: *And finally, what we all really want to know—how long does this battery last when fully charged? In how many ways can you find that? And also, how realistic is this? Turn and quickly discuss with your partner.*

The teacher elicits strategies such as doubling from the 50% (shown), but also scaling up from the 10%, the 5%, even doubling the 45% and adding 10%.

Hours left	% of battery
21	70%
10.5	35%
3	10%
13.5	45%
1.5	5%
15	50%
16.8	56%
30	100

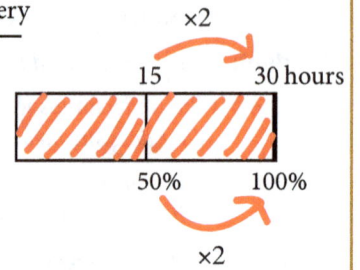

×2

15   30 hours

50%   100%

×2

Teacher: *How would you summarize some of the things that came up in this string today?*

Elicit the following:

- *Benchmark percents, like 10%, 5%, 50% are helpful.*

- *What you do to the percent, you do to the number of hours left. If you halve the percent, you halve the corresponding number of hours left.*

- *You can use what you know to figure other friendly percents.*

- *You can do your thinking with a ratio table or percent bar.*

Algebra Problem Strings
©2017 Kendall Hunt Publishing

Sample Final Display

Your display could look like this at the end of the problem string:

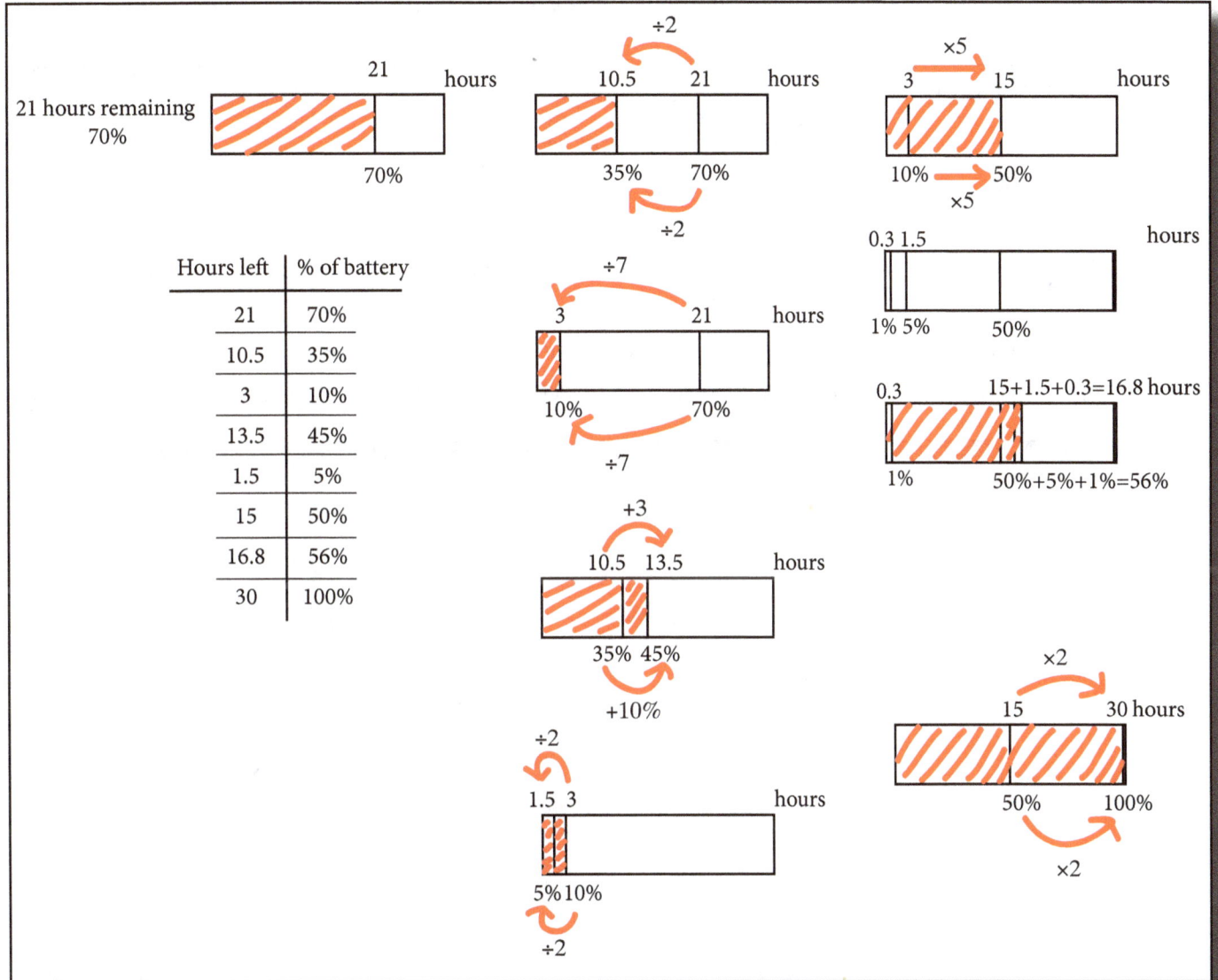

Hours left	% of battery
21	70%
10.5	35%
3	10%
13.5	45%
1.5	5%
15	50%
16.8	56%
30	100%

Facilitation Notes

This version of the problem string lists short notes for important teacher moves during the string. After you've done the string yourself and studied the relationships involved, you might make similar notes for the things you want a reminder of or deem important.

Hours left	% of battery	
21	70%	Invite kids into the story then imagining a laptop with 21 hours or 70% remaining. Put numbers on battery percent bar. How many hours are left if 35% battery left?
10.5	35%	Model on percent bar and ratio table.
3	10%	Repeat.
13.5	45%	Ask 45% and 5% together. Elicit smart combining of existing percent/hour ratios. 35 + 10%, 10% scaled to 40% + 5% or 50% – 5%.
1.5	5%	
16.8	56%	56% not a typical benchmark. What are some nice combinations?
30	100%	Of course, we probably want to know how long this battery lasts when fully charged.

Algebra Problem Strings
©2017 Kendall Hunt Publishing

2.3 | Conversions

At a Glance

Number of Bleeps	Number of Meeps
32	24
48	
	18
1	
	1

Create a ruler.

Objectives
The goal of this string is continue to help students construct the ratio table as a generalizable model for reasoning about equivalent ratios and encourage students to make sense of the relationships between two unknown units that vary proportionally, without the social knowledge of the actual units.

Placement
Use this string to help students make sense of using proportional reasoning in ratio tables with unit conversions and unit rates.

You can use this problem string to support textbook Lesson 2.3 Conversion Factors and Unit Rates.

Guiding the Problem String
This string purposely moves away from well-known units of measure, such as centimeters/meters or inches/feet, so that students are not reliant on the conversion factor (e.g., 12 inches in a foot). This allows them to construct new combinations of bleeps and meeps that are equivalent, that will help as they eventually construct the conversion factor (e.g., How many bleeps are equivalent to one meep? How many meeps are equivalent to one bleep?).

The construct of a mystery planet allows us to acknowledge that no one in the room has social knowledge about these units and how they are related. This forces students to use the values provided to reason about the relationship between the units and how they compare.

About the Mathematics
Students make errors when converting units that tend to fall into predictable patterns. Be on the look out for the erroneous reasoning known as the reversal error (see *Learning to Support Young Mathematicians at Work: An Early Algebra Resource for Professional Development* by Kara Louise Imm, et al, Heinemann Publishing, 2012), where students reason that the unit in the rate with the greater quantity is the larger unit, when in fact because it takes more of that unit to be equivalent to the other unit, it is the smaller unit. For example, some students incorrectly reason, "I know there are 12 inches in a foot, so I multiply the 48 inches by 12 to get the feet."

We can also see this confusion when students describe the relationship between quarters and dollars. They say "four quarters equals one dollar," and write 4Q = D. While this equation is true if Q and D stand for the money amounts, 4(0.25) = 1.00, it is not true that 4(number of quarters) = number of dollars. Parsing out these differences are important. The correct equation to describe the relationship between the number of quarters and the number of dollars is just the opposite of what many students initially think: 4D = Q. We can support students to trust and own this equation, when they construct a ratio table to capture the relationship and become clear about what each variable represents.

In dimensional analysis, you multiply by a rate. In a ratio table, you scale up and down to create equivalent ratios.

(continued)

Sample Interactions

Use the following as you plan how to elicit and model student strategies. This is not meant as a script, but as a view into the relationships involved and the intent of the problem string.

	Number of Bleeps	Number of Meeps
	32	24

Teacher: *Today we are heading to a mystery planet. And I'm saying mystery planet simply because I have never been there, I know you have never been there and I know very little about this planet, except for one thing.*

What I know about this planet is how they measure height or distance. Again, I don't know too much, but I do have this little chart that shows two of their common units and how they are related. Do you want to see the chart? Yes? Okay.

Take a look at this chart with your partner and just try to figure out which is the longer unit—a bleep or a meep?

Students turn and talk while the teacher listens in.

Teacher: *What did you decide?*

Student: *So I think the bleeps are the longer unit because they have 32 and that's more than 24.*

Teacher: *What do other people think of this?*

Student: *I actually disagree. I think, well, it's like with inches and feet. You would need 12 inches to make 1 foot, but the inches are smaller. Does anyone get what I'm saying?*

Student: *Yeah, I think you are saying if you need more of one unit to make the other, that unit must be smaller.*

Teacher: *Sounds like we don't quite agree. Quick poll—raise your hand if you think the bleeps are the longer unit. And now raise your hand if you think meeps are the longer unit. And how about those who are not sure yet? Okay.*

Student: *It would help if we knew what they looked like. We have no idea how big or small they are.*

Teacher: *That's true. All we have is this little table. Keep thinking while I ask some more questions. What if I told you I had 48 bleeps, how many meeps would you need to be equivalent?*

Think time.

	Number of Bleeps	Number of Meeps
	32	24
	48	

Teacher: *How many meeps? And how?*

Student: *36. To figure this out, I needed to find out 16 bleeps because that's how many you added. So I actually cut both the bleeps and the meeps in half and that's 16 bleeps is the same as 12 meeps. Then I used that one with the original one and added them together.*

Teacher: *Okay, I'm trying to capture your strategy here. Let me move the 48 down in the table, so then we can put the 16 here and scale it up. It could go either place, but for this string, I'll put it here. The ratio table doesn't have to go in order, right?*

Number of Bleeps	Number of Meeps
32	24
16	12
48	36

$32 + 16 \left[\div 2 \left(\right) \div 2 \right] 24 + 12$

Teacher: *Are there other ways to think about the meeps? Did anyone think of scaling up or down?*

Student: *Yeah, I thought about scale factor, kind of. Like what number times 32 would give me 48, and I knew it was 1.5. So I scaled up both the bleeps and the meeps by a factor of 1.5.*

Student: *Oh, right. One and half 32s is 48. So, one and half 24s is 24 and 12, 36. Yep.*

Teacher: *So, now we have a couple of nice strategies. Do a quick turn and talk just telling your partner how they are related and how they are different. Go!*

Number of Bleeps	Number of Meeps
32	24
16	12
48	36

$\times 1.5$ ... $\times 1.5$

Teacher: *So now we know 32 bleeps, 48 bleeps and also 16 bleeps in terms of meeps. How would you find 18 meeps, in terms of bleeps? Would previous strategies help us here?*

Some students may find 6 meeps: 8 bleeps and then add that combination to 12 meeps: 16 bleeps. Others might use the idea of a scale factor from 16:12. Here, again, scaling up both quantities by 1.5 helps us to find the equivalent for 18 meeps. Many will simply divide 36 by two and do the same with 48 bleeps.

Number of Bleeps	Number of Meeps
32	24
16	12
48	36
	18

Number of Bleeps	Number of Meeps
32	24
16	12
48	36
8	6
24	18

$16+8 \left[\div 2 \left(\right) \div 2 \right] 12+6$

Number of Bleeps	Number of Meeps
32	24
16	12
48	36
24	18

$\times 1.5 \left(\div 2 \left(\right) \div 2 \right) \times 1.5$

(continued)

Teacher: *Earlier I asked you to think about which unit—bleeps or meeps—was longer and we didn't quite agree. But some of you have some strong feelings about this. With your partner now, I'd like you to construct a ruler for this mystery planet which shows how the two units are related.*

Student: *Wait, how long do we make the ruler? How many units?*

Teacher: *I think that's for you and your partner to decide. As long as we can tell which unit is longer and how they are related, you can make it as short or as long as you need to. You are just making a sketch. Take a minute or two to do this.*

The teacher moves around the room, observing and recording what students are doing. Look for students who scale down the ratio to 4 bleeps: 3 meeps.

Teacher: *Tell us about your ruler.*

Student: *Well, we know that there are 4 bleeps for every 3 meeps.*

Teacher: *How?*

Student: *We scaled down in the table. I went from the 32:24 and my partner from the 16:12. So we drew 4 bleeps and then tried to eyeball that 3 meeps had to fit in the same space.*

Student: *Ours looked like that, but we drew it vertically.*

Teacher: *So, which unit is longer? Is anyone who revised their thinking willing to share what helped them change their mind?*

Student: *When I had to fit 3 meeps in the same space as 4 bleeps, I could see that the meeps had to be longer.*

Teacher: *It's tricky right. The bleeps are shorter so you need more of them to get an equivalent length of meeps.*

Number of Bleeps	Number of Meeps
32	24
16	12
48	36
24	18
12	9
4	3

$\div 8$ $\div 4$ $\div 2$ $\div 2$ $\div 4$ $\div 8$

Bleeps

Meeps

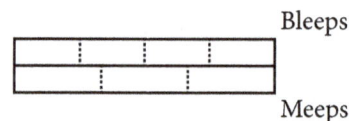

Teacher: *If we had only 1 bleep, how many meeps would we need to be the same length? A whole meep? Less than a meep? How much of a meep?*

Student: *It's got to be less than one meep. And if you divide the 4 bleeps by 4 to get one bleep, you get three fourths.*

Student: *And that looks about right, three fourths of a meeps is as long as a bleep.*

Teacher: *What if we had 1 meep? How many bleeps is that? More than one bleep? Less? How do you know?*

Student: *Well, you can look at our ruler but you can also divide the 3 meeps by 3 so the four bleeps by 3 too. That's four-thirds.*

Teacher: *Turn and talk about these unit relationships. What are you thinking?*

Number of Bleeps	Number of Meeps
32	24
16	12
48	36
24	18
12	9
4	3
1	¾
4⁄3	1

$\div 3$ $\div 4$ $\div 4$ $\div 3$

Algebra Problem Strings
©2017 Kendall Hunt Publishing

Teacher: *How would you summarize some of the things that came up in this string today?*

Elicit the following:

- *We can reason about non-unit rates in a ratio table, by scaling up and down and by combining things we know.*

- *We can convert measurements in a ratio table.*

- *You need to know if you're talking about the size of a unit or the number of units when you talk about equivalencies.*

- *Scale factors like 1.5 can be really handy.*

- *It can take a lot of a small unit to equal the length of a bigger unit.*

Sample Final Display

Your display could look like this at the end of the problem string:

Number of Bleeps	Number of Meeps
32	24
16	12
48	36
24	18
12	9
4	3
1	¾
⁴⁄₃	1

$\div 3$ $\div 4$ $\div 4$ $\div 3$

Bleeps

Meeps

Facilitation Notes

This version of the problem string lists short notes for important teacher moves during the string. After you've done the string yourself and studied the relationships involved, you might make similar notes for the things you want a reminder of or deem important.

Bleeps	Meeps	
32	24	Establish context, show ratio table. Which is longer? Who could convince us?
48		What if we had 48 bleeps, how many meeps? Elicit scaling by 1.5 and others.
	18	What if we had 18 meeps, how many bleeps? Elicit scaling by 1.5 and others.
Create a ruler.		What would a meep/bleep ruler look like? Why? Which is longer? Which do you need more of?
1		How many meeps to make the length of one bleep?
	1	How many bleeps to make the length of one meep?

Using Ratio Tables versus Dimensional Analysis

At a Glance

Convert:

40 feet to inches

52 centimeters to meters

60 inches to centimeters

70 mph to miles per minute

5 meters per second to miles per hour

Objectives

The goal of this string is to help students parse out the difference between converting units by scaling up or down using a ratio table versus using dimensional analysis. Both are fine strategies with different advantages.

Placement

Use this string to help students know when to use scaling up or down on a ratio table and when to use dimensional analysis to convert units. This string should come after students have been introduced to both strategies.

You can use this problem string after textbook Lesson 2.3 Conversion Factors and Unit Rates and before textbook Lesson 2.4 Direct Variation.

Guiding the Problem String

This string has four unit conversion problems that can be solved using scaling on a ratio table or dimensional analysis. For the first three problems, elicit both scaling on a ratio table and dimensional analysis strategies. Ask students to compare strategies. The fourth and fifth problems are converting a rate to another rate and require several conversions. Elicit the dimensional analysis strategy and discuss efficiency.

About the Mathematics

In dimensional analysis, you are multiplying by a rate to find the new unit. For example, to convert 40 feet to inches, you start with 40 feet and multiply by the rate of 12 inches: 1 foot. At the end of dimensional analysis, you have the answer as the new unit, in this case 480 inches.

$$40 \text{ feet} \times \frac{12 \text{ inches}}{1 \text{ foot}} = 480 \text{ inches}$$

In a ratio table, you are scaling up and down to create equivalent ratios. For example, to convert 40 feet to inches, you start with what you know about inches and feet, 12 inches: 1 foot, and scale up to 40 feet. At the end of the ratio table, you have a ratio of the new unit to the starting unit.

		× 40 →
feet	1	40
inches	12	480
		← × 40

For conversions that require several steps, like conversion of rates, dimensional analysis is often more efficient.

Sample Interactions

Use the following as you plan how to elicit and model student strategies. This is not meant as a script, but as a view into the relationships involved and the intent of the problem string.

Teacher: *We've just been talking about dimensional analysis. When might you want to use dimensional analysis and when might you want to scale up or down on a ratio table? Let's work on that in today's string.* *The first problem today is to convert 40 feet to inches. First, can you picture 40 feet? Where are the inches? Tell me about the inches?* *Will that be more inches or less than 40?* *Go ahead and find the number of inches in 40 feet.*	Convert 40 feet to inches.
The teacher elicits both a scaling on a ratio table strategy and a dimensional analysis strategy, modeling them both on the board. **Teacher:** *Turn to your partner and discuss the strategy that you used and the other one. How are they alike? How are they different?* The teacher does not yet start a group conversation.	$40 \text{ feet} \times \dfrac{12 \text{ inches}}{1 \text{ foot}} = 480 \text{ inches}$ <table><tr><td></td><td>$\times\,40$</td></tr><tr><td>feet</td><td>1</td><td>40</td></tr><tr><td>inches</td><td>12</td><td>480</td></tr><tr><td></td><td>$\times\,40$</td></tr></table>
Teacher: *Here's today's second problem. Convert 52 centimeters to meters. First, can you picture those 52 centimeters? Where are meters in relationship?* *I wonder which of those strategies you might use here? Go!*	Convert 52 centimeters to meters
The teacher elicits both a scaling on a ratio table strategy and a dimensional analysis strategy, modeling them both on the board. **Teacher:** *Which do you prefer to solve this problem? Why?* The teacher crafts a conversation around strategy.	$52 \text{ cm} \times \dfrac{1 \text{ meter}}{100 \text{ cm}} = \dfrac{52}{100} = 0.52 \text{ meters}$ $\div\,100 \quad \times\,52$ <table><tr><td>cm</td><td>100</td><td>1</td><td>52</td></tr><tr><td>m</td><td>1</td><td>0.01</td><td>0.52</td></tr></table>$\div\,100 \quad \times\,52$
Teacher: *The third problem today is to convert 60 inches to centimeters. The rate is 1 inch to 2.54 cm. Which strategy might be more efficient with this problem?*	Convert 60 inches to centimeters (1 in: 2.54 cm)

(continued)

The teacher elicits both a ratio table strategy and a dimensional analysis strategy, modeling them both on the board.

Teacher: *Which do you prefer to solve this problem? Why?*

How did each strategy use the unit rate of 1 inch to 2.54 cm?

What operation do you end up doing in each strategy?

Where do the answers show up in each strategy?

If you have only been using one of these strategies, do you understand the other? Can you use either? What questions do you have about the one that you don't do as readily?

$$60 \text{ inches} \times \frac{2.54 \text{ cm}}{1 \text{ inch}} = 152.4 \text{ cm}$$

	$\times\ 60$	
in	1	60
cm	2.54	152.4
	$\times\ 60$	

Teacher: *Our next problem is to convert 70 miles per hour to miles per minute. Let's see what you are thinking.*

70 miles per hour to miles per minute

The teacher elicits both a ratio table strategy and a dimensional analysis strategy, modeling them both on the board and crafting a conversation about efficiency and which strategy students prefer and why.

$$\frac{70 \text{ miles}}{1 \text{ hour}} \times \frac{1 \text{ hour}}{60 \text{ minutes}} = \frac{70 \text{ miles}}{60 \text{ minutes}} = \frac{7 \text{ miles}}{6 \text{ minutes}} \approx 1.17 \text{ miles per minute}$$

		$\div\ 6$	$\times\ 7$	
mph	60	10	70	
mpm	1	$\frac{1}{6}$	$\frac{7}{6} = 1\frac{1}{6}$	
		$\div\ 6$	$\times\ 7$	

The teacher repeats for the last problem, converting 5 meters per second to miles per hour, by having students solve, eliciting a dimensional analysis strategy, modeling it on the board, and discussing efficiency.

$$5 \text{ meters/second} \times \frac{1 \text{ mile}}{1609.34 \text{ meters}} \times \frac{3600 \text{ seconds}}{1 \text{ hour}} \approx 11.18 \text{ miles per hour}$$

Teacher: *How would you summarize some of the things that came up in this string today?*

Elicit the following:

- *We can convert measurements in a ratio table, by scaling up and down, and by combining things we know.*

- *We can convert measurements using dimensional analysis by multiplying by helpful rates.*

- *It can be efficient to solve multiple step conversions using dimensional analysis.*

Algebra Problem Strings
©2017 Kendall Hunt Publishing

Sample Final Display

Your display could look like this at the end of the problem string:

40 feet to inches

480 inches

$$40 \text{ feet} \cdot \frac{12 \text{ inches}}{1 \text{ foot}} = 480 \text{ inches}$$

		$\times 40$
feet	1	40
inches	12	480

$\times 40$

52 cm to meters

0.52 meters

$$52 \text{ cm} \cdot \frac{1 \text{ meter}}{100 \text{ cm}} = \frac{52}{100} = 0.52 \text{ meters}$$

		$\div 100$	$\times 52$
cm	100	1	52
m	1	0.01	0.52

$\div 100$ $\times 52$

60 in to centimeters

152.4 cm

$$60 \text{ inches} \cdot \frac{2.54 \text{ cm}}{1 \text{ inch}} = 152.4 \text{ cm}$$

		$\times 60$
in	1	60
cm	2.54	152.4

$\times 60$

70 mph to miles per minute

≈ 1.17 miles per minute

$$\frac{70 \text{ miles}}{1 \text{ hour}} \cdot \frac{1 \text{ hour}}{60 \text{ minutes}} = \frac{70 \text{ miles}}{60 \text{ minutes}} = \frac{7 \text{ miles}}{6 \text{ minutes}} \approx 1.17 \text{ miles per minute}$$

		$\div 6$	$\times 7$
mph	60	10	70
mpm	1	$\frac{1}{6}$	$\frac{7}{6} = 1\frac{1}{6}$

$\div 6$ $\times 7$

9.8 meters per second to miles per hour

≈ 21.92 miles per hour

$$5 \text{ meters/second} \cdot \frac{1 \text{ mile}}{1609.34 \text{ meters}} \cdot \frac{3600 \text{ seconds}}{1 \text{ hour}} \approx 11.18 \text{ miles per hour}$$

(continued)

Facilitation Notes

This version of the problem string lists short notes for important teacher moves during the string. After you've done the string yourself and studied the relationships involved, you might make similar notes for the things you want a reminder of or deem important.

40 feet to inches	Can you picture 40 feet? Where are the inches? Will there be more or fewer inches than 40? Elicit both ratio table and dimensional analysis. Partners compare.
52 cm to meters	Tell me about the centimeters and meters? Can you picture them? Elicit both ratio table and dimensional analysis. Group discussion about strategy preference.
60 in to centimeters	Elicit both ratio table and dimensional analysis. How did each strategy use the 1 inch: 2.54 cm? What operation do you end up doing in each strategy? Where do the answers show up in each strategy? Do you understand both strategies? Can use both?
70 mph to miles per minute	Elicit both. Compare.
9.8 meters/sec to miles/hour	Elicit dimensional analysis. Why not a ratio table? Too many steps, messy numbers. When is it just fine to use either strategy? When dimensional analysis?

Algebra Problem Strings
©2017 Kendall Hunt Publishing

2.5 | Reasoning About Inverse Variation 1

At a Glance

Workers	Hours
20	6
10	
40	
	24
15	
1	
	1

Objectives
The goal of this string is to reason about a situation in which the quantities relate inversely—as one increases the other decreases in a multiplicative way.

Placement
This string is the first of two inverse strings and follows a series of problem strings where proportional reasoning has played a central role. Use this problem string to introduce inverse variation.

You can use this problem string to introduce textbook Lesson 2.5 Inverse Variation.

Guiding the Problem String
This problem string draws upon intuition that suggests for any given task: the more people working on it simultaneously, the faster they will complete it. While there may be a myriad of exceptions or constraints to this idea in practice, most students will trust that this idea is generally true.

Help students trust this core idea so they can tap into the inverse relationship at play here—as the number of workers decreases the time to complete the task increases proportionally and vice versa. Throughout this problem string there is an "invisible 120" at play. How long it would take one worker to complete the job of picking the entire apple orchard is a constant that holds the other values together mathematically. Give students opportunities to notice this value, and determine where it comes from and what it means in this context. As students start to realize it, tag the 120 as the "constant of variation."

About the Mathematics
Doubling and halving is a relationship that can be used in multiplication and area—when one dimension of a rectangular array doubles and the other dimension halves, the area stays the same. Of course, it is not just doubling and halving that makes this true, but also for any other common factor of n, when one factor is multiplied by a and the other is divided by a, or multiplied by $\frac{1}{a}$.

Tables that model only proportional relationships are referred to as ratio tables. The tables in this problem string are not ratio tables—the pairs of values on the table are not equivalent ratios, $\frac{20}{6} \neq \frac{10}{12} \neq \frac{40}{3} \neq \frac{5}{24} \neq \frac{15}{8} \neq \frac{1}{120}$.

It's important to be clear about the way direct variation and inverse variation situations are related and are different. Having a clear understanding of this yourself helps you to support students as they construct these critical ideas for themselves.

In direction variation contexts, the ratios of the two quantities are constant.

In inverse variation, the product of the two values is the constant of variation. In this problem string, 120 will emerge as a significant value since it ties together all of the value pairs in the situation and helps to explain three critical relationships:

$$(\text{number of workers})(\text{number of hours}) = 120$$

$$\frac{120}{\text{number of workers}} = \text{number of hours}$$

$$\frac{120}{\text{number of hours}} = \text{number of workers}$$

(continued)

Sample Interactions

Use the following as you plan how to elicit and model student strategies. This is not meant as a script, but as a view into the relationships involved and the intent of the problem string.

Teacher: *Alright, everyone. Let's get started. I know at least some of us have been apple picking or maybe been to a farm or orchard before. So today I'd like us to envision an apple orchard where workers come in the fall to pick apples to sell to local markets. This is a pretty big place, with quite a few trees. I know one thing about this orchard, which is that it takes 20 workers a total of 6 hours to clear the orchard of apples for the market. I'll write that here on this table.* **Student:** *Are they working all at the same time or is everyone just putting in their own 6-hour shift?* **Teacher:** *Good question. Thoughts about this?* **Student:** *I think it doesn't matter, but in my mind they were all working together—like the owner called in 20 employees and 6 hours later all the apples had been picked.*		**Workers** | **Hours** 20 | 6
Teacher: *So, if it takes 20 workers 6 hours to clear the orchard, how long would it take 10 workers to do the same job?* Brief think time. **Student:** *It's either 12 or 3, I'm not really sure.* **Teacher:** *Interesting. Were other people thinking that either 3 hours or 12 hours made some sense? Yeah? Let's explore that a little.* **Student:** *So 12 hours makes more sense to me because if you have half as many workers I think it would take them longer. It's like everyone lost their partner and their workload doubled.*		**Workers** | **Hours** 20 | 6 10 | 12? 3?
Teacher: *Okay, and some of you thought 3 hours made sense. Anyone want to explain that reasoning?* **Student:** *Well, I thought this was a ratio table, where if you cut the number of workers in half, you cut the number of hours in half. Is this not a ratio table?* **Student:** *No, it looks like one, but it's just a table showing the apple-picking. 10 workers in 3 hours wouldn't make sense. Unless they are like really speedy workers.* **Teacher:** *What are the rest of you thinking now? Check in with your partner for a minute and just describe what you used to think and what you are thinking now. If that thinking changed, say why.* Students turn and talk with partners while the teacher listens in. **Teacher:** *So, is this a ratio table?* **Students:** *No!*		×½ (**Workers** | **Hours** 20 | 6) ×2 10 | 12 Not a ratio table!

Teacher: *Seems like from your conversations we are saying 10 workers would need 12 hours to pick the apples, yes? Okay. Let's keep going. If we had 40 apple pickers, how long would it take them and why?*

Workers	Hours
20	6
10	12
40	

× ½ (20 → 10) × 2 (6 → 12)

Student: *So it seems like if you double the workers you halve the time—that's what we're basically saying, so I'm going with 3 hours.*

Student: *I agree with that.*

Teacher: *And if we halved the workers instead of doubling them?*

Student: *Oh, yeah, like what happened in the last one when we went from 20 to 10. Same idea, just the reverse. If you halve the workers, you double the time.*

Teacher: *Can someone else who agrees with this say why it makes sense to them?*

Student: *It's like what we said before about losing your work partner. If there were twice as many workers and suddenly they disappeared, the rest of us would have to do our work and their work, which is like the double of our work.*

Workers	Hours
20	6
10	12
40	3

× ½ (20 → 10) × 2 (6 → 12)
× 4 (10 → 40) × ¼ (12 → 3)

If you scale the workers up, you scale the hours down and vice versa.

Teacher: *Okay, so now let's stretch this situation a bit more. What if the owners of the orchard know they have 24 hours to get the apples picked before say, a sudden freeze sets in? How many workers would they need to clear the orchard? And how do you know?*

Think time.

Workers	Hours
20	6
10	12
40	3
	24

× ½ (20 → 10) × 2 (6 → 12)
× 4 (10 → 40) × ¼ (12 → 3)

Student: *I'm thinking 5 workers would be enough.*

Teacher: *Say more, where did the 5 come from?*

Student: *I went back to the original 20 workers and 6 hours and I scaled up the hours by 4, so I scaled down the workers by 4.*

Student: *Well, I guess I multiplied the hours by 4 and divided the workers by 4?*

Teacher: *Are we okay if I write it like this?*

Students: *Yes.*

Student: *But you could go from any of the others. Double the 12 so multiply the 10 by ½.*

Student: *Or multiply the 3 hours by 8 and divide the 40 workers by 8.*

Workers	Hours
20	6
10	12
40	3
5	24

× ¼ (20 → 5) × 4 (6 → 24)
× ½ × ⅛ × 2 × 8

(continued)

Teacher: *Here's another twist. What if on a Saturday 15 apple pickers show up for the job and one of them asks, "When will we be done?" In other words, how long will we be here? What would you say now? Think about that and then turn and talk with your partner.*	**Workers | Hours**

Workers	Hours
20	6
10	12
40	3
5	24
15	

Students think, then turn and talk, while the teacher listens in.

Teacher: *I heard a lot of ideas here. Will you share your conversation with us?*

Student: *Sure. So my partner and I are thinking it's either 8 or 9.*

Teacher: *And you don't agree? Is that right?*

Student: *Exactly. So my partner said 15 is right in the middle of 10 and 20 so then the hours would also be right in the middle of 6 and 12. And the number right in the middle of 6 and 12—like you could average them—is 9. But I'm thinking that if the workers tripled—5 times 3 is 15—then the hours must be cut by ⅓, or divided by three….*

Teacher: *We have a dilemma. Two interesting ideas here. Which one makes sense to you? Let's get some more voices in the conversation.*

Student: *I think Georgia's idea fits what we've been doing so far— doubling one and halving the other, that idea.*

Student: *Yes, except this time it's tripling and multiplying by ⅓.*

Workers	Hours
20	6
10	12
40	3
5	24
15	8

×3 ( ) ×⅓   ~~Average?~~

Student: *Also, I don't know if this matters, but all of the numbers on our chart multiply to 120.*

Teacher: *What do you mean?*

Student: *Well, if you take the number of workers times the hours in every single case, it makes 120.*

Student: *Yeah, so I like my 8 hours now. It fits the pattern here.*

Teacher: *Sure, there's a pattern, but what is this 120 about anyway? Why does it keep coming up? How do we know we can trust it? Turn and talk!*

$20 \times 6 = 120$

$10 \times 12 = 120$

$40 \times 3 = 120$

$5 \times 24 = 120$

$15 \times \underline{} = 120$

Algebra Problem Strings
©2017 Kendall Hunt Publishing

Workers	Hours
20	6
10	12
40	3
5	24
15	8
1	

Student: It seems like 120 is like the total number of hours to clear this orchard, basically. And you can have one person do all of that or you could shorten the time for each person by inviting more workers.

Teacher: What do you all think of this idea?

Student: Yeah, the 120 is like the invisible number in this story that you forgot to mention.

Teacher: That's a nice way of saying it—the invisible number. In business you might refer to it is the total number of worker hours. So how long would it take the owner to pick all of these apples herself?

Student: I think we just answered that... 120 hours.

Workers	Hours
20	6
10	12
40	3
5	24
15	8
1	120

$\times \frac{1}{10}$ ... $\times 10$

Teacher: Could you defend that using another combination from the table?

Student: Sure, take 10 workers and 12 hours. Cut the 10 workers by ¹⁄₁₀ and multiply the hours by 10. Like the reciprocal: one-tenth and ten.

Teacher: Anyone else know what Dana means "like the reciprocal?"

Student: Oh yeah, that's what we've been doing this whole time—doubling and halving are reciprocals, quadrupling and fourthing, too. Is that even a word?

Teacher: Maybe you could say quartering?

Workers	Hours
20	6
10	12
40	3
5	24
15	8
1	120
120	1
x	y

120 is "constant of variation"

$x \cdot y = 120$

Teacher: For the last question, I'm wondering how many workers you'd need to clear the orchard in just one hour?

Student: That's 120. A lot of workers.

Teacher: How do you know?

Student: Lots of ways. All of the numbers of workers' times their counterpart number of hours is 120.

Student: Or you could go from the one worker and multiply and divide by 120.

Teacher: I'm going to write what you said about all of the numbers of workers times their counterpart number of hours is 120 like this: these x's time these y's equal 120, $x \cdot y = 120$.

Teacher: How would you summarize some of the things that came up in the string today?

Elicit the following:

- This was not proportional—when one variable increased, the other did not also proportionally increased.
- The table is not a ratio table.
- If you multiply one number by a, you divide the other by a or multiply the other by the reciprocal, $\frac{1}{a}$.
- The product of the number of workers and number of hours is 120, called the constant of variation, $x \cdot y = 120$.

(continued)

Sample Final Display

Your display could look like this at the end of the problem string:

Workers	Hours
20	6
10	12
40	3
5	24
15	8
1	120
120	1
x	y

Not a ratio table!

If you scale the workers up, you scale the hours down and vice versa.

$20 \times 6 = 120$

$10 \times 12 = 120$

$40 \times 3 = 120$

$5 \times 24 = 120$

$15 \times \underline{} = 120$

120 is "constant of variation"

$x \cdot y = 120$

Facilitation Notes

This version of the problem string lists short notes for important teacher moves during the string. After you've done the string yourself and studied the relationships involved, you might make similar notes for the things you want a reminder of or deem important.

Workers	Hours	
20	6	Establish context, 20 workers 6 hours to clear the orchard, put in table.
10		How long would it take 10 workers? Twice as long? Half as long? Partner talk. Discussion. Press for justification. Is this a ratio table? How do you know?
40		How long for 40 workers? Justify. Connect to doubling/halving, quadrupling/quartering.
	24	24 hours (about to freeze), how many workers needed?
15		Go slower, encourage different ways to reason about this one.
1		How long would it take just 1 worker? (May come up earlier) If not yet, bring up: reciprocal, constant of variation (120).
	1	Only one hour? How many workers? (May come up earlier) Generalize $x \cdot y = 120$.

Algebra Problem Strings
©2017 Kendall Hunt Publishing

2.6 | Reasoning About Inverse Variation 2

At a Glance

# of Volunteers	Time (hours)
250	3
500	
	2
	6
	25
10	
v	h

Objectives

The goal of this problem string is to continue to build students' reasoning about situations in which the quantities relate inversely—as one increases the other decreases in an inversely proportional way.

Placement

This problem string is the second of two inverse variation strings and follows a series of problem strings where proportional reasoning has played a central role. Use this problem string to strengthen students' sense of inverse variation behavior and equations.

You can use this problem string to support students' work with inverse variation in Lesson 2.6 Variation with a Bicycle.

Guiding the Problem String

Like the previous one, this problem string draws upon the idea that for any given task, the more people working simultaneously, the faster they will complete it. As in the previous problem string, we are asking students to shift from directly proportional relationships to inversely proportional relationships, and we are using a table to capture both.

Unlike the previous string, however, the quantities can be a bit less friendly. Because students have done a string like this, expect the first problems to go quickly, but use these to draw students who might typically be on the edges of your class community into the conversation.

Some of the problems in the string are readily solved using the relationships in the table, some are readily solved using the 750 (the constant of variation) and some are accessible with either strategy. Use this string to help students parse out which strategy they want to use when. The first problem is a straightforward double, but the second problem, how many volunteers for two hours, may surface a strategy from ratio tables where students try to find the number of volunteers for one hour in order to get the volunteers for two hours. Bring it to light and let students make sense of why that ratio table strategy does not work in a table that is not a ratio table. Help them by staying in context: if the whole job can be done in one hour with some volunteers and the whole job can be done in two hours with a different number of volunteers, what would it mean to add the one and two hours together? This problem is easily solved with the 750, the constant of variation, but can also be found by multiplying the 3 hours by ⅔ to get 2 hours. The next problem can be solved using nice relationships in the table. The fourth problem, 25 hours, cannot be solved readily with the numbers in the table, while the fifth problem, 10 volunteers, is accessible using either strategy.

About the Mathematics

Inverse variation is often represented by the division equation below, but should also be understood as related to the multiplication equation, where k is the constant of variation.

$$y = \frac{k}{x}$$
$$xy = k$$

(continued)

Sample Interactions

Use the following as you plan how to elicit and model student strategies. This is not meant as a script, but as a view into the relationships involved and the intent of the problem string.

Teacher: *We're going to build on the work we did with the apple orchard to test out some of the ideas we developed in that problem string. Today I'd like you to imagine someone new running for political office who has to get a bunch of materials out to the community so constituents begin to know the candidates' name, face, and positions on key issues. Yes, marketing, basically.*

There are lots of ways to do this, of course, but one reliable way is to send brochures or fliers right to people's apartments and homes. It takes some time to address and put stamps on the materials, but with enough volunteers you can get lots of them ready to mail in not too much time. So that's our story. Can you picture it?

Teacher: *The candidate knows that when 250 people show up to work, they are done with the work 3 hours later. I'll record that here to remember that fact. So, what happens if the candidate suddenly gets popular and 500 people show up?*

Double the volunteers, double the time?

Double the volunteers, half the time?

What was this called again—a ratio table modeling direct variation or a table modeling inverse variation?

# of Volunteers	Time (hours)
250	3
500	

# of Volunteers	Time (hours)
250	3
500	1.5

$\times 2$ (250 → 500) $\div 2, \times \frac{1}{2}$ (3 → 1.5)

Teacher: *Let's say that people start to ask for a shorter shift—3 hours feels too long, but 2 hours would be ideal. Now what? How many people would it take to get the same work done in only 2 hours? And how do you know?*

Did anyone use the 3 hours to 250 to help? How?

Did anyone think about the 250 times 3, the constant of variation? How did that help?

Did anyone try to find one hour and use that by doubling the one hour to get the two hours? Why doesn't that work? Tell us what you found. Ahhh, so if the whole job is done in one hour would you want to double that? No, this is not a directly proportional situation!

Talk with your partner about what you could pay attention to in order to make a problem like this accessible in the future.

# of Volunteers	Time (hours)
250	3
500	1.5
375	2

$\times \frac{3}{2}$ (250 → 500) ... $\times \frac{2}{3}$

$250 \times 3 = 750$ constant of variation

$\dfrac{750}{2} = 375$

Algebra Problem Strings
©2017 Kendall Hunt Publishing

Teacher: *Okay, so what if you have a six-hour shift coming up and you need to know how many volunteers to call? What do you think?*

Did anyone use the beginning information, the 3 hours? How?

Did anyone use the 2 hours? How?

Did anyone use the 750? How?

For this situation, which do you want your brain to be inclined to do next time—using the information on the table or using the 750?

I heard some of you say that it might be easier to use the table when the time is given; but when the number of volunteers is given, it might be easier to use the 750. I heard others saying that it depends on the numbers and what you already have. Let's keep thinking about this.

# of Volunteers	Time (hours)
250	3
500	1.5
375	2
125	6

$\times \frac{1}{2}$ $\times 2$ $\times \frac{1}{3}$ $\times 3$

$$\frac{750}{6} = \frac{375}{3} = 125$$

Teacher: *The next problem in the string is that the campaign had an order that they did a while ago that took 25 hours. How many volunteers did they have if they were working at the same 250 volunteers to 3 hours rate?*

Did anyone use relationships in the table? Not so much?

Why did so many of you use the 750, the constant of variation?

How did you divide 750 by 25? Did anyone think quarters?

What about the numbers this time had so many of you using the constant of variation, the 750 and not relationships in the table?

So, sometimes it's nice to use the relationships in the table and sometimes it's nice to use the constant of variation?

# of Volunteers	Time (hours)
250	3
500	1.5
375	2
125	6
30	25

$$\frac{750}{25} = \frac{700}{25} + \frac{50}{25} = 28 + 2 = 30$$

Think quarters!

Teacher: *The next problem in the string is that one day only 10 volunteers could make it. How long will those 10 volunteers take to finish the job? Look for a clever strategy!*

What did you think was clever?

Did anyone use the 30 volunteers? How?

Did anyone use the 500 volunteers? How?

Did anyone use the 250 volunteers? How?

Did anyone use the constant of variation, the 750? How?

What about the numbers this time made it easy to use either the constant of variation, the 750, or several relationships in the table?

# of Volunteers	Time (hours)
250	3
500	1.5
375	2
125	6
30	25
10	75

$\div 25$ $\div 50$ $\times 50$ $\times 25$ $\times \frac{1}{3}$ $\times 3$

$$\frac{750}{10} = 75$$

(continued)

Teacher: *Okay, great. This is going to help the campaign. But let's really bring it home with a rule. Please think about and then write down a rule—in this situation—for finding the number of hours, h, needed if you know the number of volunteers, v.*	$v \times h = 750$ $$h = \frac{750}{v}$$ $$v = \frac{750}{h}$$

And then also write a rule if you know how much time you have and are trying to recruit the right number of volunteers.

And be thinking about how these rules are related, if at all. First a minute to think and then you can jot down your rules so that we can discuss them.

Teacher: *How would you summarize some of the things that came up in this string today?*

Elicit the following:

- *Tables can be helpful in inverse variation situations.*

- *In inverse variation, if you multiply one variable by a, you divide the other variable by a to get another data point in the same scenario.*

- *In inverse variation, the product of the variables is the constant of variation. You can find one variable by dividing the constant of variation by the other.*

Sample Final Display

Your display could look like this at the end of the problem string:

Facilitation Notes

This version of the problem string lists short notes for important teacher moves during the string. After you've done the string yourself and studied the relationships involved, you might make similar notes for the things you want a reminder of or deem important.

# of Volunteers	Time (hours)	
		Establish context, put in table.
250	3	250 volunteers can get the job done in 3 hours.
500		What if the candidate gets popular and 500 volunteers come? Double/double or double/halve? Why?
	2	What if need to get the job done in 2 hours? Can you double 1 hour? Why not? Times 2/3? Use relationships in table or constant of variation, 750?
	6	Repeat with 6 hours for the job. What do you want your brain to be inclined to do next time? Use relationships in table or constant of variation, 750? Why?
10		What if 10 volunteers show up? Relationships in table or 750? Why?
	25	What if the job was done in 25 hours? Why not relationships in the table? Why use 750?
v	h	Write equations that describe the relationship between v, h, 750.

2.7 Solving Equations 1

At a Glance

$x = 3$

$x = -2$

$-x = 5$

$-x = -4$

$x - 4 = 6$

$x + 4 = -6$

$x - 4 = -10$

Follow-up String:

$x = -6$

$-x = -2$

$2x = 12$

$2x = -12$

$-2x = 10$

$-2x = -8$

$\frac{1}{2}x = 3$

$\frac{1}{3}x = -2$

$-\frac{1}{4}x = -5$

Objectives

The goal of this string is to begin to build the double open number line as a tool for reasoning about algebraic equivalence and solving equations.

Placement

This string is the first in a two string series for solving equations. It should happen early, before you are ready to have students solve equations in your course, as an introduction to using the double open number line. If your students already have such experience, you may be able to skip it or it should go very quickly.

You can use this problem string after textbook Lesson 2.7 Evaluating Expressions to get students ready for Lesson 2.8 Undoing Operations.

Guiding the Problem String

If your students have limited or no experience using a double open number line as a tool, or even an open number line at all, none of them will think to use it. Your job is to use the double open number line to model student thinking so that students can make sense of the model as they see their thinking outside of their head. This is necessary before students can use the model for thinking (as a tool). Do not expect that students will begin to use the model as a tool by the end of this string—they will probably need to experience it more as a model of their thinking before they will be ready to make the transition to using it as a tool for solving equations.

You might notice that the last three problems all deal with the same 4, 6, 10 relationships. That is purposeful—using the same relationships can help not bog down students in the midst of learning something else.

You can use the follow-up string to work with other relationships.

About the Mathematics

The open number line, or "empty" number line as the Dutch call it, is a model to represent only the quantities or expressions that the students are reasoning about. That is, it is not meant to replicate a closed number line or resemble a ruler. Only place the quantities and expressions that students are talking about on the number line. The open double number line represents equivalent expressions because the expressions are on the top and bottom at the same location.

Many thanks to Andrew Stadel for his clothes line math inspiration.

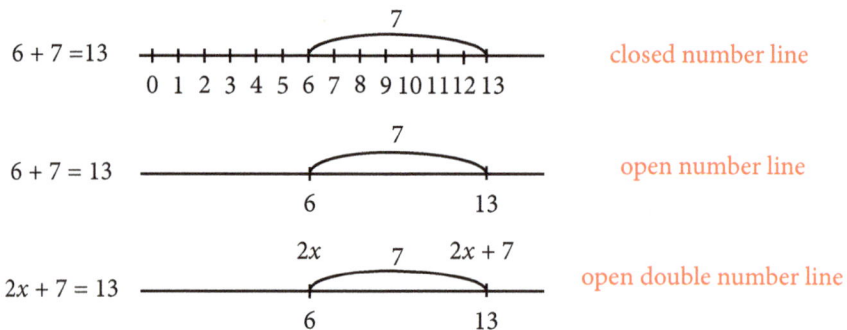

Sample Interactions

Use the following as you plan how to elicit and model student strategies. This is not meant as a script, but as a view into the relationships involved and the intent of the problem string.

Teacher: *Today's problem string starts with an easy one, x equals 3. If I tell you that x equals 3, right here on this open number line, where would 0 be?*	$x = 3$ (number line showing x above 3)
Student: *It would be that way.* **Student:** *To the left.* **Student:** *Three to the left.* **Teacher:** *Okay, I'll put it over here.*	$x = 3$ (number line showing x above 3, with 0 to the left)
Teacher: *So what if I told you that x is −2? Where is 0?*	$x = -2$ (number line showing x above −2)
Student: *Now 0 is to the right 2 units.* **Teacher:** *So 0 would be to the right, like this?*	$x = -2$ (number line showing x above −2, 0 to the right)
Teacher: *What if the opposite of x is equal to 5. Where is 0?* **Student:** *Zero has to be to the left 5.* **Teacher:** *How do you know?* **Student:** *Because zero is always to the left of 5.* **Teacher:** *So, you're not really even paying attention to the opposite of x? You're just using that zero is always to the left of 5?* **Student:** *Yes.*	$-x = 5$ (number line showing $-x$ above 5, 0 to the left)
Teacher: *If the opposite of x is at 5, where is x?* **Student:** *Does that mean that it's back 5 to the left of 0?* **Student:** *Yeah, I think so. The opposites go on either side of 0.* **Teacher:** *Does that make sense to everyone? Turn to your partner and talk about that.* Brief partner talk.	$-x = 5$ $x = -5$ (number line showing x at −5, 0, $-x$ at 5)
Teacher: *What if the opposite of x is equal to the opposite of 4? Where is 0 now?*	$-x = -4$ (number line showing $-x$ above −4)
Student: *I think 0 is to the right of −4.* **Teacher:** *Everyone agree? I'll put 0 to the right of negative 4.*	$-x = -4$ (number line showing $-x$ above −4, 0 to the right)

(continued)

Teacher: *If the opposite of x is here at negative 4, where is x?* **Student:** *So, x would have to be at 4.* **Teacher:** *Why?* **Student:** *It helped when you said opposite of x, that means that x has to be on the other side of 0.*	$-x = -4$ $x = 4$ number line showing $-x$ at -4, 0, and x at 4
Teacher: *Great. Now this next one is really easy but humor me anyway. If I tell you that a number minus 4 is 6, I bet you could tell me what that number is, right?* *Since they are equivalent, I'm going to put the x minus 4 and the 6 at the same location. If we know x – 4 is here, where do we know that number has to be?*	$x - 4 = 6$ number line showing $x - 4$ at 6
Student: *The number has to be to the right.* **Student:** *It is 4 to the right of x – 4.*	$x - 4 = 6$ number line showing $x - 4$ and x, span of 4, at 6
Teacher: *What is 4 to the right of 6?* **Student:** 10. **Teacher:** *So, the number, x, is 10. We knew that anyway from the start, but now I can see the relationship your brain was using. Nice.*	$x - 4 = 6$ $x = 10$ number line showing $x - 4$ at 6 and x at 10, span of 4
Teacher: *The next problem is x plus 4 is negative 6. What do you think x is?* **Student:** 2. **Student:** –2. **Student:** –10. **Teacher:** *So, lots of different answers. I'm going to just put this equation on the number line before I get your thinking.*	$x + 4 = -6$ number line showing $x + 4$ at -6
Teacher: *How did you get –2?* **Student:** *Well, I don't think it's –2 anymore. It's gotta be 4 to the left of –6.* **Teacher:** *What is 4 to the left of –6?* **Student:** *It's got to be more negative, so –10.*	$x + 4 = -6$ $x = -10$ number line showing x at -10 and $x + 4$ at -6, span of 4
The teacher repeats with the last problem, $x - 4 = -10$.	$x - 4 = -10$ $x = -6$ number line showing $x - 4$ at -10 and x at -6, span of 4

Algebra Problem Strings
 ©2017 Kendall Hunt Publishing

Teacher: *What did you think in this string today?*

Elicit the following:

- *It can help you to decide what's going on if you put stuff on a number line. You can see where things fit.*

- *I don't draw a line in my head, but I think about the numbers the same way.*

- *I don't think about the numbers that way, but when you put them on the number line, I can.*

Sample Final Display

Your display could look like this at the end of the problem string:

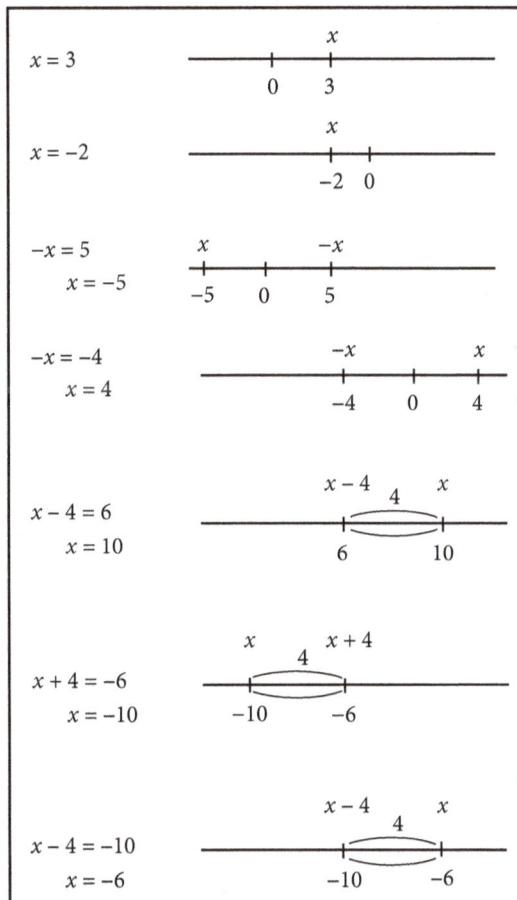

(continued)

Facilitation Notes

This version of the problem string lists short notes for important teacher moves during the string. After you've done the string yourself and studied the relationships involved, you might make similar notes for the things you want a reminder of or deem important.

$x = 3$	Put on open number line. "Where is 0?" Quick.
$x = -2$	Repeat. New open number line for each problem.
$-x = 5$	Use "opposite of x." Now where is 0? Where is x? Quick.
$-x = -4$	"Opposite of x" Where is 0? Where is x?
$x - 4 = 6$	Humor me. Equivalent so same location. What if x − 4 is here, at 6? Where is x?
$x + 4 = -6$	Equivalent so same location. What if x + 4 is here, at −6? What is 4 back from −6?
$x - 4 = -10$	What is 4 to the right of −10? It can be helpful to see your brain outside of your head.

Algebra Problem Strings
©2017 Kendall Hunt Publishing

2.8 Solving Equations 2

At a Glance

$3x + 7 = 22$

Where is $3x + 17$?

Where is $3x$?

Where is $3x - 2$?

Where is x?

$\dfrac{x}{6} - 20 = -19$

Where is 0?

Where is $\dfrac{x}{6}$?

Where is x?

Follow-up string:

$\dfrac{x+9}{3} - 1 = 4$

Where is $\dfrac{x+9}{3} + 5$?

Where is $\dfrac{x+9}{3} - 7$?

Where is $\dfrac{x+9}{3}$?

Where is $x + 9$?

Where is x?

$3(x - 5) + 7 = -14$

Where is $3(x - 5)$?

Where is $x - 5$?

Where is x?

Objectives

The goal of this string is to use a model, the open double number line, as a tool for reasoning about algebraic equivalence and solving algebraic equations.

Placement

This string is the second in a two string series and is designed to help students envision expressions as they relate to each other and to use those relationships to develop new, equivalent expressions, eventually solving the related equations. Use this problem string to give students a relationship-oriented perspective on equivalent expressions and solving equations.

This problem string could follow textbook Lesson 2.7 Evaluating Expressions.

Guiding the Problem String

Depending on their prior experiences, students may have varied experiences using the open double number line as a model for thinking. For some, the model has been a helpful tool where big ideas and efficient strategies for operations have been constructed. For others, the model might be new and, therefore, the kinds of questions here may feel unfamiliar.

This problem string (re)introduces the double open number line as a way to reason about related quantities. The string begins by posing two related quantities and instead of asking students to solve for x, you use the number line to show the equivalence by placing one expression on the top of a number line and the other on the bottom and reason about other relationships. The idea of two equivalent expressions landing in the same place is central to the mathematics of this string.

Ask students about other, related expressions, pressing students for relational justification. The questioning follows the general pattern of, "If this is true, what else is true, and how do you know?" This places the work of reasoning on students.

You could facilitate the follow-up string similarly to continue the work.

About the Mathematics

When using the open double number line, students simultaneously think about a quantity or number as a location (relative to other locations) as well as a distance. This idea will play prominently in this problem string—students will be encouraged to think flexibly about both location and distance in order to reason about unknown expressions. The example below, where 8 is a distance and 28 is a location, illustrates this distinction:

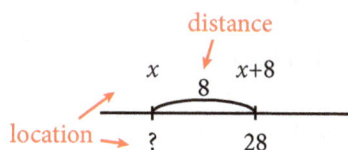

The relational actions that make sense to do on one side, go to the left 8 from $x + 8$ to find x, also imply that you subtract 8 from 28 to find the unknown. These actions mirror the balancing steps to solve the same equation and can be used to remediate students who have not been successful balancing and to extend students who have.

(continued)

Sample Interactions

Use the following as you plan how to elicit and model student strategies. This is not meant as a script, but as a view into the relationships involved and the intent of the problem string.

Teacher: *Let's say I told you that some number times 3x + 7 was at the same location on a line as 22, like this. Can we also say that 3 times some number and then add 7 is equivalent to 22, 3x + 7 = 22? Yes?* *And right now, we are not interested in what x equals— that's not our goal right now. Instead, we're going to look for 3x +17. If 3x plus 7 is here at 22, where is 3x plus 17? Turn and talk to your partner briefly.* Students turn and talk while the teacher listens in.	$3x+7$ at 22 on a line $3x + 7 = 22$
Teacher: *As I was listening in, I heard some good thinking. Where is 3x + 17?* **Student:** *It would have to be over there, to the right, 10 more.* **Teacher:** *You think it's over here, 10 to the right? Why?* **Student:** *Because if 3x plus 7 is there, then plus 17 has to be 10 to the right.* **Teacher:** *So, if 3x plus 17 is over here, 10 to the right, what is 10 to the right of 22?* **Student:** *32.* **Teacher:** *32? I'll mark that. And now I'll write an equation that represents this relationship, that 3x plus 17 is equivalent to 32.*	$3x+7$ — 10 — $3x+17$; 22 $3x+7$ — 10 — $3x+17$; 22 ... 32; 10 $3x + 17 = 32$
Teacher: *Okay, and now what about 3x? Where would we find 3x on this number line and how far away would it be from the other values and expressions we have here? Tell your partner and see if you agree. Go!* Brief partner talk. **Student:** *3x would be 7 back that way, to the left, because you had to add 7 to it to get to 3x + 7.* **Teacher:** *Nice. And so if 3x is back here 7, what's down here? What is 7 to the left of 22?* **Student:** *That's 15.* **Teacher:** *And so now we can record the equation 3x = 15.*	$3x$ — 7 — $3x+7$ — 10 — $3x+17$; 22, 10, 32 $3x$ — 7 — $3x+7$ — 10 — $3x+17$; 15, 7, 22, 10, 32 $3x = 15$
Teacher: *Now I'm interested to know where 3x minus 2 is? Where would that be?* **Student:** *That would be 2 to the left of 3x, so it would be 13 on the bottom.* **Teacher:** *So, if 3x minus 2 is 2 to the left of 3x, then 13 is 2 to the left of 15? Yes?*	$3x-2$, 2, $3x$ — 7 — $3x+7$ — 10 — $3x+17$; 13, 2, 15, 7, 22, 10, 32 $3x - 2 = 13$

Algebra Problem Strings
©2017 Kendall Hunt Publishing

Teacher: *I'm going to erase the $3x - 2$ bit for right now. So, if all of this is true, do we know enough to find zero? Where would zero be?*

Student: *We know 3 times x is 15, and we know 0 is to the left of 15.*

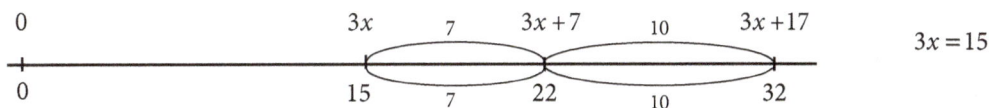

$3x = 15$

Teacher: *I said earlier that we were not trying to find x, but could you find x? If all of these are true, where would x be? Do you have enough information or do you need more? Think for a minute about that and maybe someone who is sure, one way or another, could explain to the rest of us.*

Student: *It's just 5. 3 times 5 is 15.*

Teacher: *So, if we took this distance from 0 to 3x and divided it by 3, x would land on 5? Yes?*

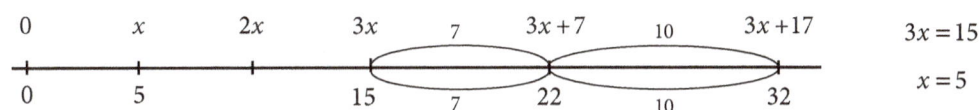

$3x = 15$

$x = 5$

Teacher: *So if we know one relationship, we can reason about several other relationships. Interesting.*

Teacher: *Here's another one to think about. If you take a certain number, divide it by 6, then subtract 20, you get −18. Got a picture of what that looks like? I'll sketch it here.*	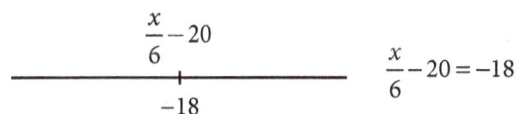 $\dfrac{x}{6} - 20 = -18$

Teacher: *If this is true, where is zero? How do you know? When you're ready, turn and tell you partner.*

Partners talk while the teacher listens in.

Teacher: *Where is zero?*

Student: *We really just ignored the top and just thought about the negative 18. That's got to be to the left of zero, so zero has to be to the right 18.*

Teacher: *Okay, so zero is over here 18. What else do we know?*

Student: *I think it would be x over 6 minus 2?*

Teacher: *You think it's x divided by 6 minus 2? Why?*

Student: *Because whatever it is minus 20 is there, so 18 that way would only be minus 2.*

Teacher: *Agree? Disagree? I'll write that down as an equation too.*

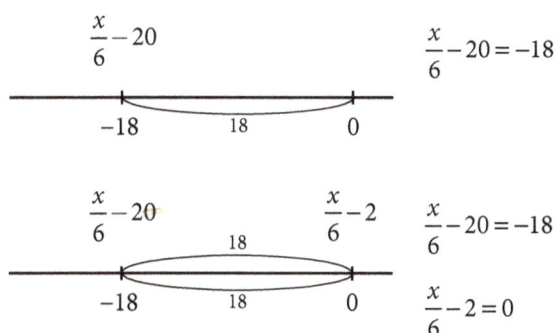

$\dfrac{x}{6} - 20 = -18$

$\dfrac{x}{6} - 20 = -18$

$\dfrac{x}{6} - 2 = 0$

(continued)

Teacher: *Okay, let's use what we just did to figure out where x divided by 6 would go. Think for a few seconds first, then you can turn and tell your partner. If you've been sketching, go ahead and add it to your model if you like. Then we'll come back to see if we agree.*	$\dfrac{x}{6}-20$ $\qquad\qquad$ $\dfrac{x}{6}$ $\qquad$ $\dfrac{x}{6}-20=-18$
Partners work and the teacher circulates, noting any misconceptions to address.	$-18 \qquad 18 \qquad 0\ 2$ $\qquad \dfrac{x}{6}-2=0$
Teacher: *What are you thinking? Where is x divided by 6?*	$\dfrac{x}{6}=2$
Student: *It would be two more to the right. And that's at 2.*	
Teacher: *So now we know that x divided by 6 is equivalent to 2.*	
Teacher: *So, a number divided by 6 is 2. My last question today is where do you think x is, and what makes you say this? Is x a really big number, a really small number, perhaps even a negative number? What is that number? I'm going to put this on a new number line.*	? $\qquad$ $\dfrac{x}{6}$ $\qquad$? $\qquad\qquad$ 2
Student: *Well, if something divided by 6 is 2, the number has to be bigger than 2.*	? $\qquad$ $\dfrac{x}{6}$ $\qquad$ x
Student: *It's 12 because 12 divided by 6 is 2.*	$\qquad\quad$ 0 2
Student: *And 2 times 6 is 12.*	
Student: *Yes, it's 12 because if you had a number and when you divided it by 6 you got 2, that number would be to the right of the result.*	? $\qquad$ $\dfrac{x}{6}$ $\qquad$ x
Teacher: *I've recorded your thinking here. I'll also put it on our original number line.*	$\qquad\quad$ 0 2 $\qquad\qquad$ 12

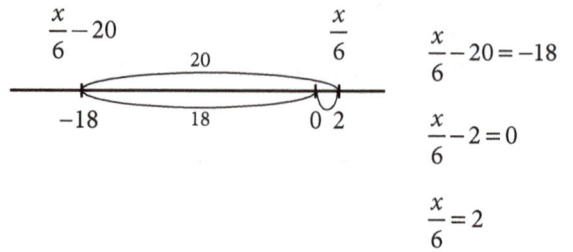

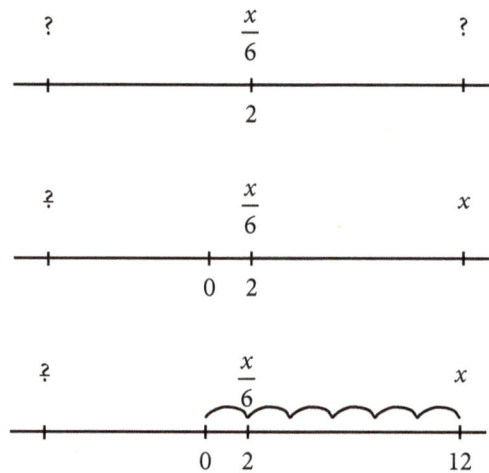

Teacher: *How would you summarize some of the things that came up in this string today?*

Elicit the following:

- *We can use what we know to find other relationships.*

- *Finding where zero is can be helpful.*

- *What you do to the top, you do to the bottom and vice versa.*

- *We can record equivalencies on a double open number line and with equations.*

Sample Final Display

Your display could look like this at the end of the problem string:

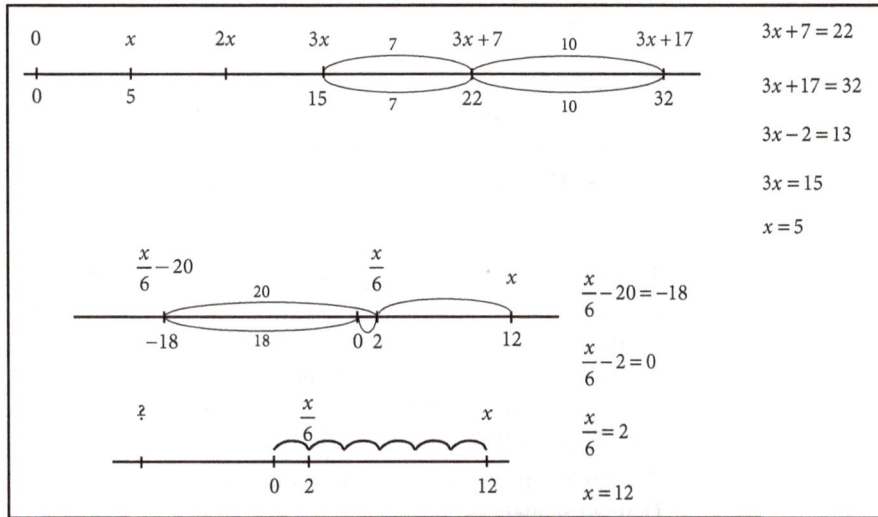

Facilitation Notes

This version of the problem string lists short notes for important teacher moves during the string. After you've done the string yourself and studied the relationships involved, you might make similar notes for the things you want a reminder of or deem important.

$3x + 7 = 22$	Say "Some number times 3 plus 7 is at the same location on a line as 22". Draw the number line then write equation.
Where is $3x + 17$?	Not finding x right now, actually want to know where 3x + 17 is? How do you know? If 3x + 17 is 10 to the right, what is 10 to the right of 22? Model on number line and equation.
Where is $3x$?	Repeat.
Where is $3x - 2$?	Repeat. Erase this part when done.
Where is x?	Where is zero? Why? Where is x? How do you know?
$\dfrac{x}{6} - 20 = -18$	New scenario. A certain number divided by 6 then subtract 20 is negative 18. Draw the number line then write equation.
Where is 0?	How do you know? What else do we know? Establish x/6 -2 = 0.
Where is $\dfrac{x}{6}$?	How do you know? Quick.
Where is x?	What do you know about x? Small, big, positive, negative? What is x?

3.1 Arithmetic Sequences

At a Glance

1, 3, 5, 7
−2, −5, −8
1, ___, 13
3, ___, ___, 15
17, ___, ___, ___, ___, 32
−10, ___, ___, 23
17, ___, ___, ___, 5

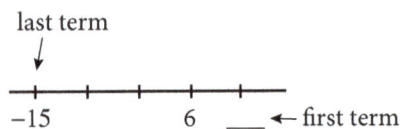

last term
↓

−15   6   ___ ← first term

Objectives

The goal of this string is to help students continue to build understanding of the additive nature of arithmetic sequences by finding the common difference when consecutive terms are missing, and prepare students for writing equations in the slope-intercept form.

Placement

Students have likely seen sequences before as ordered lists of numbers and have experienced recursive rules in which the starting value and the common difference are provided and students are asked to find missing values of the sequence. This string is designed to extend their previous work helping students to reason about the repeated addition (of positive or negative values) involved in linear relationships by providing sequences with consecutive missing terms.

You can use this problem string to introduce textbook Lesson 3.1 Recursively Defined Sequences.

Guiding the Problem String

This problem string uses an open number line as a way to visualize and make sense of arithmetic sequences. The first two problems are designed to get students naming the pattern as a sequence and linking two forms of notation: the sequence notation (listed in order separated by commas) and the open number line as a model emphasizing the distance between values. These should be quick.

The rest of the problems are missing value sequences, tinkering each time towards more complex problems so that students begin to solidify a strategy for finding the missing values as well as the common difference between values. As the problem string progresses there are more missing values to determine—nudging students away from guess and check strategies and towards efficient strategies such as finding the distance between the given values and dividing by the number of "jumps." Here we expect students to talk both about repeated addition or subtraction and multiplication or division.

The choice of the open number line is deliberate—if students can envision the space between successive terms, we suspect that they will be more likely to develop the recursive rule and/or fill in missing values. Your modeling and questioning needs to be precise so as not to create confusion. Specifically, when illustrating students' strategies, be sure to mark the distance between each value using a positive number, since distance by definition is a positive measure. But then, when you move from the open number line model to the sequence model be sure to ask whether we added or subtracted successively. This will help students realize how to transition smoothly from the model of distance towards an accurate recursive rule. The sample interactions provide examples of the teacher using precise language.

About the Mathematics

Solving the problems in this string can lead to intuition about the relationship between arithmetic sequences and the slope y-intercept form of a line. The initial term corresponds to the y-intercept of the line and the common difference (quantity being added or subtracted to each term) corresponds to the rate of change, where x is the term number and y is the term, (term number, term).

Algebra Problem Strings
©2017 Kendall Hunt Publishing

Sample Interactions

Use the following as you plan how to elicit and model student strategies. This is not meant as a script, but as a view into the relationships involved and the intent of the problem string.

Teacher: *Here we go! I'm going to put up a sequence of numbers. I'd like you to envision what they would look like as places on a number line and be prepared to describe this pattern to someone who isn't in our class.* **Student:** *You started at 1 and you went up 2 every time. Basically going by odd numbers.* **Teacher:** *And what do you envision?* **Students:** *Starting at 1 on the number line and making jumps of 2 until you get to 7.* **Teacher:** *Okay, here's a quick sketch of what you just described.*	1, 3, 5, 7
Teacher: *And what about this sequence. How do you envision it? How would you describe it?* **Student:** *You started at −2 and you jumped back twice, each time by three.* **Student:** *It's like you were adding −3 each time.* **Student:** *Or subtracting three.* **Teacher:** *Here's a model of this one. Notice that you all talked about subtracting three or negative three but in this model I'm using a positive three. Any idea why?* **Student:** *Yeah, because in the model you are just writing how far apart those numbers are, not what you did to compute them.* **Teacher:** *Are we okay if I also write the sequence in order, adding −3 each time?* **Student:** *Yeah, that helps me. Thank you.*	−2, −5, −8

(continued)

Teacher: *Now I'm going to get a bit trickier, leaving some parts out of my sequence, not telling you how I jumped or where I landed. That's your job. You should assume that whatever I added or subtracted was the same—that's what makes this an arithmetic sequence.*

Student: *I think the middle term is 7 because 1 plus 6 is 7 and 7 plus 6 is 13.*

Teacher: *And where did the 7 or the 6 come from?*

Student: *I'm not sure, I just tried a few and that one worked. I was looking for the middle, basically.*

Teacher: *Anyone else think about how far apart 1 and 13 are?*

Student: *Yeah, so I saw the space between 1 and 13 was 12 and then I just divided it by 2.*

Student: *Wait, why 2 if there is only one missing number?*

Student: *Well, I was looking for the jumps, the space between the numbers, which was 12 ÷ 2 or 6. Then I used the 6 to find the missing one, which was 7.*

1, __, 13

+6 +6

1, 7, 13

12 / 6 6

1   7   13   12 ÷ 2 = 6

Teacher: *Okay, some nice strategies are emerging. What about now? Let's think for a few seconds and then turn and talk to our partner about this one. See if the model or the strategies we've been talking about might help you here.*

Student: *So my partner and I used that idea of distance between the first and the last number, which is 12, and then we saw there would be 3 jumps. So we divided 12 by 3 and got 4. This means the sequence is 3, 7, 11, 15 because each time we added 4.*

Teacher: *Here's a model of your thinking so that we can all make sense of it.*

3, __, __, 15

+4 +4 +4

3, 7, 11, 15

?

3   ?   ?   15

12

4 4 4   12 ÷ 3 jumps = 4

3   15

Student: *Let's keep going because I'm sensing these are still pretty friendly for you. How are you thinking about this one? What's a nice way to find those missing values?*

Student: *So I kind of have a rule now. It goes like this: find the difference between the first and last numbers. 32 – 17 is 15, right? Okay. Then look at the number of missing numbers. Yes, it is four, but that means there are actually five spaces, five jumps between those numbers. So 5 is what we are dividing by, not 4. Don't be fooled, people!*

Teacher: *Everyone turn and tell your partner what she just said. Does this make sense to you?*

17, __, __, __, __, 32

+3 +3 +3 +3 +3

17, 20, 23, 26, 29, 32

15

3 3 3 3 3   32 – 17 = 15

17   32   15 ÷ 5 jumps = 3

Algebra Problem Strings
©2017 Kendall Hunt Publishing

Teacher: *Okay, that was nice. Let's try out her strategy with this one. I hear groans—was that the −10? We've got this. I wonder if it would help to draw a model of what these relationships look like. Let's give each other some time to think about this one.*

Student: *I didn't like this at first, but now I see it's not too bad. The distance between those numbers is 33. And if I use our earlier idea, I have three jumps between the four numbers. So 33 divided by 3 is 11. Each time you are going up by 11.*

Teacher: *While I make a model of that, can someone put this strategy in their own words for us?*

$-10, \_\_, \_\_, 23$

$+11 +11 +11$
$-10, \underline{1}, \underline{12}, 23$

33
11 11 11
−10   23

$10 + 23 = 33$
$33 \div 3 \text{ jumps} = 11$

Teacher: *Just a few more. Let me give you some language now to describe what we are doing. The first term in this sequence is 17 and the last term in this sequence is 5. Notice I said "term" and not "number." There are five terms altogether. With your partner will you a) make a model of this and b) find the common difference between the terms. Did that make sense? Okay, go!*

The teacher circulates, looking for ideas from students who have not spoken thus far, drawing them into the whole group conversation by honoring their thinking for the whole class to see.

$17, \_\_, \_\_, \_\_, 5$

$-3 -3 -3 -3$
$17, \underline{14}, \underline{11}, \underline{8}, 5$

12
3 3 3 3
5   17

$12 \div 4 \text{ jumps} = 3$

Teacher: *Here's the last sequence. This time I'm just going to draw a number line to represent the sequence and tell you that the last term is −15 and this term is 6, but not give you the first term. You need to find the sequence.*

last term
↓
−15   6   __ ← first term

The teacher crafts a conversation by questioning students to label the terms and the common difference. Also identifying the sequence and the terms in the sequence in order of magnitude on the number line.

last term
↓
−15   6   __ ← first term
21
7 7 7
−15 −8 −1 6 13 ←

$15 + 6 = 21$
$21 \div 3 \text{ jumps} = 7$

-7 ← common difference
$13, \underline{6}, \underline{-1}, \underline{-8}, -15$ → the sequence

the terms in the sequence in magnitude order on the number line

Teacher: *How would you summarize some of the things that came up in this string today?*

Elicit the following:

- *Arithmetic sequences are built with repeated addition, with a positive or negative common difference.*
- *The number of missing terms and the number of times you add the common difference (the number of "jumps") are not the same.*
- *To find the common difference in a sequence that has consecutive missing terms, we can find the difference between any two known terms and divide by the number of "jumps."*

(continued)

Sample Final Display

Your display could look like this at the end of the problem string:

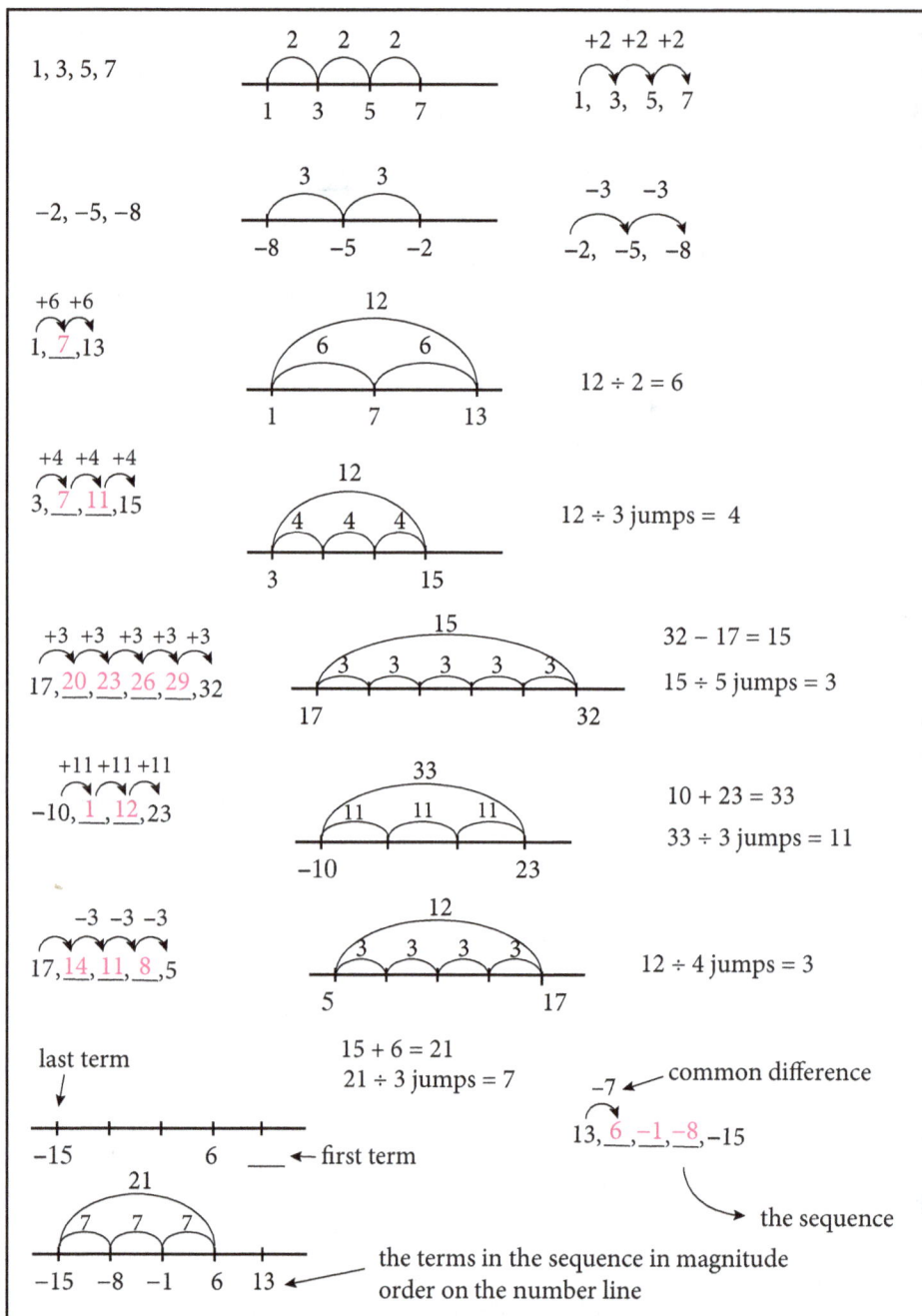

1, 3, 5, 7

2 2 2

1 3 5 7

+2 +2 +2

1, 3, 5, 7

−2, −5, −8

3 3

−8 −5 −2

−3 −3

−2, −5, −8

+6 +6

1, 7, 13

12

6 6

1 7 13

$12 \div 2 = 6$

+4 +4 +4

3, 7, 11, 15

12

4 4 4

3 15

$12 \div 3 \text{ jumps} = 4$

+3 +3 +3 +3 +3

17, 20, 23, 26, 29, 32

15

3 3 3 3 3

17 32

$32 - 17 = 15$

$15 \div 5 \text{ jumps} = 3$

+11 +11 +11

−10, 1, 12, 23

33

11 11 11

−10 23

$10 + 23 = 33$

$33 \div 3 \text{ jumps} = 11$

−3 −3 −3

17, 14, 11, 8, 5

12

3 3 3 3

5 17

$12 \div 4 \text{ jumps} = 3$

last term

↓

−15   6 ___ ← first term

$15 + 6 = 21$

$21 \div 3 \text{ jumps} = 7$

−7 ← common difference

13, 6, −1, −8, −15

the sequence

21

7 7 7

−15 −8 −1 6 13 ← the terms in the sequence in magnitude order on the number line

Algebra Problem Strings
©2017 Kendall Hunt Publishing

Facilitation Notes

This version of the problem string lists short notes for important teacher moves during the string. After you've done the string yourself and studied the relationships involved, you might make similar notes for the things you want a reminder of or deem important.

1, 3, 5, 7	Describe it. Envision it. Draw open number line. Quick.
−2, −5, −8	Repeat. Quick. Distinguish between 3 in the model and the −3 in the sequence.
1, __, 13	A missing term in this arithmetic sequence. Begin to nudge away from guess and check towards using distance and number of jumps as a strategy.
3, __, __, 15	Look to highlight the efficient, generalizable strategy of finding the difference and dividing by the number of jumps.
17, __, __, __, __, 32	How many blanks? How many jumps?
−10, __, __, 23	Repeat. Wonder aloud if a model might help.
17, __, __, __, 5	Repeat. Name "terms" and "common difference."

last term

Draw it. What is the sequence being represented?

−15   6   ___ ← first term

3.2 | Linear Plots

At a Glance

Mackinac Bridge

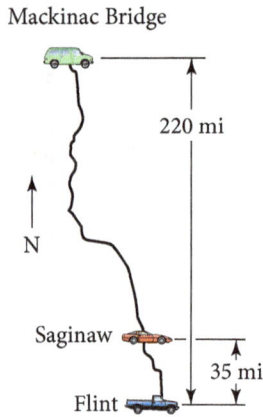

220 mi

N

Saginaw

35 mi

Flint

Green minivan,

started 220 miles away,

going toward Flint at 72 mph

Purple motorcycle,

started 110 miles away,

going toward Flint at 72 mph

Pink SUV,

started 220 miles away,

going toward Flint at 60 mph

Red sports car,

started 35 miles away,

going away from Flint at 48 mph

Light blue bicycle,

started 35 miles away,

going away from Flint at 20 mph

Green bus,

started 50 miles away,

going away from Flint at 48 mph

Objectives

The goal of this problem string is to solidify the learning about the graphs of linear plots that began in the Investigation On the Road Again about vehicles traveling at different constant rates in different directions.

Placement

This problem string can be used after students have begun to connect arithmetic sequences and plots of linear data to solidify the learning about linear plots, rates of change, and starting points.

This string is intended to support the Investigation On the Road Again in textbook Lesson 3.2 Linear Plots.

Guiding the Problem String

This problem string is based on an investigation where students create recursive rules and graphs to represent vehicles traveling north or south in Michigan. If they have not already, students will need to determine recursive rules and graphs of the green minivan and red sports car as described. Then use this problem string to reinforce and solidify that learning.

The string has two sections. The first section is based on traveling toward Flint, where students compare the green minivan's rule and graph to two new vehicles, where the starting points and rates change. The second section is based on traveling away from Flint, again comparing three new vehicles.

As you facilitate, stay in context. Use technology by entering the recursive rules to figure distances so students do not get bogged down in the calculations, then just estimate positions on the graph. As you present each problem, ask students to predict using their arms what the plot will look like and how it will compare to the other(s) previously graphed. Use color to help differentiate the different vehicles.

About the Mathematics

If the points associated with the terms of an arithmetic sequence are plotted as (term number, term), the result is a linear plot where the starting term is the y-intercept and the common difference is the rate of change of the line that models the sequence.

You can use various graphing calculators, apps, or spreadsheets to enter the recursive rules to allow the technology to figure the sequences. The numbers in the problems are purposefully cranky so that students will be nudged to either estimate or take advantage of technology.

The context in this string is a continuous distance versus time scenario. We are choosing to model the situation by using discrete data points as sequences.

Important Questions

Use the following as you plan how to elicit and model student strategies.

- How did you find the recursive rule? Why?
- Predict, how do you think the line will behave? Show me with your arms.
- Where does the starting point show up in the rule for this scenario? In the graph?
- Where does the driving rate show up in the rule for this scenario? In the graph?
- What does it look like in the rules and graphs when vehicles are traveling at the same rate?
- What does it look like in the rules and graphs when vehicles start at the same distance from Flint?
- What does it mean in the scenario when the line is increasing from left to right on the graph?
- What does it mean in the scenario when the line is decreasing from left to right on the graph?
- Consider the vehicle traveling toward Flint. If they all leave at the same time, which vehicle will arrive in Flint first? How do you know?

How would you summarize some of the things that came up in this string today?
- *The starting distance from Flint is the starting value in the recursive rule and the y-intercept on the graph.*
- *The driving rate is the common difference in the rule and affects the slope of the line in the graph.*
- *If the vehicle is traveling toward Flint, the distance to Flint is decreasing and the line is decreasing from left to right.*
- *If the vehicle is traveling away from Flint, the distance from Flint is increasing and the line is increasing from left to right.*
- *When vehicles are traveling at the same rate, the lines are parallel.*
- *When vehicles start at the same distance, the lines have the same y-intercept.*

(continued)

Sample Final Display

Your display could look like this at the end of the problem string:

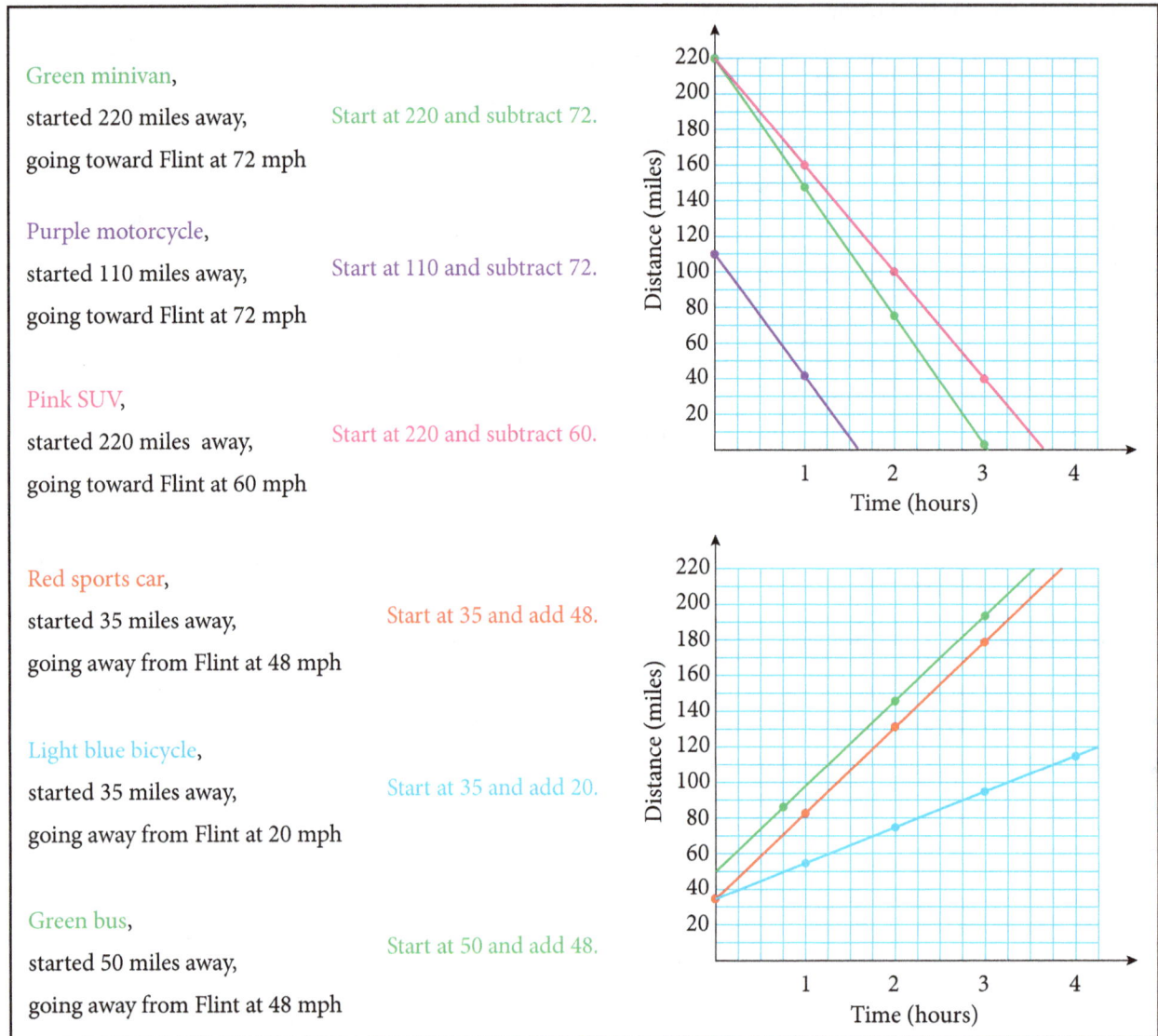

Green minivan,

started 220 miles away,

going toward Flint at 72 mph

Start at 220 and subtract 72.

Purple motorcycle,

started 110 miles away,

going toward Flint at 72 mph

Start at 110 and subtract 72.

Pink SUV,

started 220 miles away,

going toward Flint at 60 mph

Start at 220 and subtract 60.

Red sports car,

started 35 miles away,

going away from Flint at 48 mph

Start at 35 and add 48.

Light blue bicycle,

started 35 miles away,

going away from Flint at 20 mph

Start at 35 and add 20.

Green bus,

started 50 miles away,

going away from Flint at 48 mph

Start at 50 and add 48.

Algebra Problem Strings
©2017 Kendall Hunt Publishing

Facilitation Notes

This version of the problem string lists short notes for important teacher moves during the string. After you've done the string yourself and studied the relationships involved, you might make similar notes for the things you want a reminder of or deem important.

Green minivan, started 220 miles, toward Flint, 72 mph	We've got cars driving north–south or vice versa in Michigan. Stay in context throughout string. What is this car's recursive rule? Sketch a graph. Connected? Where does the starting point show up in the rule? The graph? Where does the rate show up in the rule? The graph?
Purple motorcycle, started 110 miles, toward Flint, 72 mph	New vehicle. What is this bike's recursive rule? Sketch a graph on same grid as previous. Where does the starting point show up in the rule? The graph? Where does the rate show up in the rule? The graph? How does the graph compare to the green minivan? Why?
Pink SUV, started 220 miles, toward Flint, 60 mph	New vehicle. What is this SUV's recursive rule? Sketch a graph on same grid as previous two vehicles. Where does the starting point show up in the rule? The graph? Where does the rate show up in the rule? The graph? How does the graph compare to the green minivan and the purple motorcycle? Why?
Red sports car, started 35 miles, away from Flint, 48 mph	New vehicle and direction. What is this car's recursive rule? Sketch a graph on new grid. Where does the starting point show up in the rule? The graph? Where does the rate show up in the rule? The graph? How does the graph compare to the green minivan? Why?
Light blue bicycle, started 35 miles, away from Flint, 20 mph	New vehicle, same direction. What is this bike's recursive rule? Sketch a graph on same grid as previous. Where does the starting point show up in the rule? The graph? Where does the rate show up in the rule? The graph? How does the graph compare to the red sports car? Why?
Green bus, started 50 miles, away from Flint, 48 mph	New vehicle. What is this bus' recursive rule? Sketch a graph on same grid as previous two vehicles. Where does the starting point show up in the rule? The graph? Where does the rate show up in the rule? The graph? How does the graph compare to the red sports car and the blue bike? Why?

Time-Distance Rates

At a Glance

Time (s)	Dist (ft)
0	5
3	11
6	17

Time (s)	Dist (ft)
0	10
2	6
4	2

Time (s)	Dist (ft)
2	12
4	6
6	0

Time (s)	Dist (ft)
3	30
6	45
9	60

Time (s)	Dist (ft)
0	5
4	7
8	9

Time (s)	Dist (ft)
2	9
6	6
12	3

Objectives

This problem string is designed to give students experience finding rates of change that are not whole number ratios.

Placement

This problem string supports the work of graphing sequences and finding the rates of time-distance plots that represent walks in front of a motion detector. Ideally, students will have had prior experiences actually walking in front of a motion detector or using a simulator and realizing what happens in the time-distance plots as they change their walks.

You can use this problem string to support textbook Lesson 3.3 Time-Distance Relationships.

Guiding the Problem String

The problems in this string consist of data points that represent the distance at certain times for walkers walking at constant rates in front of a motion detector. The first two problems give the starting point, but students have to find the rate. The next two problem do not give the starting point, so students must find the rate and work to find the starting point. These first four problems have rates greater than 1 foot per second. The last two problems give the starting point so students can grapple with rates between 0 and 1 foot per second.

Stay in context throughout the string. Use both the table and graph to model students' thinking, except for problem four which uses big numbers to nudge students to envision the relationships instead of graphing. Talk about the rates as ratios of distance to time. Refrain from talking about a rate of 2 feet per second as a rate of 2.

About the Mathematics

This problem string is based on the scenario of a person walking away or toward a motion detector at a constant rate. Because the rate is constant, the resulting plot of (elapsed time, distance) is linear. The plot is increasing if the rate is positive (walking away/gaining distance) and decreasing if the rate is negative (walking toward/losing distance.)

The starting point (distance from the motion detector) of each walker in these sequences corresponds to the y-intercept in linear equations. The walking rate corresponds to the rate of change (slope) of linear equations. This problem string is a precursor for students who will eventually write functions to model linear data.

Students record or display their thinking in a *model*—in this case, in a table, graph, or with words. Later students will write linear function models. This problem string helps to connect the representations, the *models,* of the table and the graph.

A *strategy* describes how students use the numbers and the way they use the relationships.

Keep your language about models and strategies consistent. For example, students can use the strategy of working backward to find the starting point in either model, the table or the graph. The strategy is working backward, the model is the table or the graph.

Sample Interactions

Use the following as you plan how to elicit and model student strategies. This is not meant as a script, but as a view into the relationships involved and the intent of the problem string.

Teacher: *So, we've been walking in front of those motion detectors. Cool, huh? Today we're going to do a problem string based on that. So, there was a student walking in front of the motion detector and this is what we know. When the detector started, he was 5 feet in front of the sensor. Then 3 seconds later he was at 11 feet and at 6 seconds, at 17 feet.*

The teacher writes the information on the board in a table.

Time (s)	Dist (ft)
0	5
3	11
6	17

Teacher: *First, tell me a little about this walk.*

Student: *Walking away.*

Student: *In three seconds, you went 6 feet.*

Student: *Started 5 feet in front.*

Teacher: *What walking instructions would you give to a different student who needs to replicate this walk? Try to find instructions that are just enough to get it right. Tell your partner.*

Students turn and talk while the teacher listens in, looking for students who need help getting started.

Teacher: *If I were to walk this right now, what instructions would you give me? What do I need to know? Someone start us off by telling me where I need to start and how you know.*

Student: *Start 5 feet in front because at time 0, the guy is at 5 feet.*

Time (s)	Dist (ft)
0	5 → start 5 feet in front
3	11
6	17

Teacher: *I'm going to put this data on a graph and let's mark that starting point. Someone start us off and tell me what to graph.*

The teacher elicits instructions about graphing the data and marking the starting point.

Teacher: *Then what? Away or toward the motion detector and how do you know?*

Student: *Away because the numbers are getting bigger.*

Teacher: *Which numbers? Does it matter?*

Student: *As time is happening, the numbers in distance are getting bigger. That means you're getting farther away.*

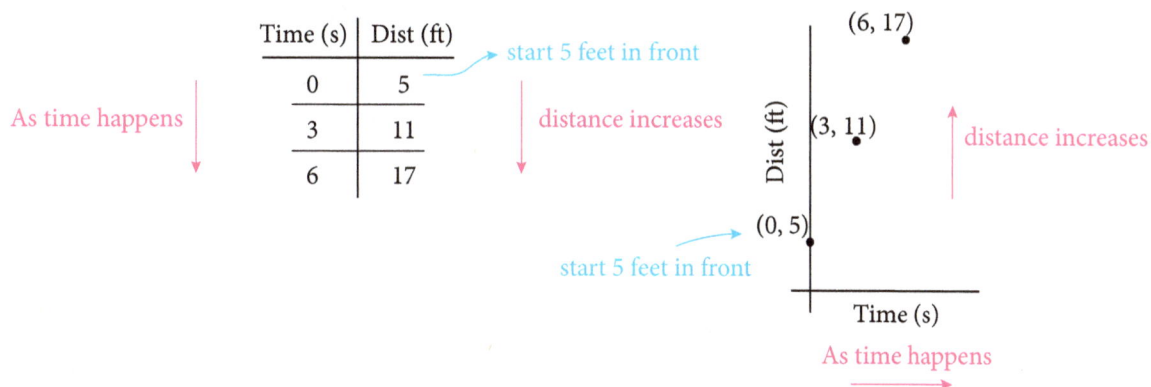

Teacher: *What else do I need to know? I know where to start and to walk away. What else?*

Student: *You need to know how fast.*

Teacher: *How fast?*

Student: 6 *feet every 3 seconds.*

Student: *Or 2 feet per second.*

Teacher: *How do you know?*

Student: *The table says that for every 6 feet he walked, 3 seconds went by. So 6 feet in 3 seconds.*

Student: *And that's like 2 feet over 1 second.*

$$\frac{6 \text{ feet}}{3 \text{ sec}} = \frac{2 \text{ feet}}{1 \text{ sec}} = 2 \text{ ft per sec}$$

Teacher: *Tell me where the those 6 feet and 3 seconds come from. I'll model what you say by marking both on the table and the graph.*

As students describe, the teacher records on both the table and the graph.

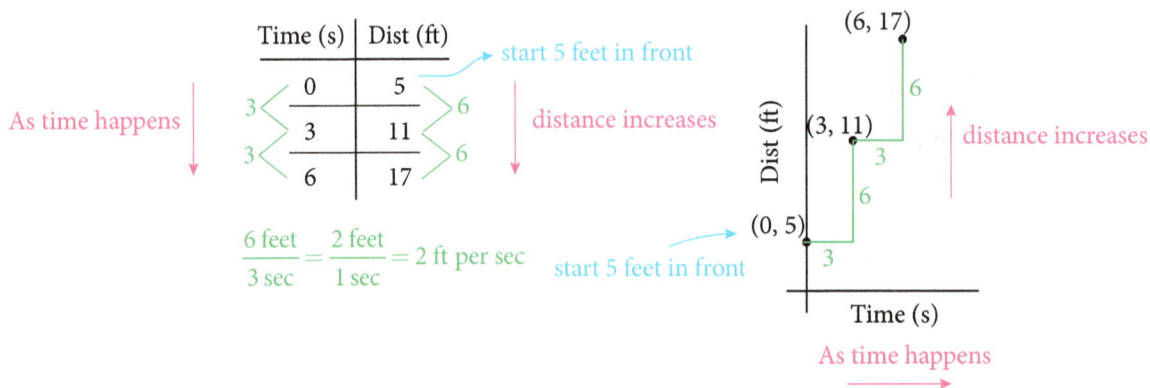

Time (s)	Dist (ft)
0	5
3	11
6	17

As time happens
distance increases
start 5 feet in front

$$\frac{6 \text{ feet}}{3 \text{ sec}} = \frac{2 \text{ feet}}{1 \text{ sec}} = 2 \text{ ft per sec}$$

start 5 feet in front

(6, 17)
(3, 11)
(0, 5)
Dist (ft)
Time (s)
distance increases
As time happens

Teacher: *So, put it all together. What instructions do we need?*

Student: *Start 5 feet in front and walk away at 2 feet per second.*

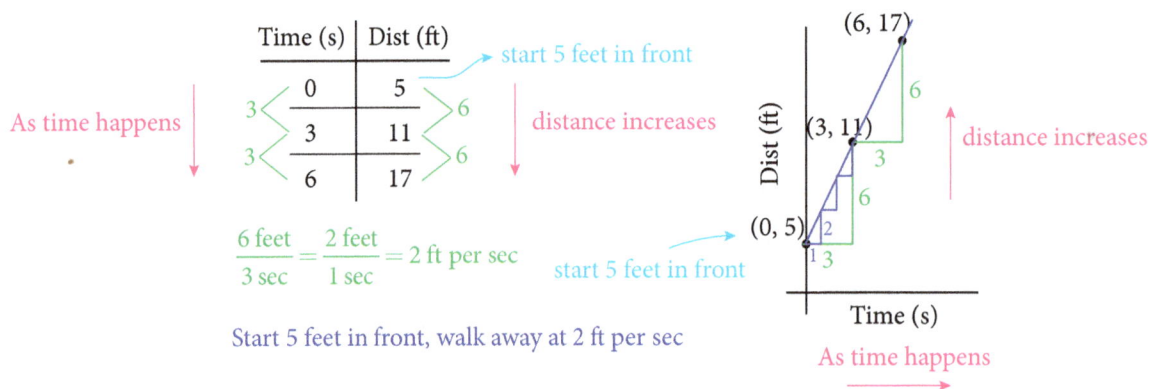

Time (s)	Dist (ft)
0	5
3	11
6	17

As time happens
distance increases
start 5 feet in front

$$\frac{6 \text{ feet}}{3 \text{ sec}} = \frac{2 \text{ feet}}{1 \text{ sec}} = 2 \text{ ft per sec}$$

Start 5 feet in front, walk away at 2 ft per sec

start 5 feet in front

(6, 17)
(3, 11)
(0, 5)
Dist (ft)
Time (s)
distance increases
As time happens

Algebra Problem Strings
©2017 Kendall Hunt Publishing

Time (s)	Dist (ft)
0	10
2	6
4	2

Teacher: *The next problem in our string is similar. We know that the walker this time is at the 10-foot mark in front of the motion detector at time 0. After 2 seconds, she is 6 feet away, and at 4 seconds she is at 2 feet. What walking instruction would you write so someone could replicate this walk? Turn and work this out with your partner.*

Students turn and talk while the teacher listens in.

The teacher crafts a conversation around creating a graph, finding the rate, and eliciting the walking directions from students.

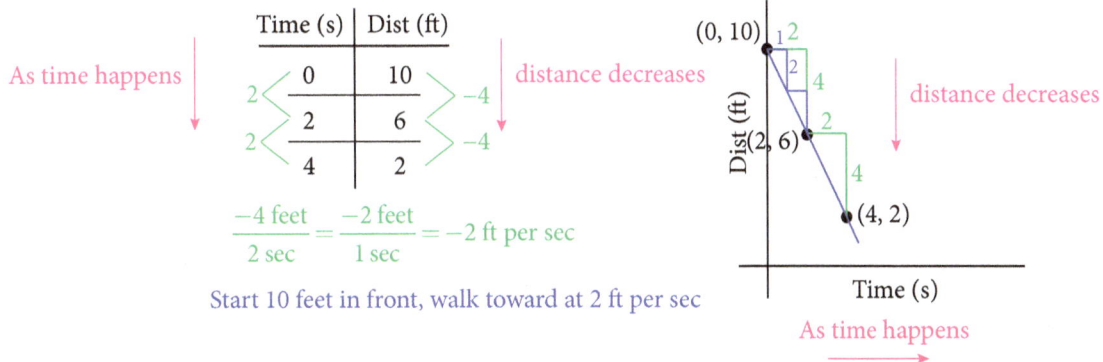

As time happens | distance decreases | distance decreases

Time (s)	Dist (ft)
0	10
2	6
4	2

$$\frac{-4 \text{ feet}}{2 \text{ sec}} = \frac{-2 \text{ feet}}{1 \text{ sec}} = -2 \text{ ft per sec}$$

Start 10 feet in front, walk toward at 2 ft per sec

(0, 10) (2, 6) (4, 2)

Dist (ft) / Time (s) / As time happens

Teacher: *The next problem in our string is a little different. We weren't paying attention at the beginning of the walk, but we know that the walker is at the 12-foot mark in front of the motion detector at time 2. At 4 seconds, he is 6 feet away, and at time 6 he is right at the motion detector. What walking instruction would you write so someone could replicate this walk? Think about this, sketch down your ideas, and then turn and share with your partner.*

Students work, then turn and talk while the teacher listens in.

Time (s)	Dist (ft)
2	12
4	6
6	0

Teacher: *I noticed that you had scooted down the values in the table. Why?*

Student: *We know we are going to have to find the starting point at time zero.*

Teacher: *Does everyone agree with that? Are we going to have to find the distance at time zero? Yes? Okay, I'll erase the titles and scoot them up so we have room. Keep telling us about your thinking?*

Student: *I found that he was walking toward the thing at 6 feet in 2 seconds. That's the same as 3 feet per second. And it's negative 3 because he's walking toward it.*

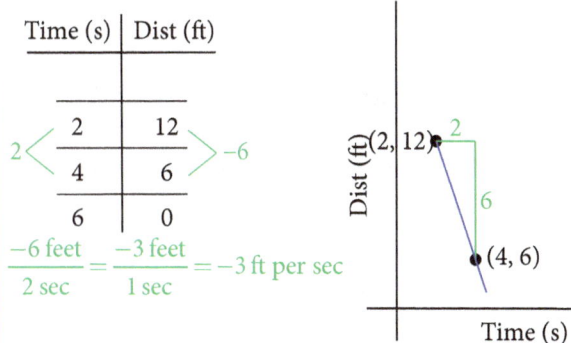

Time (s)	Dist (ft)
2	12
4	6
6	0

$$\frac{-6 \text{ feet}}{2 \text{ sec}} = \frac{-3 \text{ feet}}{1 \text{ sec}} = -3 \text{ ft per sec}$$

(2, 12) (4, 6)

Dist (ft) / Time (s)

(continued)

Teacher: *Someone else pick it up. How did you find the starting point?*

Student: *We just walked backward in the table.*

Student: *We used the graph.*

Teacher: *Does it matter? Either way, you are backing up 2 seconds and?*

Student: *And 6 feet. So that puts you at 18 feet at time 0.*

Teacher: *So, I'll model that strategy, backing up by the rate of 6 feet for every 2 seconds, on both the table and the graph. What are your walking instructions?*

Student: *Start at 18 feet and walk towards the motion detector at 3 feet per second.*

Time (s)	Dist (ft)
0	18
2	12
4	6
6	0

$$\frac{-6 \text{ feet}}{2 \text{ sec}} = \frac{-3 \text{ feet}}{1 \text{ sec}} = -3 \text{ ft per sec}$$

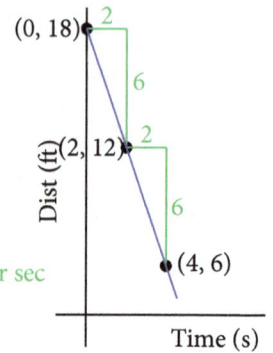

Start at 18 feet, walk toward at 3 ft per sec

Teacher: *The next problem in our string is similar. Here's the information we know. What walking instruction would you write so someone could replicate this walk? Turn and work this out with your partner.*

Students turn and talk while the teacher listens in. The numbers are such that students who were only thinking using the graph might be more inclined to use just the table.

Time (s)	Dist (ft)
3	30
6	45
9	60

The teacher crafts a conversation around finding the rate, finding the starting point, and eliciting the walking directions from students. Then the teacher discusses the efficiency of using the table because of the numbers.

Teacher: *I don't think I saw very many of you drawing a graph. Why not?*

Student: *The numbers are so big. That would a be big graph.*

Student: *We didn't need it. We got this by just dealing with the numbers.*

Teacher: *So, we've got graphs to understand what is going on when we need them, but if we don't need them, we can reason about the relationships using the table.*

Time (s)	Dist (ft)
0	15
3	30
6	45
9	60

$$\frac{15 \text{ feet}}{3 \text{ sec}} = \frac{5 \text{ feet}}{1 \text{ sec}} = 5 \text{ ft per sec}$$

Start 15 feet in front, walk away at 5 ft per sec.

Teacher: *Here's the information we know for the next walker. What walking instruction would you write so someone could replicate this walk? Turn and work this out with your partner.*

Students turn and talk while the teacher listens in, looking for students who are grappling with the magnitude of the rate.

Time (s)	Dist (ft)
0	5
4	7
8	9

Algebra Problem Strings
©2017 Kendall Hunt Publishing

Teacher: *What can we say about the rate of this walker?*

Student: *It's slow. Slower than any of the other ones we've just done.*

Teacher: *How are you thinking about the rate?*

Student: *Well, we know that she is covering 2 feet in 4 seconds. So that's like 1 foot in 2 seconds. Does that work?*

Teacher: *What do you all think? What do you think about a rate of 1 foot in 2 seconds?*

Student: *It's right, but it's not for 1 second.*

Teacher: *Can we find the rate for 1 second? How far would she walk in just 1 second if she's going 2 feet in 4 seconds and 1 foot in 2 seconds?*

Student: *That's just ½ of a foot in 1 second.*

Teacher: *I'll sketch that in the table and the graph. So, how many of those little half-foot per 1 second would fit in this 2 feet per 4 seconds triangle?*

Student: *Four of them.*

Teacher: *What walking instructions work for this walk?*

Student: *Start at 5 feet and walk away at ½ foot per second.*

Time (s)	Dist (ft)
0	5
4	7
8	9

4 ... 2 ; 4 ... 2

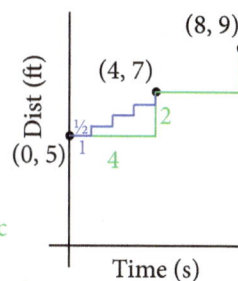

(8, 9)   (4, 7)   (0, 5)

$$\frac{2\ \text{feet}}{4\ \text{sec}} = \frac{1\ \text{foot}}{2\ \text{sec}} = \tfrac{1}{2}\ \text{ft per sec}$$

Start at 5 feet, walk away at ½ ft per sec

Teacher: *Here's the information we know for the next walker. What walking instruction would you write so someone could replicate this walk? Turn and work this out with your partner.*

Students turn and talk while the teacher listens in, looking for students who are grappling with the sign of the rate.

Time (s)	Dist (ft)
0	9
6	6
12	3

The teacher crafts a conversation around finding the rate, finding the starting point, and eliciting the walking directions from students.

Time (s)	Dist (ft)
0	9
6	6
12	3

6 ... -3 ; 6 ... -3

$$\frac{-3\ \text{feet}}{6\ \text{sec}} = \frac{-1\ \text{foot}}{2\ \text{sec}} = -\tfrac{1}{2}\ \text{ft per sec}$$

Start at 9 feet, walk toward at ½ ft per sec

Teacher: *What information did we always need when we found the walking directions?*

Student: *The starting point, the rate, and the direction.*

Student: *You can say the direction by the sign of the rate.*

(continued)

Teacher: *How would you summarize some of the things that came up in this string today?*

Elicit the following:

- *The starting point is at time 0.*

- *The rate in this scenario is the distance traveled in the time. The unit rate is the distance traveled in 1 second.*

- *If we don't know the starting point, we can find it by working backwards.*

- *The sign of the rate tells you the direction of the walker, away (gaining distance) is positive and toward (losing distance) is negative.*

- *You can use the table and the graph to help you think about rate.*

Algebra Problem Strings
©2017 Kendall Hunt Publishing

Sample Final Display

Your display could look like this at the end of the problem string:

Time (s)	Dist (ft)
0	5
3	11
6	17

As time happens ↓   start 5 feet in front   distance increases ↓

$$\frac{6 \text{ feet}}{3 \text{ sec}} = \frac{2 \text{ feet}}{1 \text{ sec}} = 2 \text{ ft per sec}$$

start 5 feet in front

Start 5 feet in front, walk away at 2 ft per sec.

(6, 17)   (3, 11)   (0, 5)

distance increases ↑

Dist (ft) / Time (s)

As time happens →

Time (s)	Dist (ft)
0	10
2	6
4	2

As time happens ↓   distance decreases ↓

$$\frac{-4 \text{ feet}}{2 \text{ sec}} = \frac{-2 \text{ feet}}{1 \text{ sec}} = -2 \text{ ft per sec}$$

Start 10 feet in front, walk toward at 2 ft per sec.

(0, 10)   (2, 6)   (4, 2)

distance decreases ↓

Dist (ft) / Time (s)

As time happens →

Time (s)	Dist (ft)
0	18
2	12
4	6
6	0

$$\frac{-6 \text{ feet}}{2 \text{ sec}} = \frac{-3 \text{ feet}}{1 \text{ sec}} = -3 \text{ ft per sec}$$

Start at 18 feet, walk toward at 3 ft per sec

(0, 18)   (2, 12)   (4, 6)

Dist (ft) / Time (s)

Time (s)	Dist (ft)
0	15
3	30
6	45
9	60

$$\frac{15 \text{ feet}}{3 \text{ sec}} = \frac{5 \text{ feet}}{1 \text{ sec}} = 5 \text{ ft per sec}$$

Start 15 feet in front, walk away at 5 ft per sec

Time (s)	Dist (ft)
0	5
4	7
8	9

$$\frac{2 \text{ feet}}{4 \text{ sec}} = \frac{1 \text{ foot}}{2 \text{ sec}} = \tfrac{1}{2} \text{ ft per sec}$$

Start at 5 feet, walk away at ½ ft per sec

(8, 9)   (4, 7)   (0, 5)

Dist (ft) / Time (s)

Time (s)	Dist (ft)
0	9
6	6
12	3

$$\frac{-3 \text{ feet}}{6 \text{ sec}} = \frac{-1 \text{ foot}}{2 \text{ sec}} = -\tfrac{1}{2} \text{ ft per sec}$$

Start at 9 feet, walk toward at ½ ft per sec

(continued)

Facilitation Notes

This version of the problem string lists short notes for important teacher moves during the string. After you've done the string yourself and studied the relationships involved, you might make similar notes for the things you want a reminder of or deem important.

Time (s)	Dist (ft)
0	5
3	11
6	17

We've been walking in front of motion detectors. Here's what we know about a walk. What walking directions would replicate this walk?
First, tell me a little.
Draw graph. Elicit as time increases, distance increases.
Elicit starting point, non-unit rate, unit rate, direction.

Time (s)	Dist (ft)
0	10
2	6
4	2

Repeat.
Work with partner.
Discuss negative rate corresponding to "toward the sensor."

Time (s)	Dist (ft)
2	12
4	6
6	0

No starting point given.
Scoot up/down to make room for starting point.
Elicit rate.
Elicit using rate to find starting point.

Time (s)	Dist (ft)
3	30
6	45
9	60

Repeat. No graph.
Why didn't anyone use graph?

Time (s)	Dist (ft)
0	5
4	7
8	9

Rate of 1/2 ft/sec.
Elicit relationship between non-unit rate 2 ft/4 sec and 1/2 ft per second.

Time (s)	Dist (ft)
0	9
6	6
12	3

Repeat.
Rate of -1/2 ft/sec.
What information do we need to give good walking instructions?
Only graph if needed.

Algebra Problem Strings
©2017 Kendall Hunt Publishing

3.4 | Linear Equations 1

At a Glance

$y = 215 + 3.8x$

$y = 150 + 3.8x$

$y = 3.8x$

$y = -100 + 3.8x$

$y = 215$

$y = 215 + 4.5x$

$y = 215 + 2x$

$y = 350 + 2.5x$

Objectives

The goal of this problem string is to solidify the learning about linear equations that took place in the Working Out with Equations Investigation in which students used linear data to write the slope y-intercept form of the equation of a line, $y = a + bx$.

Placement

This is the first of two strings where students find an equation and corresponding graph of an exerciser who has burned an initial 215 calories then rides a stationary bike at the gym, burning an additional 3.8 calories per minute, this problem string can help students solidify how changing parts of the scenario affect the parts of the equation and graph.

You can use this problem string to follow up after the Investigation: Working Out with Equations in textbook Lesson 3.4 Linear Equations and the Intercept Form.

Guiding the Problem String

This problem string is designed to get students thinking about the role of the y-intercept or "starting value," and rate of change in linear relationships. We begin by introducing, or re-introducing, the character of Manisha, a fitness-loving and calorie-counting character. We purposely move between representations in this string—from context to equation to graph and back again—so that students are making sense of a changing y-intercept and rate, and how those models connect. It's ideal if each pair of students has one grapher between them so that they can verify their predictions and compare them with your displays as the problem string progresses.

At first Manisha jogs to the gym and then works out on a stationary bike at the rate of 3.8 calories per minute. This establishes our story with the graph and equation. As you facilitate the rest of this string, present each problem as a scenario. The next three problems tinker with her starting value, the y-intercept, so that students realize how the y-intercept appears in the graph and the equation, and what it means for her total calories burned over time. The last three problems tinker with the rate of calorie burning once Manisha gets to the gym so that students realize how the rate of change appears in the graph and equation, and what it means for her total calories burned over time.

The last problem changes both the rate and y-intercept simultaneously. This can be a good moment for you to assess students' understanding and decide how you will proceed.

About the Mathematics

The slope y-intercept form of the equation of a line is sometimes written as $y = mx + b$. We use the form $y = a + bx$ because it more closely fits the scenarios where the y-intercept is the starting value (the y-value at $x = 0$) and then the rate describes the slope of the line from the initial point. In this scenario, you could think about this form, $y = b + mx$ as "the y-values are equal to the *beginning value* plus how you *move* times the variable."

(continued)

Sample Interactions

Use the following as you plan how to elicit and model student strategies. This is not meant as a script, but as a view into the relationships involved and the intent of the problem string.

The teacher has previously set up the scenario of Manisha running to the gym burning 215 calories and then biking to burn an additional 3.8 calories per minute.

Teacher: *Remember this equation that represents Manisha. What does the y, the 215, the 3.8, and the x represent?*

Brief think time.

Teacher: *Turn and tell your partner what all of this means. See if you agree.*

Students turn and talk while the teacher listens in.

Teacher: *What do each of these represent? Where is the 215 on the graph? The 3.8? Convince us!*

The teacher leads a discussion where students identify the relevant parts and connect the equation and the graph. The teacher has set a window in which students can reasonably find the unit rate.

$$y = 215 + 3.8x$$

215 calories running to gym
3.8 calories per minute on bike

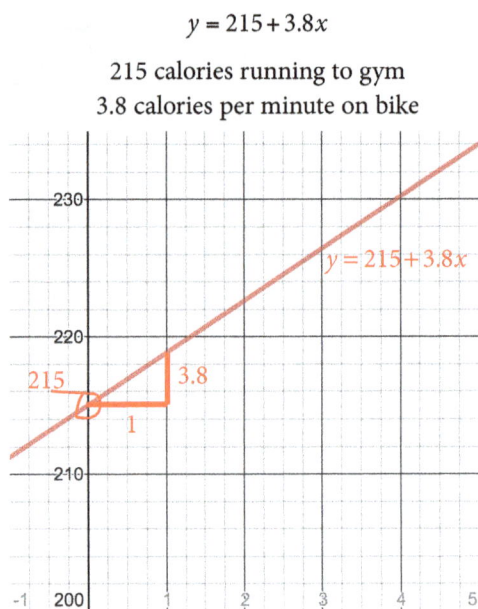

Teacher: *Okay, take out your grapher. With your partner enter this equation and change the window together so that you get a good picture of what's happening for the first 20 minutes or so of Manisha's workout.*

The teacher circulates, helping students as needed set the window parameters. The teacher elicits suggestions and changes the window for the display graph.

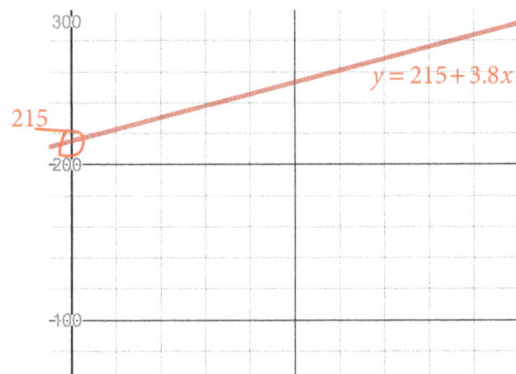

Algebra Problem Strings
©2017 Kendall Hunt Publishing

Teacher: *Well, sometimes Manisha doesn't jog to the gym. Sometimes she bikes to the gym. And she's learned that this burns exactly 150 calories from her house to the gym. So, what changed? What didn't change? Talk to your partner about what changed in the equation, and how this will look in the graph. Keep the graph of the original and add the graph of this 150 calorie bike ride to the gym.*

When students are ready, the teacher projects the new graph for students to compare to their own.

Teacher: *So what happened here? What's different and what's the same? Does that make sense? Why?*

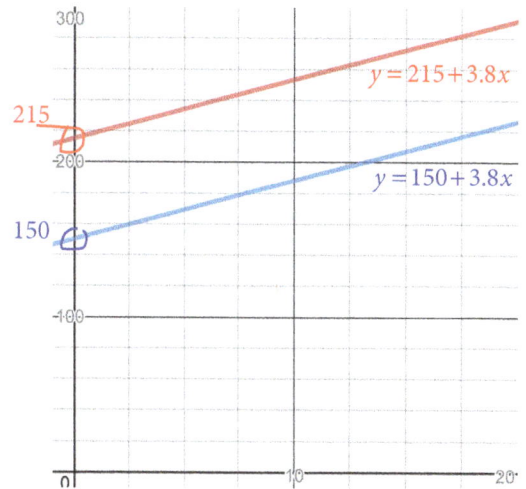

Teacher: *What would happen if Manisha got a ride to the gym from a friend? Essentially burning no calories (by exercising) as she sat in the car and got dropped off at the gym. Think for a moment about what would happen to the equation and the graph.*

Think time.

Okay, briefly tell your partner how to represent this new situation. Go!

Brief partner talk.

Will you describe what you and your partner did to represent this situation as an equation and graph? Everyone else, be listening for whether you did the same, or something different.

The teacher displays the graph as students discuss.

If it doesn't come up, the teacher puzzles aloud about why the graph goes through (0, 0) and why the lines have been shifting down the y-axis each time—why does that make sense?

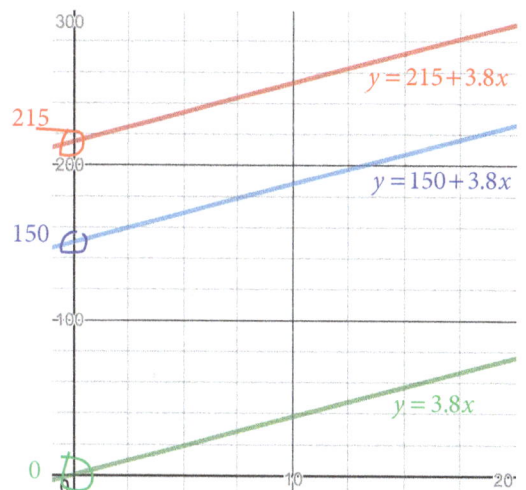

(continued)

Teacher: *Sometimes, anticipating a big workout at the gym, Manisha stops for a snack. Her favorite is a peanut butter cookie. She just gets one, and it's only 100 calories, but she eats it right before her gym workout. How does this affect how many calories she will burn during this workout? How might this change the graph and the equation?*

Think for a moment, then get ready to talk to your partner and graph this situation.

What's going on in the story at x-intercept? What does it have to do with burning off the 100-calorie cookie?

The teacher could open up the window to show the *x*-intercept.

Teacher: *Let's take a minute to look at all four situations.*

What is going on here? What do you notice about these four workouts for Manisha?

What's changing across these four graphs? And what's staying the same? What explains this?

If needed:

Why do the graphs each start in a different place—what explains this?

Why do they look parallel—what explains this?

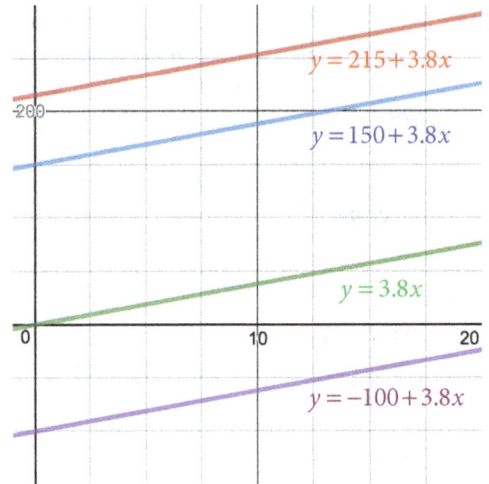

$y = 215 + 3.8x$

$y = 150 + 3.8x$

$y = 3.8x$

$y = -100 + 3.8x$

Teacher: *Now, one day Manisha ran to the gym, turned her ankle and had to be driven home. What would that situation look like, in the equation and in the graph? Why?*

Why is the line horizontal?

What does the y-intercept represent?

What changed? What didn't change?

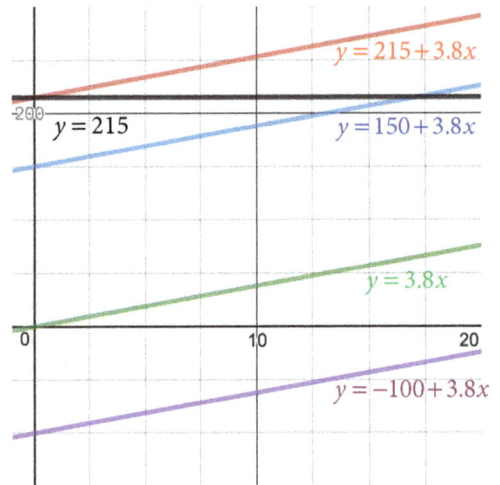

$y = 215 + 3.8x$

$y = 215$

$y = 150 + 3.8x$

$y = 3.8x$

$y = -100 + 3.8x$

Algebra Problem Strings
©2017 Kendall Hunt Publishing

Teacher: *On a different day, Manisha still ran to the gym, but this time she decided to try a new elliptical machine and she was able to burn 4.5 calories per minute. How will that affect the equation and the graph? Why?*

What changed? What didn't change?

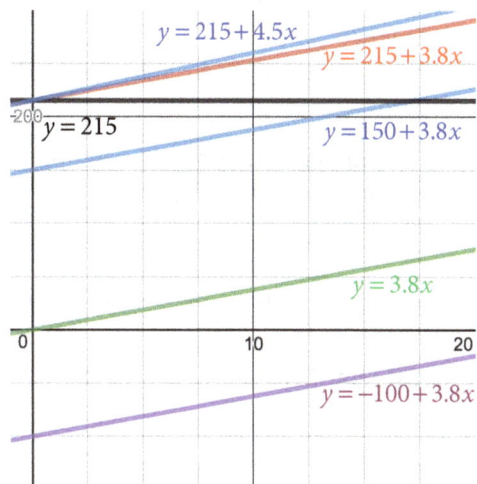

$y = 215 + 4.5x$
$y = 215 + 3.8x$
$y = 215$
$y = 150 + 3.8x$
$y = 3.8x$
$y = -100 + 3.8x$

Teacher: *Manisha decided that she was tired so instead of biking after she ran to the gym, she swam laps instead, burning only 2 calories per minute. How will that affect the equation and the graph? Why?*

What changed? What didn't change?

Look across the last 3 graphs. Why are they not parallel?

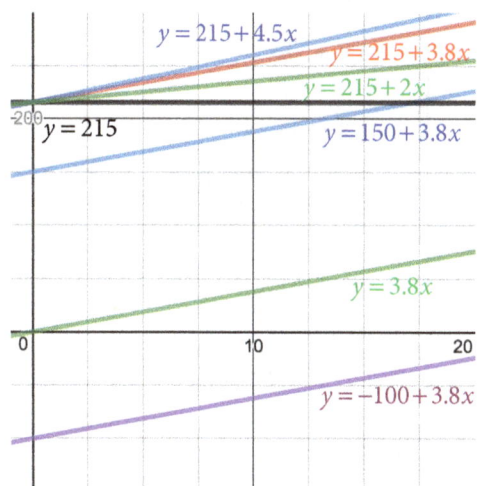

$y = 215 + 4.5x$
$y = 215 + 3.8x$
$y = 215 + 2x$
$y = 215$
$y = 150 + 3.8x$
$y = 3.8x$
$y = -100 + 3.8x$

Teacher: *Here's the last scenario. What would happen if Manisha walked to the gym burning 100 calories and then tried a new rowing machine that burned 5.4 calories per hour? What would that graph look like? How might it be related to anything we've talked about already?*

After you've got your equation graphed in your grapher, let me know and I'll come check in with you. Then we'll discuss.

Let's compare only the first and last scenarios. I'll turn off all the others.

What changed? What didn't change? Why?

What does the y stand for in this situation? The x?

In the equation $y = a + bx$, what does the a mean in the graph and the situation? What does the b mean in the graph and the situation?

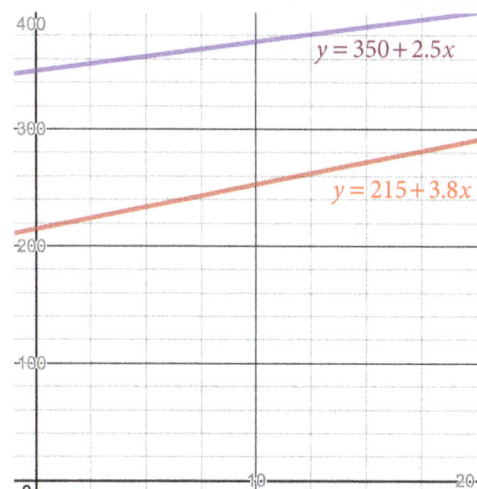

$y = 350 + 2.5x$
$y = 215 + 3.8x$

(continued)

Teacher: *How would you summarize some of the things that came up in this string today?*

Elicit the following:

- *In the equation y = a + bx, the constant "a" is the y-intercept and represents the initial calories burned in this scenario.*

- *In the equation y = a + bx, the constant "b" is the rate of change and represents the calories per minute burned in the gym over time in this scenario.*

- *When the y-intercept increases or decreases, the graph shifts up or down.*

- *When the rate increases or decreases, the graph rotates, increasing (in this case) at a different rate.*

Sample Final Display

Your display could look like this at the end of the problem string:

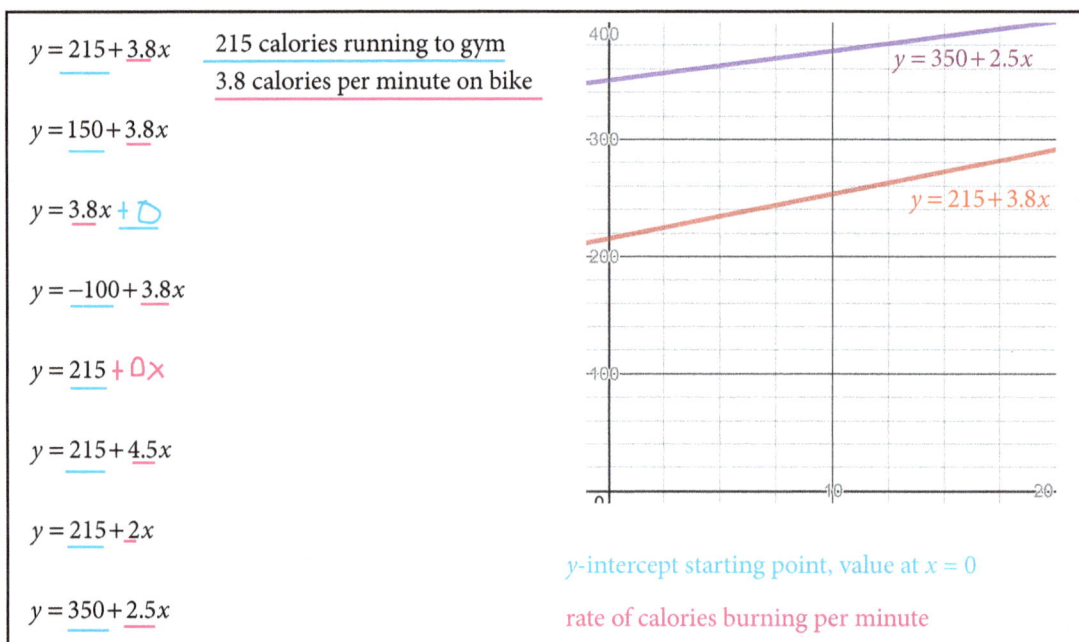

$y = 215 + 3.8x$ 215 calories running to gym
               3.8 calories per minute on bike

$y = 150 + 3.8x$

$y = 3.8x + \square$

$y = -100 + 3.8x$

$y = 215 + \square x$

$y = 215 + 4.5x$

$y = 215 + 2x$

$y = 350 + 2.5x$

$y = 350 + 2.5x$

$y = 215 + 3.8x$

y-intercept starting point, value at $x = 0$

rate of calories burning per minute

Facilitation Notes

This version of the problem string lists short notes for important teacher moves during the string. After you've done the string yourself and studied the relationships involved, you might make similar notes for the things you want a reminder of or deem important.

$y = 215 + 3.8x$	Remember Manisha? Runs to the gym and burns 215 calories, then bikes, burning additional 3.8 calories per minute. What does graph look like? Where do you see the 215? The 3.8? Graph in small window and draw in a unit slope triangle. Have students graph a 20 minute window and change display graph accordingly.
$y = 150 + 3.8x$	Present all the rest as scenarios. Students write and graph equations. What would happen if Manisha bikes to gym, burning 150 calories? Partners predict. Display both graphs together. What changed? What didn't change? Partner talk.
$y = 3.8x$	This day Manisha got dropped off at the gym. Repeat. Display all three graphs together. Why (0, 0)?
$y = -100 + 3.8x$	A different day, she ate a snack while getting dropped off. Repeat, but change window. Why negative y-intercept? Look at 4 graphs. What do you notice? What's changing across the four graphs? What stays the same? Why parallel?
$y = 215$	Oh no! After running to the gym, she turned her ankle and got a ride home. Repeat. Why horizontal?
$y = 215 + 4.5x$	What if she ran to the gym like normal, but worked out extra hard, 4.5 calories per min? Repeat.
$y = 215 + 2x$	Another day is a tired day—after running, she barely burned 2 calories per min. Repeat. Look at last 3 graphs. Why not parallel?
$y = 350 + 2.5x$	Predict! What will change and how, in the equation? In the graph? Write the equation and graph. Display the original graph and this graph. Compare. Why?

At a Glance

Start at 2 feet, walk away at 1.5 feet per second

$y = -2 + 1.5x$

$y = 10 - x$

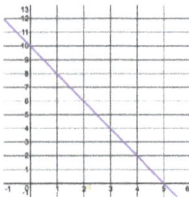

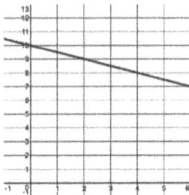

Time (s)	Dist (ft)
0	12
1	11

Time (s)	Dist (ft)
0	5
2	3

$y = 3 + 2.5x$

Objectives

The goal of this problem string is to continue to solidify the parameters in the slope y-intercept form of the equation of a line, $y = a + bx$. In this problem string, the rates and y-intercepts are positive and negative.

Placement

This problem string is the second in a series of two strings. This string extends the work of problem string 3.4 where students continue to construct and/or solidify ideas about a changing y-intercept and rate of change, this time set in the context of walking in front of a motion detector, making sense of both positive and negative values.

This problem string supports the work of textbook Lesson 3.5 Linear Equations and Rate of Change.

Guiding the Problem String

This problem string is designed to help students connect the parameters of a linear situation with the representations—from context to equation to table to graph and back again—so that students are making sense of the meaning of a changing y-intercept and rate of change. It's ideal if each pair of students has one grapher between them. You might assign one student to tell the other student what to enter and the other student to do the entering. This can encourage communication between partners.

Ideally students would have experience walking in front of a motion detector or playing with a motion detector simulator. The first problem starts with the scenario that a walker starts 2 feet in front of the detector and is moving away at a constant rate of 1.5 feet per second. Ask students to enter the equation and get a graph in a suitable viewing window. The second problem is to make sense of the given equation—that the walker started 2 feet behind the detector—and create the graph. After each problem discuss the meaning of the y-intercept and rate and where they appear in the equation, table, and graph. The third question is also a given equation, but the fourth and fifth questions are graphs for which students are to write the equation and describe the context. The sixth and seventh equations are given in tables. The last question is given as an equation.

About the Mathematics

The purpose of a problem string like this is not just to help students create different representations of a scenario, but rather to help students connect the different representations. We all understand linearity better when we connect the parts of scenarios, graphs, equations, and tables together.

Algebra Problem Strings
©2017 Kendall Hunt Publishing

Important Questions

Use the following as you plan how to elicit and model student strategies.

- What information can help to write the equation of a line?

- Where can you find the y-intercept in an equation? A graph? A motion detector scenario? A table?

- Where can you find the rate of change in an equation? A graph? A motion detector scenario? A table?

- What is true about equations of lines that have graphs that are parallel?

- What is true about equations of lines that have graphs that have the same y-intercept?

How would you summarize some of the things that came up in this string today?

- *The y-intercept is the y-value at $x = 0$, when the time starts on a motion detector, and the a in $y = a + bx$.*

- *The rate of change is the slope of the line, the speed and direction of the walker, and the b in $y = a + bx$.*

- *If the rates are equivalent, the lines are parallel.*

- *If the y-intercepts are equivalent, the lines intersect at the y-intercept.*

Sample Final Display

Your display could look like this at the end of the problem string. Note that the graph showing is just for the last problem as the display grapher is updated for each problem.

Start at 2 feet, walk away at 1.5 feet per second.	$y = 2 + 1.5x$
$y = -2 + 1.5x$	Start at 2 feet behind the sensor, walk away at 1.5 ft/sec.
$y = 10 - x$	Start 10 feet in front of the sensor, walk toward at 1 ft/sec.

$y = 10 - 2x$
Start 10 feet in front of the sensor, walk toward at 2 ft/sec.

$y = 10 - 0.5x$
Start 10 feet in front of the sensor, walk toward at ½ ft/sec.

Time (s)	Dist (ft)
0	12
1	11

$y = 12 - x$
Start 12 feet in front of the sensor, walk toward at 1 ft/sec.

Time (s)	Dist (ft)
0	5
2	3

$y = 5 - x$
Start 5 feet in front of the sensor, walk toward at 1 ft/sec.

$y = 3 + 2.5x$

Start 3 feet in front of the sensor, walk away at 2.5 ft/sec.

$y = 3 + 2.5x$

(continued)

Facilitation Notes

This version of the problem string lists short notes for important teacher moves during the string. After you've done the string yourself and studied the relationships involved, you might make similar notes for the things you want a reminder of or deem important.

Start at 2 feet, walk away at 1.5 feet per second	Remember motion detectors? What would the graph and equation look like for this walker? One partner enter on the grapher, the other partner dictate as you both decide. Display equation, graph. Compare.
$y = -2 + 1.5x$	What does this equation mean if it represents a walker? Partners predict. Display both graphs together. What changed? What didn't change? Partner talk. Walker started behind the motion detector.
$y = 10 - x$	Repeat.
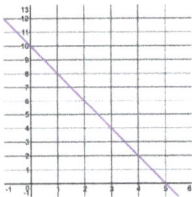	Present graph of y = 10 − 2x. What is the scenario? What is the equation? Enter your prediction. Check graph, table. Display last graph and this one together. What is the same? Different? Why?
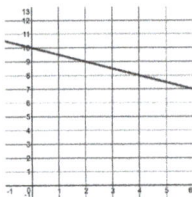	Present graph of y = 10 − 0.5x. Repeat. How did you find the rate? Draw in a rate unit triangle.
Time (s) Dist (ft) 0 — 12 1 — 11	Present table of values for y = 12 − x. Predict equation! Enter your prediction. Check graph, table. Display last four graphs and this one together. What is the same? Different? Why?
Time (s) Dist (ft) 0 — 5 2 — 3	Present table of values for y = 5 − x. Predict equation! Enter your prediction. Check graph, table. Display last five graphs and this one together. What is the same? Different? Why?
$y = 3 + 2.5x$	Clear grapher. Predict graph! Enter your prediction. Check graph, table. What is the same? Different? Why?

Algebra Problem Strings
©2017 Kendall Hunt Publishing

Solving Equations Using Relational Thinking

At a Glance	Objectives

Objectives
This problem string helps students solve equations using relational thinking on an open double number line and connect those relational moves to the recording of balancing moves.

At a Glance

$2x + 3$

$$\frac{\quad\quad\quad|\quad\quad\quad}{17}$$

$2x - 3 = 17$

$2x + 3$

$$\frac{\quad\quad\quad|\quad\quad\quad}{-17}$$

$2x - 3 = -17$

$2 + 4x = x + 8$

Follow-up String

$14x = 28$
$14x + 10 = 66$
$-14x + 10 = -46$
$2x + 4 = -22$
$2(x + 4) = -18$
$2x + 4 = 3x - 5$

Placement
In the two string series in 2.7 and 2.8, the open double number line was introduced as an equivalence model for supporting students to compare and relate expressions. This problem string continues the work of finding equivalent equations toward solving equations but also connects what is happening in the open double number line to the balancing moves in equations.

You can use this problem string to support the work in textbook Lesson 3.6 Solving Equations Using the Balancing Method.

Guiding the Problem String
Present the first problem of this string by drawing the double open number line that represents the equation $2x + 3 = 17$, asking students to interpret it and what else they know. Ask them to think about related equations or "truths" that would help us move closer to knowing what x is, our eventual goal. Encourage students to develop equivalent and simpler equations and eventually find the value of x. The model is a visual representation of how the quantities are related and is essential to the string. As students suggest equivalencies on the model, represent their moves with equations. Try to avoid using language like "add three to both sides" and replace it with language that shows how the quantities are related and are positioned on a number line. Focus also on sense-making: "If $2x + 3$ is here on the number line, would it make sense that $2x$ is here? What do we think?" Notice that throughout the sample dialogue we use the language of location and distance—an important pedagogical shift from "what is x?" to "where is x?"

Expect that as negative quantities emerge in problems 3 and 4 students may need to slow down to verify their reasoning about integers, which often can be shaky. Get students to defend their thinking to their peers. The last problem in the string is different from the previous set, so that students can begin to make use of their relational thinking when solving new problems. If it proves to be too big a challenge, let students think about it—and identify what exactly made it hard to think about—and return to it another day.

You can use the follow up string to solve other equations using relational thinking.

About the Mathematics
For more introductory information see problem string lessons 2.7 and 2.8.

The equation solving "moves" on an open double number line can be represented both on the open double number line model and using equations as a model. The moves students choose to make and in which order are strategies. How you and students choose to represent those strategies, whether on the open double number line or with equations, are models. If you want to know how students are messing with the relationships, ask about strategy, not models. Were you thinking about finding $4x$ first? $3x$? Not, did you use a number line or equations or balancing?

(continued)

Sample Interactions

Use the following as you plan how to elicit and model student strategies. This is not meant as a script, but as a view into the relationships involved and the intent of the problem string.

Teacher: *For our problem string today, I'm going to put these two expressions, 2x + 3 and 17, in the same location on this open double number line. If these two expressions are at the same location, what else do we know? What other relationships?* As students think, the teacher records the equation.	$2x + 3$ ———————|——————— 17 $2x + 3 = 17$
Teacher: *What else do we know? For example, where would 2x be? And what else can you say about it?* **Student:** *I think it would be to the left of what we have so far.* **Teacher:** *So, if I go 3 over here, that's 2x? Yes? So, what is 3 to the left of 17?* **Student:** *It would have to be 3 less, so 14.* **Teacher:** *And what equation matches where 2x and 14 just ended up?* **Student:** *2x equals 14.*	$2x \;_3\; 2x + 3$ ————⌢———— 17 $2x \;_3\; 2x + 3$ ————⌢———— $14 \;^3\; 17$ $2x = 14$

Teacher: *I've just recorded your thinking on the open double number line. I'm also going to record it using equations. You said that we know that 2x is 3 to the left of 2x + 3. I can record that as 2x + 3 subtract 3. And you said that 3 to the left of 17 is 14. I can record that as 17 subtract 3. I can write that horizontally. Or I can also write that vertically. Could someone say in their own words how these equations can represent what we were thinking on the number line?*

Student: *From 2x + 3, you go back 3 to get to 2x. So you also go back 3 from 17 to get to 14.*

$2x \;_3\; 2x + 3$
————⌢————
$14 \;^3\; 17$

$2x + 3 = 17$
$2x = 14$

$2x + 3 - 3 = 17 - 3$
$2x = 14$

$$2x + 3 = 17$$
$$\underline{-3 \quad -3}$$
$$2x = 14$$

Algebra Problem Strings
©2017 Kendall Hunt Publishing

Teacher: *So now think about two things: Where is zero on this number line? And where is x?*	
Take a moment to think about both of those ideas and then we'll talk.	
Brief think time.	$0x \qquad x \qquad 2x_{\ 3}\ 2x+3$
Teacher: *Where is zero? Where is x?*	$0 \qquad 7 \qquad 14\ ^3\ 17$
Student: *Zero is to the left of 14, so it's also to the left of 2x. And x would be halfway between 0 and 2x. That's at 7.*	
Teacher: *Who understand what she is telling us? Let's take this in pieces. Is 0 to the left of 14?*	
Student: *Yes, on a number line, it goes from left to right, small to big.*	
Teacher: *And is x is in the middle of 0 and 2x? Is this always going to be true?*	
Student: *I don't know how to explain it, but x is half of 2x and halves go in the middle. Imagine 20 and 10 and 0, where x is 10 and 2x is 20. You can kind of picture that x is the middle, or the half of 2x.*	
Student: *Yeah, knowing where 0 is helps. Since 14 is there, then 0 is to the left. Then x is in between 0 and 2x.*	

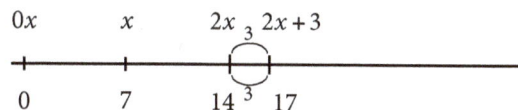

Teacher: *I'm going to record what we just did with equations. You said that you were thinking about x being in the middle of 0 and 2x, or half of 2x. We can represent that by writing 2x divided by 2. Then you said you thought about half of 14, so we can write 14 divided by 2. We can write that horizontally or vertically.*

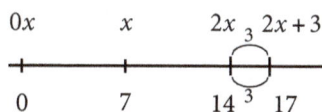

$0x \qquad x \qquad 2x_{\ 3}\ 2x+3$

$0 \qquad 7 \qquad 14\ ^3\ 17$

$2x + 3 = 17$

$2x = 14$

$x = 7$

$2x + 3 - 3 = 17 - 3$

$2x = 14$

$2x \div 2 = 14 \div 2$

$$2x + 3 = 17$$
$$\underline{-3 \quad -3}$$
$$\frac{2x}{2} = \frac{14}{2}$$

Teacher: *Let's try another one. This time I'll put up the equation and you tell me what the number line could look like.*	$2x - 3 = 17$
The teacher writes $2x - 3 = 17$.	$2x - 3$
Student: *Just put them in the middle at the same place.*	17
Teacher: *So what would help us to reason about this equation?*	
Student: *I think we should find just plain 2x.*	$2x-3_{\ 3}\ 2x$
Teacher: *Why?*	17
Student: *Because we are trying to solve these, which means that eventually we want to know where x is.*	
Teacher: *Okay—so where's 2x and what else can you say about it?*	$2x-3_{\ 3}\ 2x$
Student: *This time 2x is to the right because 2x minus 3 is less than 2x and would land there. So we need to shift up 3 to 20. I guess I am saying that 2x equals 20.*	$17\ ^3 20$

(continued)

Teacher: *I'll model what you just said with equations. If you move right 3 of 2x − 3, we can write that as 2x − 3 plus 3. And then since you moved right 3, right 3 of 17 is 20, so 17 add 3 and you get 20. We can write that horizontally or vertically.*

$$2x - 3 \quad _3 \quad 2x$$
$$17 \quad ^3 \quad 20$$

$2x - 3 = 17$

$2x = 20$

$2x - 3 + 3 = 17 + 3$

$2x = 20$

$$\begin{array}{l} 2x - 3 = 17 \\ \underline{+3 \quad +3} \\ 2x = 20 \end{array}$$

Teacher: *Where is 0 and where is x this time? Given what we've got up here, can you tell where 0 and x should be? How?*

Student: *It's like the first problem. 0 is to the left of 17.*

Student: *And x is in between 0 and 20, so x is 10.*

Teacher: *I'll model that on the number line and with equations.*

As the teacher writes the equations, the teacher describes the moves on the number line.

$$0x \qquad x \qquad 2x-3 \; _3 \; 2x$$
$$0 \qquad 10 \qquad 17 \; ^3 \; 20$$

$2x - 3 = 17$

$2x = 20$

$x = 10$

$2x - 3 + 3 = 17 + 3$

$2x = 20$

$2x \div 2 = 20 \div 2$

$$\begin{array}{l} 2x - 3 = 17 \\ \underline{+3 \quad +3} \\ \dfrac{2x}{2} = \dfrac{20}{2} \end{array}$$

Teacher: *Here's our next one.*

The teacher writes the number line shown.

Teacher: *What's the equation for this one?*

As students say the equation, the teacher writes it next to the number line.

Teacher *And now think about where 2x would be, and what is it equivalent to? Have a quick talk with your partner about this one.*

Student: *So my partner and I said that 2x is to the left of 2x + 3. We used the first problem to be sure.*

Teacher: *What do other folks think of this? Would this make sense?*

If necessary the teacher could have a different student or two try to convince the class that any quantity that is three less than another would be three spaces to the left.

Teacher: *And what is 3 to the left of −17?*

Students: *Negative 20.*

$$2x + 3$$
$$-17$$

$2x + 3 = -17$

$$2x \quad 2x + 3$$
$$-20 \; ^3 \; -17$$

Teacher: *I'll record that horizontally this time.*

$$2x \; _3 \; 2x + 3$$
$$-20 \; ^3 \; -17$$

$2x + 3 = -17$

$2x = -20$

$2x + 3 - 3 = -17 - 3$

$2x = -20$

Algebra Problem Strings
©2017 Kendall Hunt Publishing

Student: *I got it. I know where x is!*

Teacher: *Hold on. Let's give people a minute to think about where x and 0 are. And if you've already situated both of those, be thinking about this idea that x is in the middle of 0 and 2x—is that still true? Is that always true?*

Turn to your partner and discuss that.

Students turn and talk briefly while the teacher listens in.

Teacher: *Where is 0?*

Student: *It would have to be to the right of 2x.*

Teacher: *Why?*

Student: *I think it's because of the −20 and −17. Zero is to the right of them.*

Student: *Ahhh right. That makes sense.*

Teacher: *Great! Then where is x?*

Students discuss that x has to be half way between 0 and $2x$ and they know that −10 is half way between 0 and −20. The teacher models those relationships as students describe them, making the thinking visible.

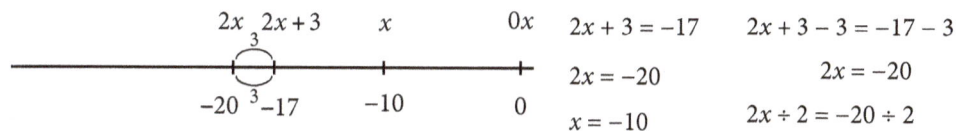

$$2x + 3 = -17$$
$$2x = -20$$
$$x = -10$$

$$2x + 3 - 3 = -17 - 3$$
$$2x = -20$$
$$2x \div 2 = -20 \div 2$$

Teacher: *Let's try another one: 2x minus 3 equals −17. Now what? Let's think for a few seconds first and then describe to our partners: Where would 2x be? Where is x? At what point can you determine where 0 is?*

You can model it any way you want, but I wonder what the open double number line would look like that models your thinking.

The teacher leads a conversation about finding $2x$, 0, and x and this time models the actions with vertical equations.

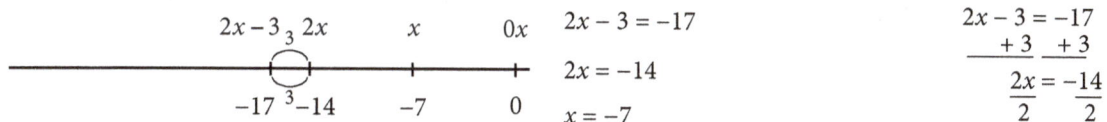

$$2x - 3 = -17$$
$$2x = -14$$
$$x = -7$$

$$2x - 3 = -17$$
$$\underline{ + 3 \quad + 3}$$
$$\frac{2x}{2} = \frac{-14}{2}$$

(continued)

Teacher: *So, here's the last problem today. Use everything you've learned so far to try to crack it. Why don't you each sketch along with me. The equation is $2 + 4x = x + 8$. Go ahead and set that up. If these two expressions are in the same place, what else do we know?*	$$2 + 4x = x + 8$$ $$2 + 4x$$ $$\rule{6cm}{0.4pt}$$ $$x + 8$$

Student: *I have two ideas—either we could find $4x$ on the top or we could find x on the bottom.*

Teacher: *Okay, why don't I give everyone a minute to think through this, using your models and your partners.*

Teacher: *Who tried finding $4x$, and could explain if that was helpful?*

Student: *Sure, so $4x$ is the same as $x + 6$ on the bottom. Everything just shifts over 2 to the left. We were gonna try to find $2x$ or x but we decided those were messy, so we found $3x$ instead.*

Teacher: *Okay, can your partner finish your thinking?*

Student: *Sure, so $3x$ is at 6.*

Student: *How did you get that?*

Student: *We shifted the whole thing back x spaces.*

Student: *But how did you know how far that was?*

Teacher: *Really good question. What do others think?*

Student: *I don't think it matters right now—we don't know how big x is right now. Later you could re-draw it if you wanted to. I think you just want to be consistent.*

Algebra Problem Strings
©2017 Kendall Hunt Publishing

The teacher circulates and looks for different starting points, nudging students who need help getting started. Then the teacher calls on students who solved the equation differently, modeling as students talk. For each strategy, the teacher elicits descriptions of where the student started.

$2 + 4x = x + 8$	$2 + 4x - 2 = x + 8 - 2$	Find $4x$ on the top first.
$4x = x + 6$	$4x = x + 6$	
$3x = 6$	$4x - x = x + 6 - x$	
$x = 2$	$3x = 6$	
	$3x \div 3 = 6 \div 3$	

$2 + 4x = x + 8$	$2 + 4x - 8 = x + 8 - 8$	Find x on the bottom first.
$4x - 6 = x$	$4x - 6 = x$	
	$4x - 6 - x = x - x$	
	$3x - 6 = 0$	
$3x - 6 = 0$	$3x - 6 + 6 = 0 + 6$	
$3x = 6$	$3x = 6$	
$x = 2$	$3x \div 3 = 6 \div 3$	

$2 + 4x = x + 8$	$2 + 4x = x + 8$	Find $3x$ on the top first.
$2 + 3x = 8$	$\underline{ -x \quad -x }$	
$3x = 6$	$2 + 3x = 8$	
$x = 2$	$2 + 3x - 2 = 8 - 2$	
	$\dfrac{3x}{3} = \dfrac{6}{3}$	

Teacher: *So, we have three different starting places where each eventually found x to be 2. Which of these do you think was efficient? Which of these do you wish your brain would lean toward naturally next time you run into problems like these?*

Student: *The middle one, finding x on the bottom first, seems like it took more steps.*

Student: *Both the second and the third ones, where you subtract x, were kind of weird to me since we didn't know how big x was at that point. But once I got over that, they weren't too bad.*

Student: *The first one seemed pretty straight forward, back 2, back x, and voila you've just got to divide by 3.*

Student: *The first and last one each moved an x and 2, just in a different order. I like both of these.*

Teacher: *How would you summarize some of the things that came up in this string today?*

Elicit the following:
- *We can use what we know.*
- *The number line goes left to right, small to big.*
- *I can add and subtract integers by realizing where they are on a number line.*
- *It can be helpful to find 0.*
- *There are alternate strategies for solving equations—I can choose how I want to start, but it makes sense to try to be efficient.*
- *We used three ways to model our thinking: open double number line, horizontal equations, and vertical equations.*

(continued)

Sample Final Display

Your display could look like this at the end of the problem string:

Row 1:
$2x + 3 = 17$
$2x = 14$
$x = 7$

$2x + 3 - 3 = 17 - 3$
$2x = 14$
$2x \div 2 = 14 \div 2$

$$2x + 3 = 17$$
$$\underline{-3 \quad -3}$$
$$\frac{2x}{2} = \frac{14}{2}$$

Row 2:
$2x - 3 = 17$
$2x = 20$
$x = 10$

$2x - 3 + 3 = 17 + 3$
$2x = 20$
$2x \div 2 = 20 \div 2$

$$2x - 3 = 17$$
$$\underline{+3 \quad +3}$$
$$\frac{2x}{2} = \frac{20}{2}$$

Row 3:
$2x + 3 = -17$
$2x = -20$
$x = -10$

$2x + 3 - 3 = -17 - 3$
$2x = -20$
$2x \div 2 = -20 \div 2$

Row 4:
$2x - 3 = -17$
$2x = -14$
$x = -7$

$$2x - 3 = -17$$
$$\underline{+3 \quad +3}$$
$$\frac{2x}{2} = \frac{-14}{2}$$

Row 5:
$2 + 4x = x + 8$
$4x = x + 6$
$3x = 6$
$x = 2$

$2 + 4x - 2 = x + 8 - 2$
$4x = x + 6$
$4x - x = x + 6 - x$
$3x \div 3 = 6 \div 3$

Find $4x$ on the top first.

Row 6:
$2 + 4x = x + 8$
$4x - 6 = x$

$2 + 4x - 8 = x + 8 - 8$
$4x - 6 = x$
$4x - 6 - x = x - x$
$3x - 6 = 0$

Find x on the bottom first.

Row 7:
$3x - 6 = 0$
$3x = 6$
$x = 2$

$3x - 6 + 6 = 0 + 6$
$3x = 6$
$3x \div 3 = 6 \div 3$

Row 8:
$2 + 4x = x + 8$
$2 + 3x = 8$
$3x = 6$
$x = 2$

$2 + 4x = x + 8$
$$\underline{-x \quad -x}$$
$2 + 3x = 8$
$2 + 3x - 2 = 8 - 2$
$$\frac{3x}{3} = \frac{6}{3}$$

Find $3x$ on the top first.

Algebra Problem Strings
©2017 Kendall Hunt Publishing

Facilitation Notes

This version of the problem string lists short notes for important teacher moves during the string. After you've done the string yourself and studied the relationships involved, you might make similar notes for the things you want a reminder of or deem important.

$2x + 3$ ——————|—————— 17	Start with the model only, then record equation. What else do we know? Where is 2x? Where is zero? Where is x? Record moves on the number line, with equations horizontally, vertically.
$2x - 3 = 17$	Start with equation. What would the double open number line look like? What would help us here? Repeat modeling on number line and with equations.
$2x + 3$ ——————|—————— −17	Start with the model only. What's the equation? Where is 2x now? So where is 0 and where is x?
$2x - 3 = -17$	Start with equation. Repeat.
$2 + 4x = x + 8$	Start with equation. What would be helpful to find first? Work with partner. Elicit different strategies: Getting to 4x on the top Getting to x on the bottom Moving back an x first. Discuss efficiency. Which one do you want your brain to lean toward next time?

3.7 Rate of Change

Time (s)	Dist (ft)
0	4
2	10

Time (s)	Dist (ft)
0	4
2	7

Time (s)	Dist (ft)
2	5
6	8

Time (s)	Dist (ft)
−2	9
1	7

Time (s)	Dist (ft)
−3	17
0	13

Time (s)	Dist (ft)
0	10
5	8

Time (s)	Dist (ft)
0	10
5	2

Objectives
The goal of this problem string to help students find and make sense of rational rates of change.

Placement
Use this problem string to help students continue to build intuition for rates of change in linear situations. In this string, students are given two points in the context of distance-time data which can help them reason about rates of change as students find the slope between the two points.

This problem string is designed to precede or support textbook Lesson 3.7 A Formula for Slope.

Guiding the Problem String
You can use the detailed Sample Interaction notes in problem string 3.3 to help guide your facilitation of this problem string. Whereas that string dealt primarily with integer rates and rates of ½, this string delves into rational rates, giving students experience with making sense of the fact that rates are ratios that can be expressed as non-unit ratios, but are usually simplified to unit rates.

Pose these problems as data collected from walkers moving in front of motion detectors at constant rates. The first problem is easy, but sets up the second problem where the walker covers half the distance in the same amount of time, therefore traveling half as fast. Similarly, in the third problem the walker covers twice the distance in the same amount of time, traveling twice as fast. The fourth problem starts a new pair, where the fifth problem again travels twice the distance, −4 feet, in the same 3 seconds as the fourth (−⁴⁄₃ of a foot per second for a rate that is twice as fast −²⁄₃ of a foot per second). The sixth and seventh problems are similarly related but by four times the rate, from −0.4 ft/sec to −1.6 ft/sec. Keep in context. Help students consider the negative rates as losing distance as a walker moves towards the motion detector and the positive rates as gaining distance as a walker moves away from the detector.

About the Mathematics
Of course not all rates are distance over time, but students can use this familiar scenario to help them make sense of rates, especially when the numbers get less friendly, like the rational number rates in this problem string.

Rate of change is a large, encompassing category that describes many attributes in many places in mathematics, from linear equations to calculus-related rate problems. Slope is one example in this large category. Slope has more of a physical connotation, such as the slope of a road or stair case, whereas rate of change is much broader. Talk about slope as an example of rate of change, not as an end in itself. Too often students and some teachers view slope in a narrow, physical sense of the slant of a line, not in context of the variables' units. Then, when we need students to broaden to the more encompassing category of rate of change, students find that transition difficult. If we can help students develop first a sophisticated sense of rate of change with linear equations, where slope is a physical manifestation of the graphed relationship, then students will not have to break out of the limited view later.

Important Questions

Use the following as you plan how to elicit and model student strategies.

- How did you find the rate every time? Why? How do you make sense of that?
- What is the meaning of the difference of the *y*-values in this distance-time scenario?
- What is the meaning of the difference of the *x*-values in this distance-time scenario?
- What is the meaning of the ratio of the difference of the *y*-values to difference of the *x*-values in this distance-time scenario?
- What happens to the rate when the walker travels twice as far in the same amount of time? Half as far in the same amount of time? How do you know?
- What does a positive rate mean in this distance-time scenario? What would it look like on a graph?
- What does a negative rate mean in this distance-time scenario? What would it look like on a graph?
- What does a rate like ⅔ feet per second mean in this distance-time scenario? What would it look like on a graph? How can a non-unit rate like 2 feet in 5 seconds help?

How would you summarize some of the things that came up in this string today?

- *To find the rate of a walker, you need to know how much distance the walker covered in an amount of time.*
- *If the rate is positive, the walker is gaining distance or walking away from the motion detector. This looks like an increasing graph.*
- *If the rate is negative, the walker is losing distance or walking toward the motion detector. This looks like a decreasing graph.*
- *Non-unit rates like 4 feet for every 3 seconds can help us make sense of ⁴⁄₃ of a foot per second and vice versa.*
- *If you cover twice the amount of distance in the same amount of time, you are traveling twice as fast.*

(continued)

Sample Final Display

Your display could look like this at the end of the problem string:

Time (s)	Dist (ft)
0	4
2	10

$2 \langle \dots \rangle 6$

$\dfrac{6 \text{ feet}}{2 \text{ sec}} = \dfrac{3 \text{ feet}}{1 \text{ sec}} = 3 \text{ ft per sec.}$

Time (s)	Dist (ft)
0	4
2	7

$2 \langle \dots \rangle 3$

$\dfrac{3 \text{ feet}}{2 \text{ sec}} = \dfrac{1.5 \text{ feet}}{1 \text{ sec}} = 1.5 \text{ ft per sec.}$ Half as far in the same amount of time.

Time (s)	Dist (ft)
2	5
6	8

$4 \langle \dots \rangle 3$

$\dfrac{3 \text{ feet}}{4 \text{ sec}} = \dfrac{0.75 \text{ feet}}{1 \text{ sec}} = \frac{3}{4} \text{ or } 0.75 \text{ ft per sec.}$ The same distance in twice the time.

Time (s)	Dist (ft)
-2	9
1	7

$3 \langle \dots \rangle -2$

$\dfrac{-2 \text{ feet}}{3 \text{ sec}} = \dfrac{-\frac{2}{3} \text{ feet}}{1 \text{ sec}} = -\frac{2}{3} \text{ ft per sec.} \approx -0.67 \text{ ft per sec.}$

Time (s)	Dist (ft)
-3	17
0	13

$3 \langle \dots \rangle -4$

$\dfrac{-4 \text{ feet}}{3 \text{ sec}} = \dfrac{-\frac{4}{3} \text{ feet}}{1 \text{ sec}} = -\frac{4}{3} \text{ ft per sec.} \approx -1.33 \text{ ft per sec.}$ Twice the distance in the same amount of time.

Time (s)	Dist (ft)
0	10
5	8

$5 \langle \dots \rangle -2$

$\dfrac{-2 \text{ feet}}{5 \text{ sec}} = \dfrac{-\frac{2}{5} \text{ feet}}{1 \text{ sec}} = -\frac{2}{5} \text{ ft per sec.} = -0.4 \text{ ft per sec.}$

Time (s)	Dist (ft)
0	10
5	2

$5 \langle \dots \rangle -8$

$\dfrac{-8 \text{ feet}}{5 \text{ sec}} = \dfrac{-\frac{8}{5} \text{ feet}}{1 \text{ sec}} = -1\frac{3}{5} \text{ ft per sec.} = -1.6 \text{ ft per sec.}$ Quadruple the distance in the same amount of time.

Algebra Problem Strings
©2017 Kendall Hunt Publishing

Facilitation Notes

This version of the problem string lists short notes for important teacher moves during the string. After you've done the string yourself and studied the relationships involved, you might make similar notes for the things you want a reminder of or deem important.

Time (s)	Dist (ft)
0	4
2	10

We've got walkers moving in front of motion detectors at constant rates.
Stay in context throughout string.
What is this walker's rate?
How do you know?
What would the graph look like? Just motion with your arm.
Increasing or decreasing (gaining ground or losing ground)?

Time (s)	Dist (ft)
0	4
2	7

What is this walker's rate?
Repeat as previous.
How does this rate relate to the previous walker's? How do you know?

Time (s)	Dist (ft)
2	5
6	8

Repeat.
How does this rate relate to the previous walker's? How do you know?
What does it mean that you cover the same distance in twice the time?

Time (s)	Dist (ft)
−2	9
1	7

Repeat.
What's going on with the −2 time? Two secs before the motion detector started.
What does 2 feet for every 3 secs mean?
What does −2 feet for every 3 secs mean? What's the unit rate?
What does negative rate mean?

Time (s)	Dist (ft)
−3	17
0	13

Repeat.
What does 4 feet for every 3 secs mean?
What does −4 feet for every 3 secs mean? What's the unit rate?
How does this rate relate to the previous walker's? How do you know?

Time (s)	Dist (ft)
0	10
5	8

Repeat.
What does 2 feet for every 5 secs mean? What's the unit rate?
What does negative rate mean?

Time (s)	Dist (ft)
0	10
5	2

Repeat.
What does 8 feet for every 5 secs mean?
What does −8 feet for every 5 secs mean? What's the unit rate?
How does this rate relate to the previous walker's? How do you know?

4.0 | **Relations and Functions 1**

At a Glance

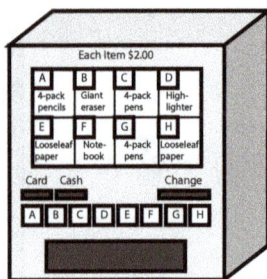

Button (x)	Item (y)
7	folder
2	graph paper
8	mechanical pencil
3	paper
4	folder
1	hand-sanitizer
3	eraser
5	mechanical pencil
6	paper
4	folder

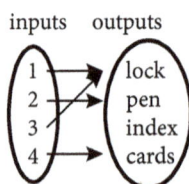

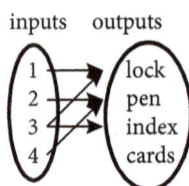

Objectives

This problem string helps students construct the big idea that a function is a relation for which there is only one output for each input.

Placement

This is the first of two problem strings based on relations and functions. Use this problem string to introduce students to the idea of relations by providing an imaginable context to anchor their ideas.

You can use this problem string to introduce textbook Lesson 4.1 Secret Codes.

Guiding the Problem String

In addition to providing students a context to make sense of functions, this short problem string also uses the ideas of consistency, predictability, and/or a broken vending machine as a way to think about the difference between functions and non-functions, and introduces mapping as a way to represent relations, including functions and non-functions.

As you move through the string, you need not worry about whether students have a working definition of a function or if they are aware of the vertical line test. This is not our focus. Instead, you are codifying language like input and output so that in the next string, and subsequent work on functions, students will have a contextual anchor by which to reason about functions. It's important for students to have the chance to do some mapping, which may be a new representation for them. Here we use their mappings to determine why some vending machines feel consistent or predictable (e.g., "working") and others do not. Later, we will formally connect this work to functions, but it's not the goal to do so here.

The first problem is the image of a school supply vending machine that has multiple inputs mapped to the same output. Students create a table and map and then discuss. The second problem is a table that represents a different vending machine with multiple outputs for the same input. Students discuss the unpredictability of such a machine. The third and fourth problems are presented as maps that represent different vending machines.

About the Mathematics

In this string we provide the context of a vending machine to help students make sense of relations conceptually before introducing functions with either a formal definition or the vertical line test. Students have likely developed some emergent ideas about relations and functions previously as a result of their experiences. Our goal here is to begin to solidify those ideas into some big ideas that students can trust with a context that may be generalizable for future work with functions.

Sample Interactions

Use the following as you plan how to elicit and model student strategies. This is not meant as a script, but as a view into the relationships involved and the intent of the problem string.

Teacher: *We are going to be thinking of vending machines today, and how we can use them as a metaphor for our work in algebra. I'm going to show you a unique vending machine that I recently saw on a visit to another high school. Take a look. Notice anything that feels important? Get ready to describe how it works to your partner. Ready?*

The teacher shows the image of the vending machine. After 30 seconds or so, the teacher asks partners to talk together about how the machine works. Then, after a minute or two, asks students to share their ideas and descriptions with the whole class.

Student: *So we noticed that everything was $2, which is convenient, and that there are 8 buttons, but only 6 items in the machine.*

Student: *Yeah, we saw the doubles, too. Like you can press C or G to get pens and E or H to get loose-leaf paper.*

Teacher: *Why do you think that is?*

Student: *Maybe kids buy those the most so they wanted a bigger supply.*

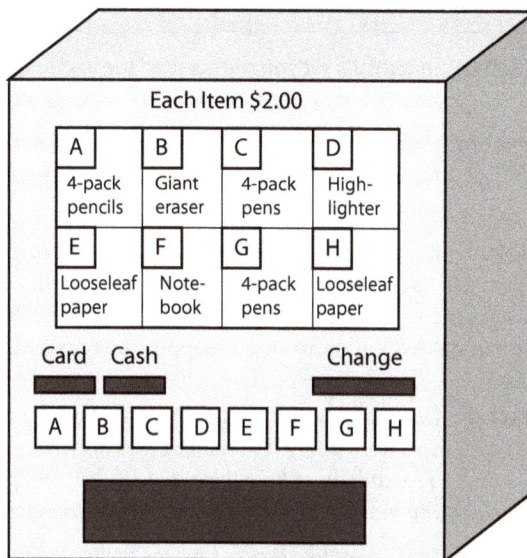

Each Item $2.00			
A 4-pack pencils	**B** Giant eraser	**C** 4-pack pens	**D** High-lighter
E Looseleaf paper	**F** Note-book	**G** 4-pack pens	**H** Looseleaf paper

Card   Cash   Change

A B C D E F G H

Teacher: *Seems like we have a good sense of how the machine works. What if you put $6 in the machine and then pressed F 3 times?*

Student: *You should get 3 notebooks. I mean, every time you press F you should get a notebook.*

Teacher: *Any exceptions to that?*

Student: *I guess if they are sold out, but usually you can tell from the front of the machine and you get your money back.*

Teacher: *Okay, now we are going to represent this vending machine in a few ways. Make a table of the data in this vending machine. Be thinking about what the two variables are and which one depends on the other. I'll give you a minute to do that now.*

The teacher circulates and nudges students who are having a hard time getting started.

Teacher: *What is dependent? What is independent?*

After a minute or two of work time, the teacher asks a student to describe his or her table (as shown) as the teacher quickly draws a version of it for the whole class to see.

depends on

Button (x)	Item (y)
A	pencils
B	eraser
C	pens
D	highlighter
E	paper
F	notebook
G	pens
H	paper

(continued)

Teacher: *Now let's do some mapping. This is just another way to visually represent what's going on in this vending machine where we use arrows to associate each input with its output. What do I mean by "input" and "output" here?* **Student:** *I think the buttons you press are the inputs and the supplies you get, like the pencils, are the output.* **Student:** *So inputs will be letters and outputs will be names of supplies.*	
Teacher: *Exactly. I'll get you started with the mapping. If we wanted to show what happens when you press A, we might draw this.* **Teacher:** *Take a minute with your partner to make a mapping of the entire vending machine, then we'll look at some of the ones you made.* The issue of how to represent outputs that repeat (e.g., pencils and paper) will likely come up. Let students first map in a way that makes sense to them—do not interrupt or fix their initial attempts. Later you can discuss and refine their mappings, as needed, as a class.	A ⟶ pencils
Teacher: *Okay, as you were working, I was taking notes about what you were doing. I made sketches of two students' maps. Take a look at how these two students mapped the vending machine, looking for anything interesting here.* **Student:** *So it seems like they are basically the same—except Imani numbered the "doubles" and Jada just drew arrows to them.* **Student:** *In a way they both work, but I like how we said before there were 8 buttons and 6 options. Jada's kind of shows that better.* **Teacher:** *Okay, and who can say that same idea another way?* **Student:** *There are 8 inputs and 6 outputs.* **Teacher:** *So, they both represent the situation, but mathematicians use Jada's notation when we have more than one input associated with the same output. You all said earlier there were two ways to get pens and two ways to get loose-leaf paper from this machine, right? Jada's mapping just gives us a slightly easier way of seeing that. From now on use that approach.*	A ⟶ pencils B ⟶ eraser C ⟶ pens 1 D ⟶ highlighter E ⟶ paper 1 F ⟶ notebook G ⟶ pens 2 H ⟶ paper 2 -Imani A ⟶ pencils B ⟶ eraser C ⟶ pens D ⟶ highlighter E ⟶ paper F ⟶ notebook G H -Jada

Algebra Problem Strings
©2017 Kendall Hunt Publishing

Button (x)	Item (y)
7	folder
2	graph paper
8	mechanical pencil
3	paper
4	folder
1	hand-sanitizer
3	eraser
5	mechanical pencil
6	paper
4	folder

Teacher: *Next, I want to tell you a story of a different vending machine. I want you to think about how you would make a mapping of this and what it would tell you about this machine.*

The principal kept records one morning to see how the kids were using the machine. Your job, with your partner, is to think about this data. What would it look like as a mapping? Does it tells us anything interesting about the machine? Ready?

Teacher: *Okay, let's talk about your mapping and also what you learned about this machine. Describe your mapping and I'll make a picture of your thinking.*

Student: *Sure. We started by putting the buttons in order. So there's an arrow from 1 to sanitizer. Then from 2 to graph paper. And this is where it got weird for us—we wrote 3 and then it had two arrows, one to loose-leaf and another to an eraser.*

Teacher: *Why is Luna saying "it got weird" here?*

Student: *Seems like the machine is inconsistent. Like, how do you know if you are getting loose-leaf or the eraser if you press 3. That's super confusing.*

Student: *Maybe it's broken.*

Teacher: *This seems like a problem, yes? Let's keep going with the mapping and see if that happens again.*

Student: *With 4 we drew one arrow to folder, even though two students bought folders by pressing 4—we weren't sure if we needed to have one arrow or two.*

Teacher: *Good question. So those arrows are meant to show the relationship between inputs and outputs, not how many times a button was pressed. Keep going.*

Student: *5 went to clicky pencil, 6 went to loose-leaf, 7 to folder and we used the same folder as we used for 4. Two arrows going to the same place, like what they did before. 8 went to clicky pencil, same as 5.*

Teacher: *Some of you are saying this machine felt inconsistent, perhaps broken—where does that idea show up in the mapping? Could you spot it? Turn and talk—go!*

The teacher circles or highlights the place on the map and the table where students see an input with more than one output.

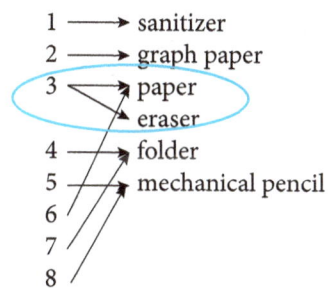

```
1 ──────▶ sanitizer
2 ──────▶ graph paper
3 ──────▶ paper
          ▲ eraser
4 ──────▶ folder
5 ──────▶ mechanical pencil
6
7
8
```

Button (x)	Item (y)
7	folder
2	graph paper
8	mechanical pencil
3	paper
4	folder
1	hand-sanitizer
3	eraser
5	mechanical pencil
6	paper
4	folder

weird inconsistent broken?

(continued)

Teacher: *Alright, last idea for today. I'm going to just show you two mappings for two other vending machines, and I'd like you and your partner to first study both for 30 seconds. I'll give us silent think time to do that. Then I'd like you to talk with your partner about these questions, which I will leave up:* *What do you notice about these two vending machines?* *What can we say about these two vending machines?*	What do you notice about these two vending machines? What can we say about these two vending machines? 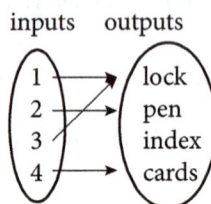 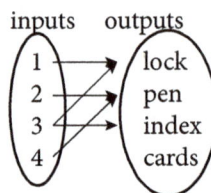
The teacher captures students' ideas and descriptions.	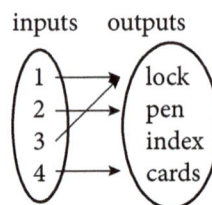 4 inputs 3 outputs not broken "you can predict what you will get" 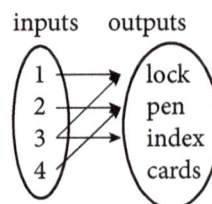 4 inputs 3 outputs broken: the 3 button gives 2 different outputs 2 ways to get a lock in each machine

Teacher: *How would you summarize some of the things that came up in this string today?*

Elicit the following:

- *Relations can be represented in tables and by mappings.*
- *If there are two inputs (buttons) that map to the same output (school supplies), it provides more opportunities to get the same output.*
- *If there is more than one output for a given input in a vending machine, the machine might be broken, unpredictable, or inconsistent.*

Algebra Problem Strings
©2017 Kendall Hunt Publishing

Sample Final Display

Your display could look like this at the end of the problem string:

depends on

Button (x)	Item (y)
A	pencils
B	eraser
C	pens
D	highlighter
E	paper
F	notebook
G	pens
H	paper

A ⟶ pencils
B ⟶ eraser
C ⟶ pens 1
D ⟶ highlighter
E ⟶ paper 1
F ⟶ notebook
G ⟶ pens 2
H ⟶ paper 2

-Imani

A ⟶ pencils
B ⟶ eraser
C ⟶ pens
D ⟶ highlighter
E ⟶ paper
F ⟶ notebook
G
H

-Jada

Button	Item
7	folder
2	graph paper
8	mechanical pencil
3	paper
4	folder
1	hand-sanitizer
3	eraser
5	mechanical pencil
6	paper
4	folder

weird
inconsistent
broken?

1 ⟶ sanitizer
2 ⟶ graph paper
3 ⟶ paper
    eraser
4 ⟶ folder
5 ⟶ mechanical pencil
6
7
8

What do you notice about these two vending machines?
What can we say about these two vending machines?

inputs   outputs

1
2   lock
3   pen
4   index
      cards

4 inputs
3 outputs
not broken
"you can predict
what you will get"

inputs   outputs

1
2   lock
3   pen
4   index
      cards

4 inputs
3 outputs
broken: the 3 button
gives 2 different outputs

2 way to get a lock in each machine

(continued)

Facilitation Notes

This version of the problem string lists short notes for important teacher moves during the string. After you've done the string yourself and studied the relationships involved, you might make similar notes for the things you want a reminder of or deem important.

Display the vending machine.
What do you see? Describe to partner.
Elicit more than one input for output: loose-leaf, pens.
Elicit representing as a table. What is dependent? Independent?
Give example of and then elicit mapping.

Button (x)	Item (y)
7	folder
2	graph paper
8	mechanical pencil
3	paper
4	folder
1	hand-sanitizer
3	eraser
5	mechanical pencil
6	paper
4	folder

Display the table of the new vending machine.
Elicit mapping. Anything interesting about this machine?
Weird! More than one output for one input.
Highlight.
Is it broken? Inconsistent? Unpredictable?

Display the mapping of two new vending machines.
Study, then discuss with partner.
What do you notice about these two vending machines?
What can we say about these two vending machines?
Elicit number of outputs for one input and vice versa.
Predictable?

For Display

Each Item $2.00

A 4-pack pencils	B Giant eraser	C 4-pack pens	D High-lighter
E Looseleaf paper	F Note-book	G 4-pack pens	H Looseleaf paper

Card   Cash   Change

A B C D E F G H

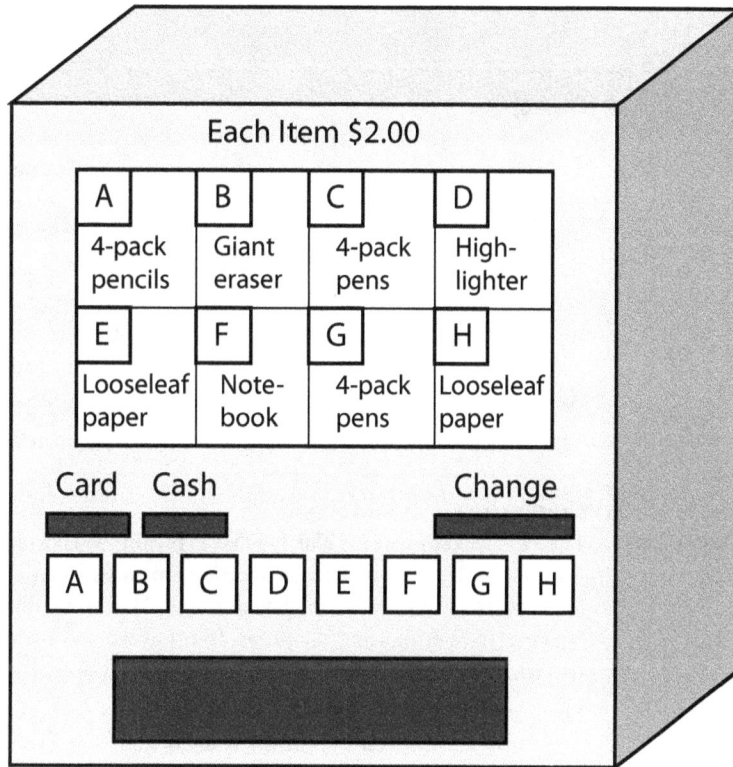

Button (x)	Item (y)
7	folder
2	graph paper
8	mechanical pencil
3	paper
4	folder
1	hand-sanitizer
3	eraser
5	mechanical pencil
6	paper
4	folder

What do you notice about these two vending machines? What can we say about these two vending machines?

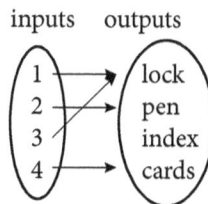

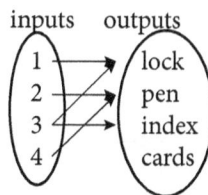

inputs   outputs

1 2 3 4 → lock pen index cards

inputs   outputs

1 2 3 4 → lock pen index cards

4.1 Relations and Functions 2

At a Glance

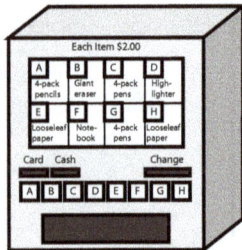

Button (x)	Item (y)
7	folder
2	graph paper
8	mechanical pencil
3	paper
4	folder
1	hand-sanitizer
3	eraser
5	mechanical pencil
6	paper
4	folder

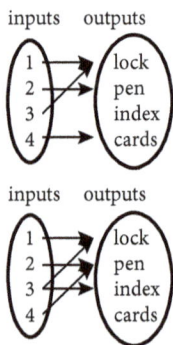

Objectives

This string extends an imaginable context so that students deepen the big idea that a function is a relation in which there exists exactly one output for each input.

Placement

In the previous problem string, students were introduced to the metaphor of a vending machine as a way to reason about mathematical relationships involving inputs and outputs. Use this string to name, explore, and formalize what is meant by function and use students' emerging ideas about functions to decide whether some relations are functions or not.

This string could precede or support textbook Lesson 4.1 Secret Codes.

Guiding the Problem String

First we help students to clarify what is meant by a "broken" vending machine. It is not that the buttons do not match the items distributed, but that there is no reasonable way to predict what one has purchased based on the input-output relationship between the buttons and the items. In other words, if you press *button one* and get a pen sometimes and a pencil other times, the vending machine is not predictable.

Once students have a sense of this idea—that a predictable, consistent relationship between inputs and outputs is what we are after—then bridge this idea to the formal definition of a function. As you are doing this, you are also beginning to move away from the vending machine context towards other relations in context, and eventually towards some non-contextual numerical values.

About the Mathematics

A relation is a mapping of inputs to outputs. A function is a relationship in which every input has exactly one output. The idea of a function is, on the surface, not deeply complex, though its implications for continued study in mathematics is critical. Yet, most students have a shallow or procedural understanding of functions at best. Students tend to not be sure what makes a function a function, what a non-function might look like, whether certain transformations of functions maintain function status, and why all of this talk of functions even matters. By focusing on the predictable nature of functions, students gain a sense of how they are different from other relationships and why they are a valuable construct to study.

Sample Interactions

Use the following as you plan how to elicit and model student strategies. This is not meant as a script, but as a view into the relationships involved and the intent of the problem string.

Teacher:	*In our previous problem string, we talked about those school supply vending machines and we left with the idea that you could spot a broken vending machine based on its mapping or its table. There was something different about the relationship between inputs and outputs, right? Someone remind us what made a machine "broken" exactly.*
Student:	*When you press a button and got one item and then your friend pressed the same button and got something else.*
Student:	*Or when you pressed the button in the morning and got one thing, but got something else in the afternoon.*

Teacher: *Who wants to add on?*

Student: *We said that you can have two or more buttons that go to the same thing—like if pencils are really popular then you have a bunch of buttons that correspond. But what you cannot have is a single button that gives you different things, but you don't know what when.*

Teacher: *And how did this idea—that each input needed only one output—show up in the mapping? I'll put up the last problem we talked about yesterday and I'd like you to explain to your partner how you'd spot a "broken" machine based on the mappings alone. What do you look for exactly?*

Students turn and talk while the teacher listens in, noting any misconception and big ideas to elicit during the problem string.

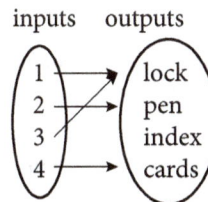

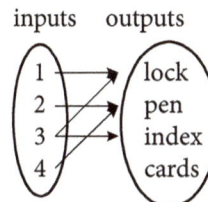

Teacher: *So this idea that there are relationships where each input has exactly one output is really important for our work in algebra and beyond. Whenever you have a relationship like this, we call it a function. And when the inputs do not associate with the outputs in this way, we say it is not a function. They are both relations, but a function is a specific kind of relation.*

Which of these is a function?

The teacher records the correct labels with student input.

Teacher: *Which vending machine would you rather use? Why?*

Student: *The function one, so I know I am going to get what I want.*

Teacher: *I hear you saying that you value the predictability of the function vending machine. Today we'll look at a few more examples of mappings, tables, and situations to try to figure out whether these are functions or not. Ready?*

Example of function

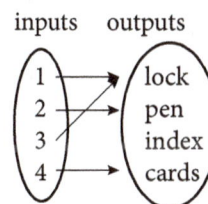

Example of non-function

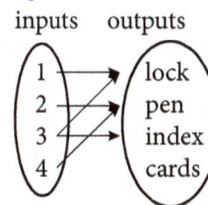

Teacher: *Here's an old picture from the side of a city taxi. First, who can explain what it means?*

Student: *I think it means you have to pay $2.50 just to get in the taxi, then for every ⅕ of a mile, they add on 40 cents.*

Teacher: *So now, with your partner, represent this taxicab situation with a table and decide whether this is a function or not.*

Students work while the teacher circulates, nudging students who need help getting started.

TAXI FARE

$2.50 -- Initial charge
$0.40 -- Per 1/5 mile

(continued)

Teacher: *I've put your table on the board. Function or not?*

Student: *So my partner and I think it is a function, because you can pretty much figure out the fare if you know how far you've gone.*

Teacher: *Are you saying that functions have rules that allow you to predict an output if you know the input?*

Student: *I think so.*

Teacher: *Let's come back to that idea. Everyone be thinking about whether this is also true of functions. Who else thinks this is a function?*

Student: *We made a table and we then turned it into a mapping, just to be sure. And it's definitely a function.*

Teacher: *I'll quickly put your mapping up here for all of us to see.*

Brief think time.

Teacher: *Are you convinced that this is a function? Anyone need some more convincing?*

Is it important to you that a taxi ride of 3 miles has a predictable cost? Is it okay if sometimes it costs $8.50 and sometimes it costs $12.50 or $20.50?

Student: *It wouldn't be fair to have different prices charged on different days for the same ride if you didn't know about it. Customers wouldn't like that.*

miles traveled	fare
0	$2.50
1	$4.50
2	$6.50
3	$8.50
$\frac{1}{5}$	$2.90

miles		fare
0	$\longrightarrow$	$2.50
$\frac{1}{5}$	$\longrightarrow$	$2.90
1	$\longrightarrow$	$4.50
2	$\longrightarrow$	$6.50
3	$\longrightarrow$	$8.50

Teacher: *Okay, here's a table of states and their capital cities. Hopefully you know a bunch of these. Is this a function? Why or why not? Let's think for 30 seconds silently, and then we'll talk.*

Student: *I think this must be a function because each state has only one capital city and there aren't two with the same name. Wait, are there?*

Student: *No, each state capital has its own name, so it's a function.*

Teacher: *If we made a mapping of this, which map, that we already have on the board, would it most look like?*

Student: *It would look like the taxicab mapping, where every input has only one output. No doubles!*

Teacher: *Yeah, mathematicians could say each input has a unique output.*

State	Capital
California	Sacramento
Texas	Austin
Colorado	Denver
Wisconsin	Madison
South Carolina	Columbia

Algebra Problem Strings
©2017 Kendall Hunt Publishing

Teacher: *Here's a table of some famous cities in the U.S. Is this a function? Again, let's think first, and then you can talk with your partner.*

Student: *It was a function until we got to Springfield, New Jersey.*

Teacher: *Say more…*

Student: *Well, up until that point each city had only one state associated with it. But with the last two on the table, you have two cities that are located in two different states.*

Teacher: *Who agrees that this is not a function, but can explain it in a different way?*

Student: *Sure! Think of it like a vending machine—if you pressed Jackson you get Mississippi sometimes and other times Michigan. Totally different places.*

Teacher: *Can anyone else deepen the argument that this is not a function? I'm not sure we are all convinced.*

Student: *If we go back to the idea of mapping, Springfield and Jackson would have two arrows from input to outputs, and that means they have more than one output. So, definitely not a function.*

Teacher: *If you were buying a ticket to Springfield, would you care if the ticket-buying is a function or not a function? Why?*

Student: *It better be a function. I need the plane to go to the correct Springfield!*

City	State
Minneapolis	Minnesota
Springfield	Missouri
Chicago	Illinois
Jackson	Mississippi
Portland	Oregon
Springfield	New Jersey
Jackson	Michigan

Teacher: *Okay, here's the last set for today. Here's two tables, this time with just numerical values. Pay really close attention to what's going on here. Try to envision the mappings of both of them. Think about whether they are functions or not and how you would convince all of us.*

Let's study them silently for 30 seconds, then a little talk with our partners.

The teacher elicits students noticing that the inputs and outputs were switched from table A to table B, and student conclusions that table A represents a function while table B does not. The teacher puzzles over this idea and encourages students to explain to each other why the same values when "flipped" go from being a function to not being a function in this case.

The teacher encourages students to use mapping, the metaphor of the vending machine, or the definition provided earlier to strengthen their arguments.

A.

x	y
1	5
4	9
7	12
15	9

B.

x	y
5	1
9	4
12	7
9	15

(continued)

Teacher: *How would you summarize some of the things that came up in this string today?*

Elicit the following:

- *Relations are mappings from inputs to outputs.*

- *Functions are examples of relations.*

- *Functions are relations where each input has exactly one output.*

- *Functions can have data where more than one input has the same output.*

- *Functions offer us predictability that non-functions do not have. That's important if you want a certain outcome.*

Sample Final Display

Your display could look like this at the end of the problem string:

Example of function

inputs   outputs

1 → lock
2 → pen
3 → index
4 → cards

Example of non-function

inputs   outputs

1 → lock
2 → pen
3 → index
4 → cards

TAXI FARE

$2.50 -- Initial charge
$0.40 -- Per 1/5 mile

miles traveled	fare
0	$2.50
1	$4.50
2	$6.50
3	$8.50
$1/5$	$2.90

miles	fare
0 →	$2.50
$1/5$ →	$2.90
1 →	$4.50
2 →	$6.50
3 →	$8.50

State	Capital
California	Sacramento
Texas	Austin
Colorado	Denver
Wisconsin	Madison
South Carolina	Columbia

City	State
Minneapolis	Minnesota
Springfield	Missouri
Chicago	Illinois
Jackson	Mississippi
Portland	Oregon
Springfield	New Jersey
Jackson	Michigan

A.

x	y
1	5
4	9
7	12
15	9

B.

x	y
5	1
9	4
12	7
9	15

Algebra Problem Strings
©2017 Kendall Hunt Publishing

Facilitation Notes

This version of the problem string lists short notes for important teacher moves during the string. After you've done the string yourself and studied the relationships involved, you might make similar notes for the things you want a reminder of or deem important.

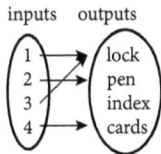

Remember the vending machines? What made one broken, unpredictable?
How would you spot a broken machine based on a mapping?
Define a function.
Which of these is a function?
Label.

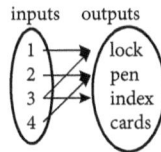

TAXI FARE

$2.50 -- Initial charge
$0.40 -- Per 1/5 mile

Represent this taxi fare situation with a table.
Function or not a function?
Why?
Elicit mapping, vending machine comparison, function defintion.

State	Capitol
California	Sacramento
Texas	Austin
Colorado	Denver
Wisconsin	Madison
South Carolina	Columbia

Function or not a function?
Each state has a unique capital city and capital city name.
Elicit mapping, vending machine comparison, function defintion.

City	State
Minneapolis	Minnesota
Springfield	Missouri
Chicago	Illinois
Jackson	Mississippi
Portland	Oregon
Springfield	New Jersey
Jackson	Michigan

Repeat.
Why not a function?

A.

x	y
1	5
4	9
7	12
15	9

Show both tables at once.
What do you notice?
Function or not a function?
When is a relation a function?
Elicit mapping, vending machine comparison, function defintion.

B.

x	y
5	1
9	4
12	7
9	15

(continued)

For Display

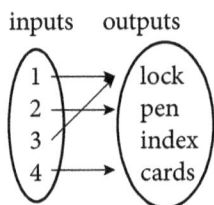

```
        TAXI FARE

$2.50 -- Initial charge
$0.40 -- Per 1/5 mile
```

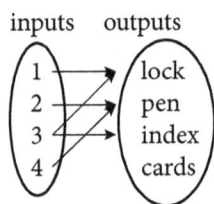

State	Capital
California	Sacramento
Texas	Austin
Colorado	Denver
Wisconsin	Madison
South Carolina	Columbia

City	State
Minneapolis	Minnesota
Springfield	Missouri
Chicago	Illinois
Jackson	Mississippi
Portland	Oregon
Springfield	New Jersey
Jackson	Michigan

A.

x	y
1	5
4	9
7	12
15	9

B.

x	y
5	1
9	4
12	7
9	15

Algebra Problem Strings
©2017 Kendall Hunt Publishing

4.2 Functions and Graphs

At a Glance

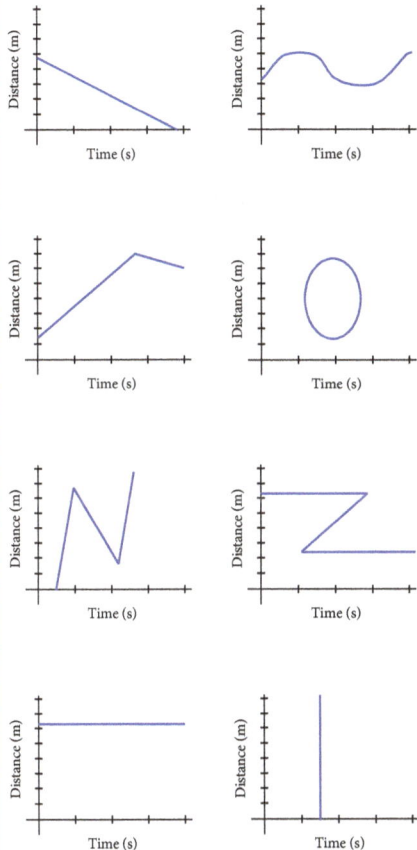

Objectives
This problem string helps students visually construct the big idea that a function is a relation for which there is only one output for each input because you can only be in one place at a time.

Placement
This problem string can be used to either introduce the vertical line test or support the work of using the vertical line test to determine if a graph represents a function. The questions and facilitation notes are written to support students after learning about the vertical line test. Ideally students would have prior experience walking in front of motion detectors.

You can use this problem string to support textbook Lesson 4.2 Functions and Graphs.

Guiding the Problem String
This problem string is based on the context of walking in front of a motion detector. To start, remind students of their prior experiences with motion detectors. You could begin the string by having a student walk at different rates and in different directions in front of a motion detector. Then display each problem (graph) in turn and ask students if the graph could represent a walk in front of a motion detector. If so, ask students to describe walking directions. As an option, you could ask a student to walk those directions in front of a motion detector. Then ask students if the graph represents a function and how they know. Connect their ideas about walkability (you can only be in one place at a time or more specifically, you can only be at one location in front of the motion detector at a given time) with the vertical line test and the definition of a function. Problem D (oval) is not walkable for several reasons: it is not a function (displays a walker in two locations during an interval of time), also because the walker would have to appear suddenly right after one second (sort of "beam me in Scotty"), and some students may talk about not being able to turn back time. The rest of the problems are either functions and could be walked (A, B, C, E, G) or they are not functions because the walker would be in multiple places at one time (F, H).

Because you are asking students to determine walking directions, this string has the added benefit of helping students read graphs, and reason about rates and y-intercepts.

About the Mathematics
A function is a relation in which every input has exactly one output. The word *exactly* in mathematics means that for an input, there is one, and only one, output. On a graph, this means that for every x-value, there is one, and only one, y-value.

The vertical line test helps you determine whether a relation is a function. If all possible vertical lines cross the graph once or not at all, then the graph represents a function. The graph does not represent a function if you can draw even one vertical line that crosses the graph two or more times.

A function can have multiple inputs that have the same outputs. On a graph, this means that there can be multiple x-values that have the same y-values. An example of this is a horizontal line.

(continued)

Important Questions

Use the following as you plan how to elicit and model student strategies.

- *Could someone create this graph by walking in front of a motion detector? How? Why not?*

- *Does the graph represent a function? Why or why not?*

- *How does the vertical line test relate to the walkability of a graph?*

- *How does the vertical line test relate to the definition of a function?*

How would you summarize some of the things that came up in this string today?

- *The vertical line test means that if all vertical lines possible to draw only cross the graph at most one time, then the graph represents a function.*

- *Graphs that represent walks in front of a motion detector (time, distance in front of a motion detector) are functions because you can only be in one place (one location in front of a motion detector) at one time.*

- *If the graph shows at walker in more than one place at a time, the graph is not walkable nor is it a function.*

Sample Final Display

Your display could look like this at the end of the problem string:

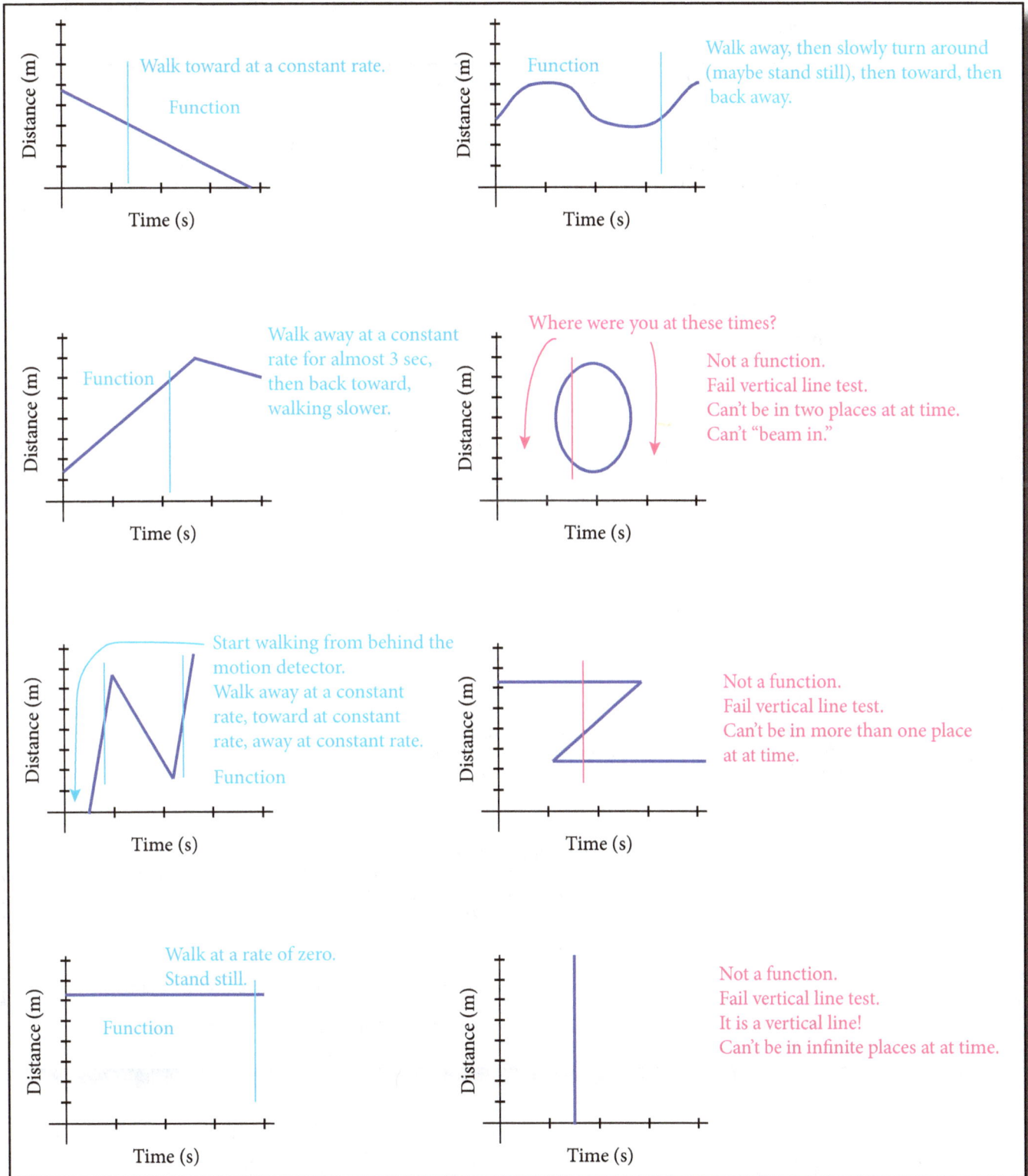

Walk toward at a constant rate.

Function

Function

Walk away, then slowly turn around (maybe stand still), then toward, then back away.

Function

Walk away at a constant rate for almost 3 sec, then back toward, walking slower.

Where were you at these times?

Not a function.
Fail vertical line test.
Can't be in two places at at time.
Can't "beam in."

Start walking from behind the motion detector.
Walk away at a constant rate, toward at constant rate, away at constant rate.

Function

Not a function.
Fail vertical line test.
Can't be in more than one place at at time.

Walk at a rate of zero.
Stand still.

Function

Not a function.
Fail vertical line test.
It is a vertical line!
Can't be in infinite places at at time.

(continued)

Facilitation Notes

This version of the problem string lists short notes for important teacher moves during the string. After you've done the string yourself and studied the relationships involved, you might make similar notes for the things you want a reminder of or deem important.

Graph (Distance (m) vs Time (s))	Notes
	Remind students of prior experiences walking in front of a motion detector. Optional: Ask a student to walk at different rates and in different directions in front of a motion detector. Display graph. Could this represent a walk in front of a motion detector? What are walking directions? Optional: walk the directions in front of a motion detector. Does this graph represent a function? Vertical line test?
	Display graph. Could this represent a walk in front of a motion detector? What are walking directions? Does this graph represent a function? Vertical line test?
	Repeat.
	Repeat. Why can't this graph represent a walk? Can't be in two places at one time. And can't "beam in". Why doesn't this graph represent a function? More than one output for an input, doesn't pass vertical line test.
	Repeat. Lines are slanted just enough—fast walks, but still walkable (runable).
	Repeat. Why can't this graph represent a walk? Can't be in two places at one time and can't turn back time. Why doesn't this graph represent a function? More than one output for an input, doesn't pass vertical line test. How does "can't be in more than one place at a time" relate to the vertical line test?
	Repeat. It's walkable, maybe a boring stand still walk, but it's a walk with a rate of zero. It's a vertical line test, not a horizontal or any other kind of line test, so yes, a function. How does the vertical line test relate to walking in front of a motion detector? How does the vertical line test relate to the definition of a function?
	Repeat. Why can't this graph represent a walk? Can't be in an infinite number of places at one time. Why doesn't this graph represent a function? More than one output for an input, doesn't pass vertical line test. It is a vertical line!

Algebra Problem Strings
©2017 Kendall Hunt Publishing

For Display

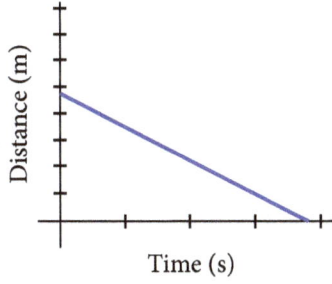

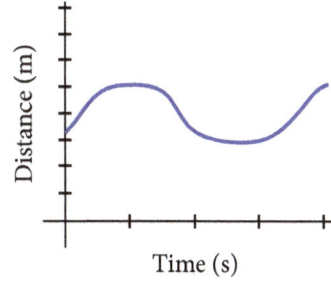

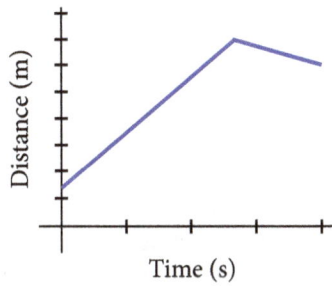

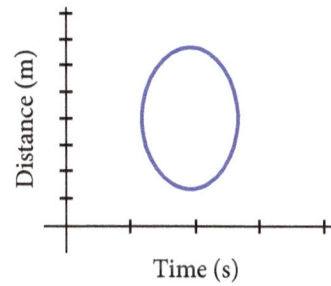

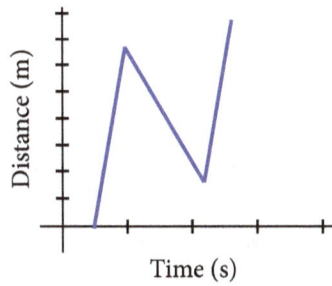

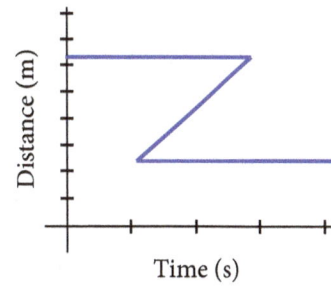

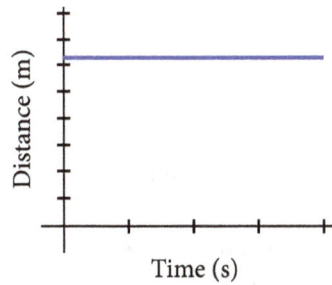

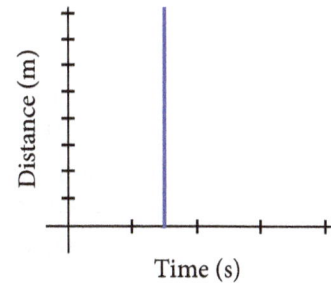

Reading Graphs

At a Glance

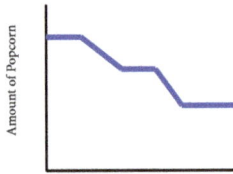

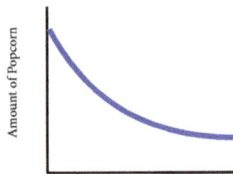

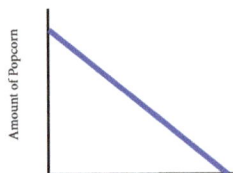

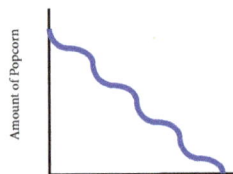

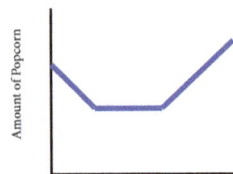

Objectives

This purpose of this problem string is to help students continue to make meaning from functions represented as graphs and to add more precision to their interpretations of these graphs in context, such as function/not a function, increasing/decreasing, and intercepts.

Placement

This problem string builds on the previous strings in lessons 1.7, 4.0, 4.1, and 4.2. Students have told the story of graphs and explored the idea of functions. Use this string to bridge these two ideas—that functions in the form of graphs can also be interpreted as a story with increasing levels of detail.

You could use this problem string to introduce or support textbook Lesson 4.3 Graphs of Real-World Situations.

Guiding the Problem String

The first problem sets the stage with a graph of a student's amount of popcorn as he watches a movie. Annotate the graph with student insights. Ask questions to prompt new ideas as necessary. Press students for justification. The string continues with family members' popcorn amounts as they watch the movie. Present each graph one at a time, but keep them up as you present a new one. Encourage students to compare the graphs, scenarios, rates, etc.

Because the independent variable time in this scenario is passing as the people are eating popcorn, it would be ideal if the graphs could feel dynamic to students, as if the story is unfolding in real time. One way to do this is the cover a graph and reveal the graph slowly from left to right, suggesting to students one way to "read" a graph on their own. This could also give the graphs a bit of suspense, since students won't be able to see the entire graph until the end and if the rate changes there will be a noticeable turn in the graph.

As your students get more experienced reading graphs, deepen your questions. It is appropriate and important for students to be able to distinguish between linear/nonlinear functions (constant rates/non-constant rates), between increasing/decreasing functions and between continuous/discrete functions. You can also bring in: input/output, independent/dependent variables, low/high rates, a rate of zero, and domain and range.

About the Mathematics

A function is increasing if the graph rises from left to right. A function is decreasing if the graph falls from left to right.

The function is a relationship between sets and the graph is a representation of that relationship.

A function is continuous when there are no breaks in the domain or range. A function with a domain or range that can be counted or measured only in whole numbers is discrete. Many discrete situations are represented by continuous graphs because it is easier to graph a continuous curve than many separate points and the continuous curve can show the pattern or trend.

Algebra Problem Strings
©2017 Kendall Hunt Publishing

Sample Interactions

Use the following as you plan how to elicit and model student strategies. This is not meant as a script, but as a view into the relationships involved and the intent of the problem string.

Teacher: *Today I'm going to show you a few graphs of a boy named Minh and his family on a day at the movies. They each decide to get popcorn—same size bag for each person so they don't have to share. Your job is to analyze the graph:*

Be able to tell the story of each person's popcorn eating.

Justify your story with facts from the graph.

Think about where the graph is increasing, decreasing, or staying the same and what that means.

Think about whether these graphs are functions or not.

Think about how realistic each of these situations is.

Ready? Okay, here's the first one.

Teacher: *Here's Minh. Take a few seconds now to really study the graph.*

Someone tell us the story of this graph. What happened to Minh's popcorn during the movie?

Who can add more detail or description to this interpretation?

If necessary, ask:

Did Minh finish his popcorn? How do you know?

What does the y-intercept mean?

What do the turning points, where the rates change, in the graph mean?

What does the steepness of the line segments mean?

What do the horizontal segments mean? With popcorn? With the movie?

Is this a function? Who could convince us of this?

Was this function increasing, decreasing, neither, or both?

Next to each graph, record students' ideas and annotate the graph accordingly. Leave each graph displayed so students can refer to them throughout the string.

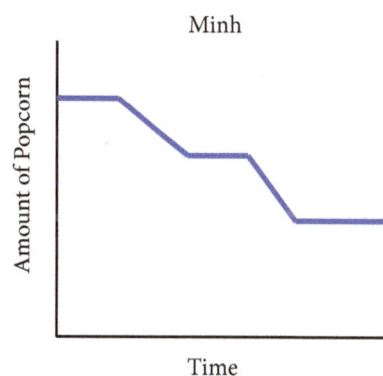

Minh

Teacher: *Alright, here's Minh's cousin. What's going on here? Get ready to tell the story. First, a few seconds to really study the graph.*

After think time, ask students to share their insights. If necessary, ask:

There are no turning points in this graph. What does this mean?

How does the rate of eating popcorn compare to Minh's?

Was the rate of popcorn eating constant or changing for Minh's cousin?

Is this a function? Who could convince us of this?

Was this function increasing, decreasing, neither, or both?

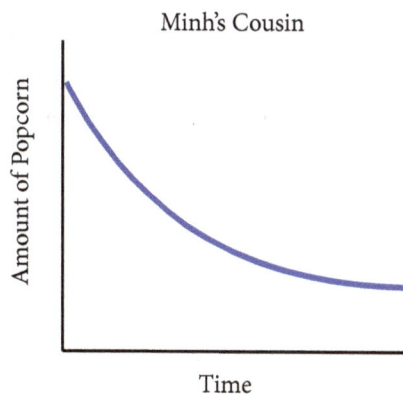

Minh's Cousin

(continued)

Teacher: *Here's our next graph. What's the story here, in terms of popcorn eating?*

Then you could ask….

What's different about this graph?

There are no turning points in this graph. What does this mean?

What can you say about how fast or slow Minh's grandfather was eating his popcorn?

Did he finish all of his popcorn? How do you know?

Is this a function—how do you know?

Was this function increasing, decreasing, neither, or both?

Minh's Grandfather

Teacher: *This one feels a little different. What's going on here?*

Then you might ask….

What's different about this graph?

There are no turning points in this graph. What does this mean?

Is this a function? How do you know?

Was this increasing, decreasing, neither, or both?

Did she finish all of her popcorn? How do you know?

Is this believable? Why or why not?

Minh's Mom

Teacher: *Here's a graph showing the popcorn eating of Minh's aunt. How would you tell the story of how she ate her popcorn during the movie?*

Follow-up questions could include….

Anything interesting or different about this graph?

What does the turning point in this graph mean? Is that realistic in the story?

Is this a function—how do you know?

Was this function increasing, decreasing, neither, or both?

Did she finish all of her popcorn? How do you know?

Is this believable? Why or why not?

Minh's Aunt

Algebra Problem Strings
©2017 Kendall Hunt Publishing

Teacher: *Here's a graph of Minh's uncle. What's going on here? What could explain the shape of this graph?*

This graph is intended to be a bit puzzling and also to give students a chance to see an increasing portion of a graph, even though it doesn't make sense in the context here.

You might consider asking:

What's interesting or different about this graph?

Is this a function—how do you know?

Was this function increasing, decreasing, neither, or both?

Did he finish all of his popcorn? How do you know?

Is this believable? Why or why not?

Are parts of the graph believable? Which parts and why?

What part of the graph doesn't make much sense?

Do these graphs represent continuous or discrete functions?

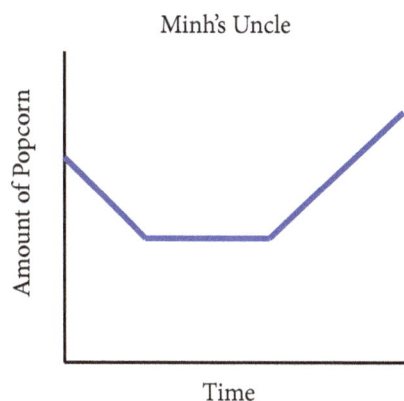

Minh's Uncle

Amount of Popcorn

Time

Teacher: *How would you summarize some of the things that came up in this string today?*

Elicit the following:

- *It can be important to look at what the dependent variable is doing as the independent variable is changing, looking at how the amount of popcorn is changing as time goes by.*

- *When the amount of popcorn is the same over time, it looks like a horizontal line on the graph of (elapsed time, amount of popcorn).*

- *When someone eats popcorn quickly the graph of (elapsed time, amount of popcorn) is steeper.*

- *You can tell the initial amount of popcorn by looking at the y-intercept of the graph of (elapsed time, amount of popcorn). You can tell when a person ate all of the popcorn by looking at the x-intercept, if there is one, of the graph of (elapsed time, amount of popcorn).*

- *Increasing means the graph goes up from left to right. Decreasing means the graph goes down from left to right.*

- *These graphs were all functions because in this scenario a person can't have two different amounts of popcorn at the same time.*

- *These graphs are continuous representations of a discrete situation. The domain of time is continuous but the range of the amount of popcorn, measured in kernels, is not continuous.*

(continued)

Sample Final Display

Your display could look like this at the end of the problem string:

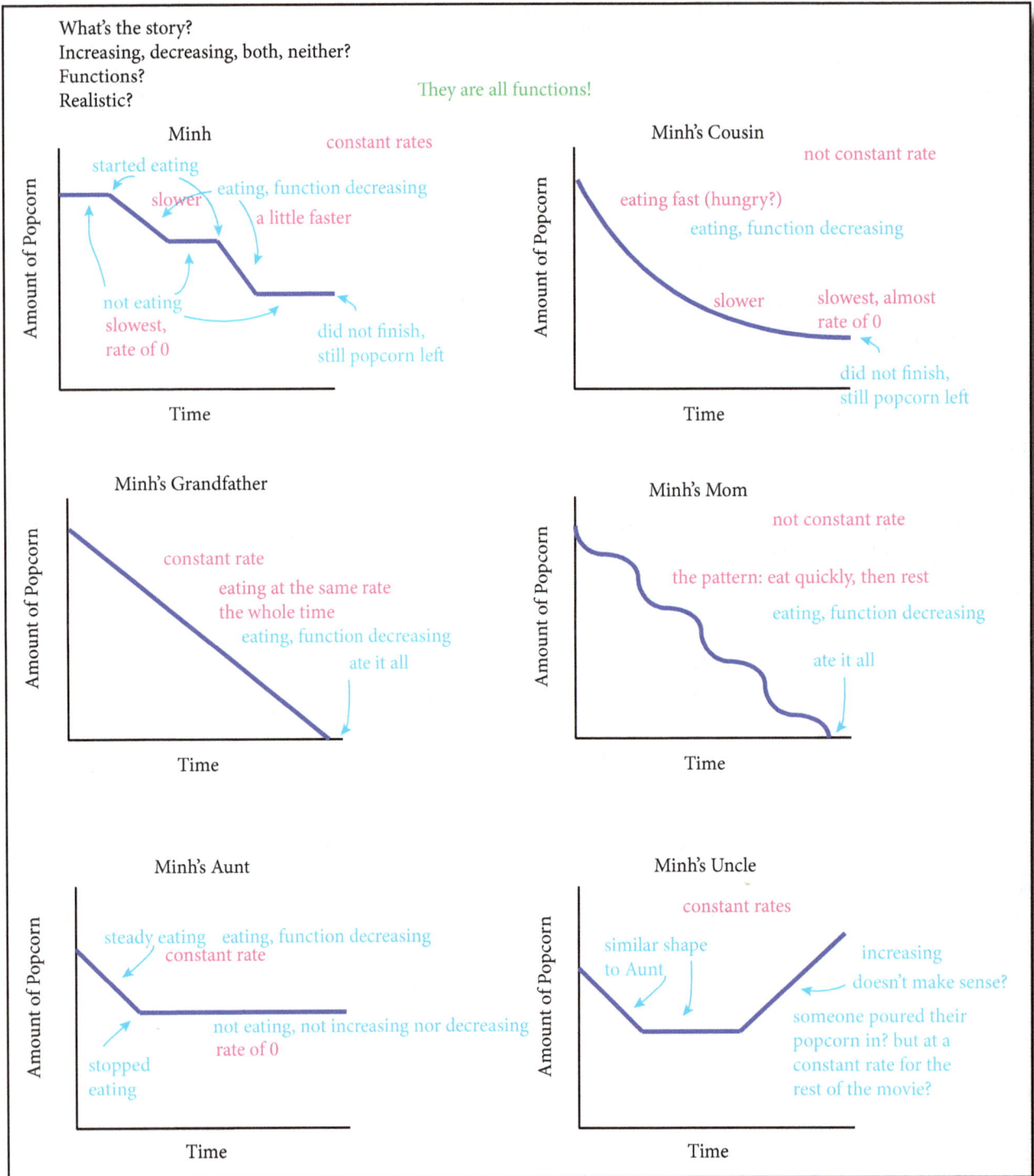

What's the story?
Increasing, decreasing, both, neither?
Functions?
Realistic?

They are all functions!

Minh — constant rates
started eating
slower
eating, function decreasing
a little faster
not eating
slowest, rate of 0
did not finish, still popcorn left

Minh's Cousin — not constant rate
eating fast (hungry?)
eating, function decreasing
slower
slowest, almost rate of 0
did not finish, still popcorn left

Minh's Grandfather
constant rate
eating at the same rate the whole time
eating, function decreasing
ate it all

Minh's Mom — not constant rate
the pattern: eat quickly, then rest
eating, function decreasing
ate it all

Minh's Aunt
steady eating   eating, function decreasing
constant rate
not eating, not increasing nor decreasing
rate of 0
stopped eating

Minh's Uncle — constant rates
similar shape to Aunt
increasing doesn't make sense?
someone poured their popcorn in? but at a constant rate for the rest of the movie?

Algebra Problem Strings
©2017 Kendall Hunt Publishing

Facilitation Notes

This version of the problem string lists short notes for important teacher moves during the string. After you've done the string yourself and studied the relationships involved, you might make similar notes for the things you want a reminder of or deem important.

Minh

Day at the movies! Student Minh and family eating popcorn.
This graph represents the amount of popcorn Minh has during the movie.
Gradually uncover the graph to simulate time passing.
What is the story?
Did Minh finish the popcorn? What does the y-intercept mean?
What do the turning points mean? Steepness of the line segments?
Increasing or decreasing?
Function or not a function?
Domain, range?
Leave the graph up.

Minh's Cousin

Repeat.
What does it mean that there's no turning points?
How does the rate for the cousin relate to the rate for Minh?
Was the rate of popcorn eating constant or changing?

Minh's Grandfather

Repeat.
What's different?
What can you say about how fast or slow the grandfather was eating popcorn?

Minh's Mom

Repeat.
What's different?
Is this believable? Why or why not?
If so, what could explain the rate changes?

Minh's Aunt

Repeat.
What's different?
What does the turning point mean?
Is this believable? Why or why not?
If so, what could explain the rate change?

Minh's Uncle

Repeat.
What's interesting or different?
Are parts of this graph believable? Parts unbelievable?
Why?
Do these graphs represent continous or discrete functions?
Are there breaks in the domain or range? If so, then discrete.
Sometimes we represent discrete situations with continuous graphs.

(continued)

For Display

What's the story?
Increasing, decreasing, both, neither?
Functions?
Realistic?

Minh

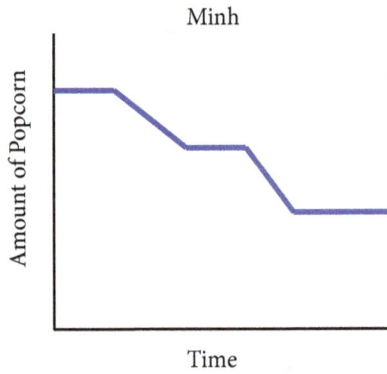

Minh's Cousin

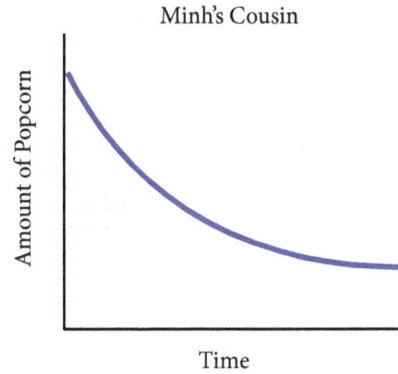

Minh's Grandfather

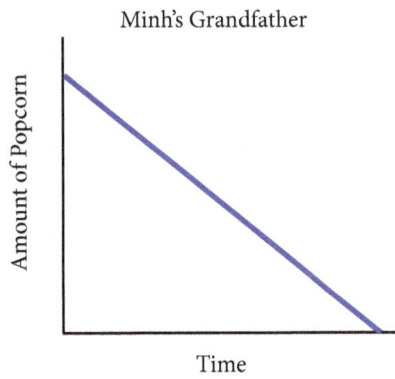

Minh's Mom

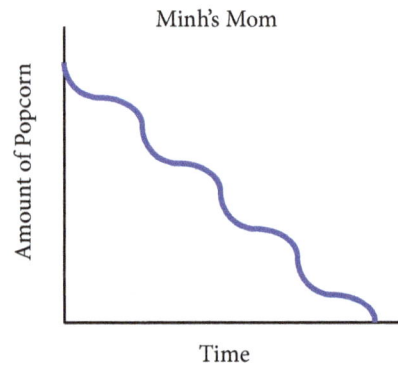

Minh's Aunt

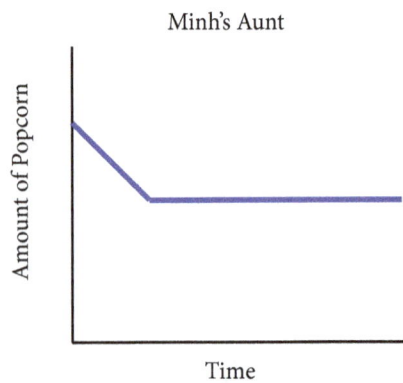

Minh's Uncle

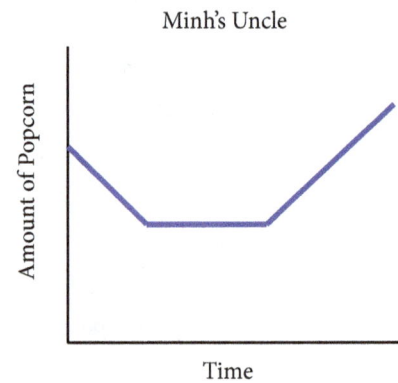

Algebra Problem Strings
©2017 Kendall Hunt Publishing

4.4 | Writing Linear Equations 1

At a Glance	**Objectives**

At a Glance

rate of 2 m/s, start at 0

rate of 2 m/s, (3, 7)

rate of 3 m/s, (4, 10)

Objectives

The goal of this problem string is to help students develop strategies for finding the slope y-intercept form of the equation of a line.

Placement

This is the first in a series of six strings on writing linear equations which builds from the strings in lessons 3.4 and 3.5. Use this problem string to draw out student thinking about writing equations of lines given a rate and a point before students are offered the point-slope form of the equation of a line. This can help strengthen students' facility with the slope y-intercept form of a line and prepare students for success with other forms.

You can use this problem string to precede the Investigation Crosstown Traffic in textbook Lesson 4.4 Slope-Intercept Form of a Linear Function in order to get the ideas percolating that students will use in the investigation.

Guiding the Problem String

The first problem should be quick. By having students briefly consider the proportional relation of $y = 2x$, there is the potential that they might use it to find the equation for the second problem. Setting the problems in the context of walking in front of a motion detector is purposeful. It suggests reasoning using rates to find the "starting point" which is the y-intercept.

Allow students ample time to work on the second problem. Circulate and listen carefully to students. Look for students using these three strategies:

- Walking back in time one second at a time. Model that unit rate strategy by drawing one jump at a time on the graph and table.

- Reasoning about a bigger jump, 3 seconds. Model that non-unit rate strategy by drawing a big jump of 3 seconds on the graph and table.

- Focusing on the rate of 2 meters per second and how that should be 6 meters but it's not, the location is 7 meters at 3 seconds. Sketch the line $y = 3x$ and plot the point (3, 7) and wonder aloud how they might move the line to include (3, 7), while keeping the rate the same. This is a transformation strategy.

If you do not find all three strategies, wonder about the missing one(s) and ask students to help make sense of them.

As students work on the last problem, encourage them to make sense of the problem first, but then to look back and wonder if they can make sense of a different strategy. Wonder which strategy might be more efficient in which cases. During the last problem, ask students to describe the three strategies.

About the Mathematics

Graphs and tables are models. On these models, we can represent strategies—how students used relationships to find the y-intercept.

Textbook lesson 4.4 introduces the slope-intercept form, $y = mx + b$, different from the intercept form, $y = a + bx$ from textbook lesson 3.4. The unit rate and non-unit rate strategies lead to $y = a + bx$ and the transformation strategy yields $y = mx + b$.

(continued)

Sample Interactions

Use the following as you plan how to elicit and model student strategies. This is not meant as a script, but as a view into the relationships involved and the intent of the problem string.

Teacher: *Today I want you to consider a walk in front of a motion detector. For this walk, we know the walker started right at the motion detector and is walking 2 meters per second. Where would the walker be at each second? I know this is easy. Let's just get some points up here quickly.* **Student:** *At time 0, the walker is at 0 meters.* **Student:** *And then 1 second, they are at 2 meters. And then 2 more meters at 2 seconds.* **Teacher:** *What's an equation that would model this walk?* **Student:** *I think y = 2x. But I don't think that's a walk. That's kind of fast!* **Student:** *Yeah, that's more like a jog.* **Teacher:** *You think? Everyone agree? I am going to use the grapher to display that along with the points. Looks like a match! Someone explain why this equation models this walk please.* **Student:** *The equation means that the distance equals 2 times the time and the walker is going at 2 meters per second. It matches.*	 $y = 2x$ (1, 2)
Teacher: *The next problem I want you to think about is still a person walking in front of a motion detector. This person is also moving away at 2 meters per second. You don't know where the person started, but at 3 seconds, the person was 7 meters away. What is an equation to model this walk?* As the teacher says this, the teacher writes the information as shown.	rate = 2 m/s, (3, 7)

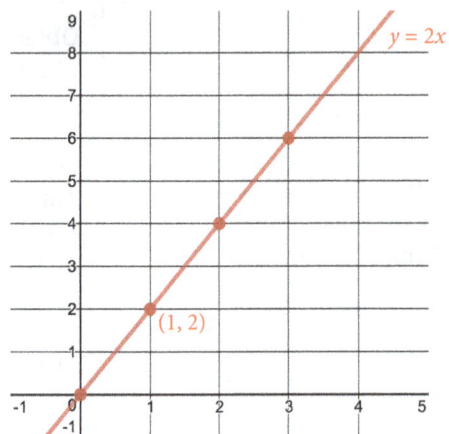

Students work and the teacher circulates, looking for a unit rate strategy, a non-unit rate strategy, and a transformation strategy. To help some students get started the teacher asks:

Teacher: *What do you know? How can you use that? What would be helpful to figure out? Can you use where the person is to work backwards? Can you use the first problem to help you?*

Algebra Problem Strings
©2017 Kendall Hunt Publishing

Teacher: *Some of you are using the rate of 2 meters per second, starting from the point (3, 7). Tell us about that.*

Student: *Well, since we know the walker was at 7 meters at 3 seconds, I went back in time to 2 seconds, 5 meters and then 1 second, 3 meters, and so they must have started at 1 meter.*

Teacher: *I am going to model that on the graph like this. Does that represent what you did?*

Student: *Yes.*

Teacher: *It could also look like this in a table. Right?*

Student: *Yes, that's how I thought about it, not in the graph, but in a table.*

Teacher: *We can model that thinking, using those relationships either on a graph or in a table. So, what is the equation?*

Student: *Now that we know the starting point of 1, I think of starting at 1 and adding 2 meters times the number of seconds, $y = 1 + 2x$.*

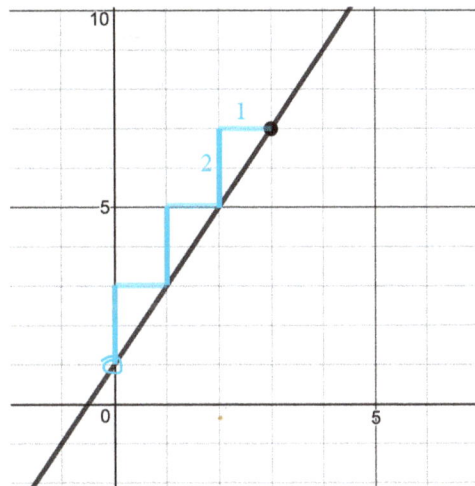

t	d
0	1
1	3
2	5
3	7

$$y = 1 + 2x$$

(continued)

Teacher: *Did anyone do something similar, starting with the (3, 7) but use the rate in a different way? By thinking about how far the walker went in all 3 seconds?*

Student: *Yeah, I started there too, but I knew that since they were walking at 2 meters per second and they were walking for 3 seconds, they covered 6 meters.*

Teacher: *That sounds important, but different than our other strategy. Who can repeat that for us?*

Student: *Since they walked for 3 seconds at 2 meters per second, they went a total of 3 times 2 is 6 meters in 3 seconds.*

Teacher: *I will model what you said on the graph like this. Go back all 3 seconds at once, and back all 6 meters at once. And check it out—we end up with the same starting point. So, you get the same explicit formula.*

Teacher: *Interesting, so some of you went back 1 unit at a time, using the unit rate. And some of you went back 3 seconds all together, so all 6 meters at a time, using the non-unit rate of 6 meters per 3 seconds. Nice.*

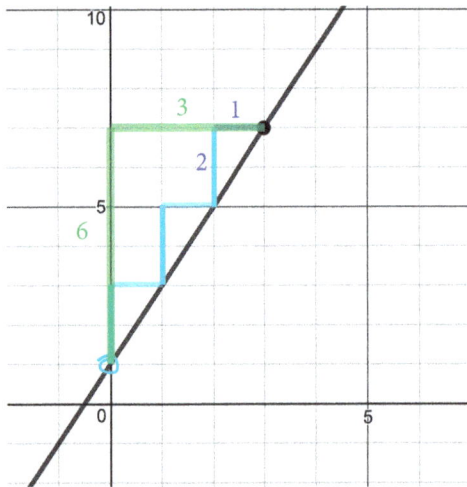

$$y = 1 + 2x$$

Teacher: *But I saw some of you using the first problem. Who can tell us about that?*

Student: *I was thinking about walking at 2 meters per second. If you did that, it would look like the first problem. But then at 3 seconds, you would be at 6 meters, not 7.*

Teacher: *I will model that by sketching the $y = 2x$ and the point (3, 7). Hmmm... that's interesting. Can anyone else see where you might go from here?*

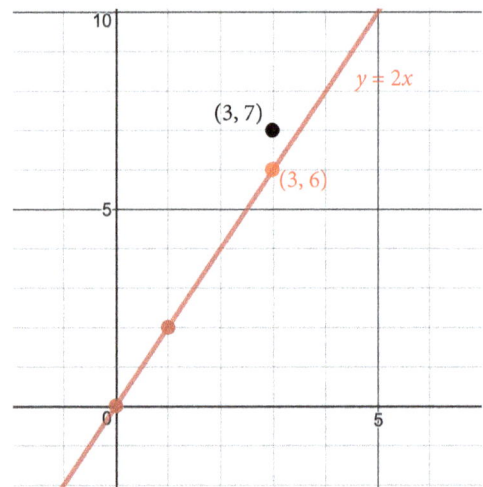

Algebra Problem Strings
©2017 Kendall Hunt Publishing

Student: *Well, since the line is just 1 unit too low, I thought about shifting it all up 1, so that would be $y = 2x + 1$.*	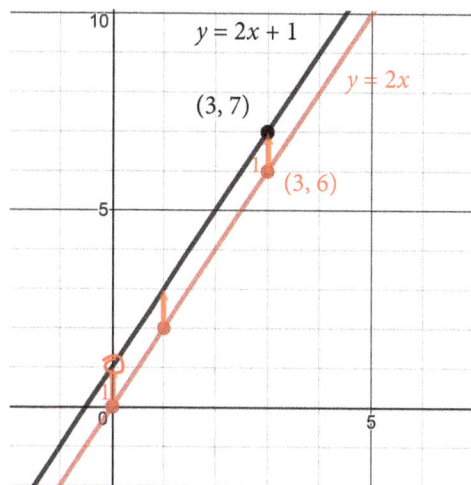
Teacher: *Are there any questions about this strategy? No? Then how might you describe it?*	
Student: *Think about the rate and where you would be if you started at 0. Then shift that line to where the walker actually was.*	
Teacher: *I noticed that if you thought about the starting point, the y-intercept first, you might have thought of the line as $y = 1 + 2x$, but if you thought about the rate first and then shifted, you may have thought of the line as $y = 2x + 1$. Interesting.*	
Teacher: *Great! Our next problem is a different walk in front of a motion detector. This person is moving away at 3 meters per second. Again we don't know where the person started, but at 4 seconds, the person was 10 feet away. What is an equation to model this walk?* As the teacher describes the scenario, the teacher records the information on the board. *Make sense of this problem, but after you have, I wonder if you can look back and consider other strategies? See if you can make sense of these other strategies.*	rate of 3 m/s, (4, 10)

(continued)

Students work and the teacher circulates, looking for a unit rate strategy, a non-unit rate strategy, and a transformation strategy.

Teacher: *A few of you were thinking about that starting point and working backwards one second at a time. Tell us about that.*

Student: *We knew that at 4 seconds, she was 10 feet away, so 3 seconds, 7 feet; 2 seconds, 4 feet; 1 second, 1 foot. And it got a little weird, but when time started, she was behind the motion detector 2 feet. So that's the y-intercept.*

Teacher: *I just recorded all of that on the graph, but we can also record that thinking in a table. Some of you used a graph and some a table, but you were all working with the same relationships, working backward by 1 second to find the y-intercept. So, what equation did you write?*

Student: $y = -2 + 3x$.

As the teacher records the student's equation, the teacher says the equation in context.

Teacher: *The distance, y, equals start at −2 and gains distance at 3 feet per second times the number of seconds.*

rate of 3, (4, 10)

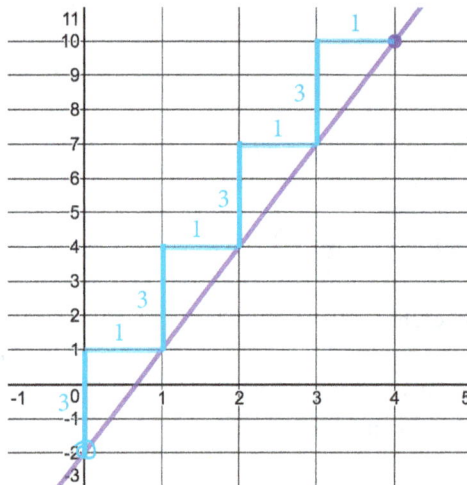

t	d
0	−2
1	1
2	4
3	7
4	10

$y = -2 + 3x$

Algebra Problem Strings
©2017 Kendall Hunt Publishing

Lesson 4.4 • Writing Linear Equations 1 (continued)

Teacher: *Okay, some of you thought about backing up in time in one big jump. Convince us about that!*

Student: *If you know that she walked four seconds at 3 meters per second, then she covered 12 feet. But she was at 10 feet, so she must have started at −2.*

As the student is describing the teacher is modeling these relationships on the graph and table.

Teacher: *Who could explain this thinking?*

Student: *Since you know she's at 10 meters at 4 seconds and you know she went 3 meters per second for 4 seconds, she went 12 feet in those 4 seconds, so back from 10 by 12 is −2.*

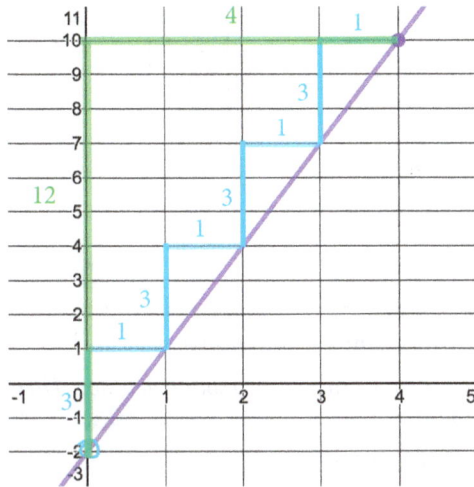

$$y = -2 + 3x$$

Teacher: *Turn to your partner and put some words to these strategies.*

After partners talk briefly, the teacher asks students to share their ideas and the teacher records some brief descriptions.

Teacher: *Turn to your partner and compare these two strategies. How are they alike? Different?*

Students turn and talk while the teacher listens in, taking note of students' emerging ideas.

The teacher crafts a short conversation with students about efficiency, helping students verbalize that bigger, fewer jumps is a goal.

Teacher: *Some of you were working on a strategy that used the rate of 3 meters per second. Help us make sense of that please.*

Student: *So, I thought about a person walking at 3 meters per second. After four seconds, she should be at 12 feet, but she's not. You said she's at 10.*

Teacher: *Let me mark that. At 4 seconds, if she started right at the motion detector, she would now be 4 seconds times 3 meters per second, 12 meters away.*

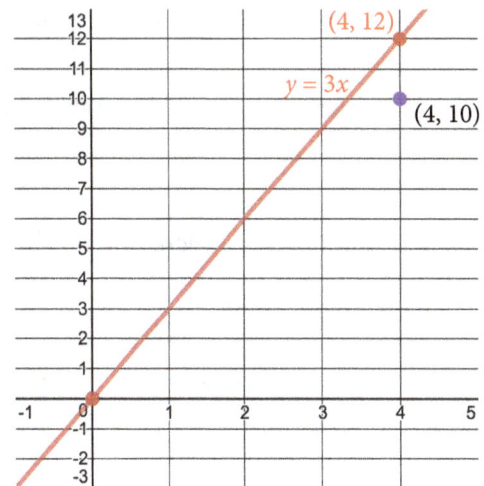

(continued)

Algebra Problem Strings
©2017 Kendall Hunt Publishing

171

Student: *So it shifts down. The line goes down 2. That way, she's still walking at 3 meters per second and she's at 10 feet in 4 seconds. Since the line shifted down 2, so did the y-intercept. So the line is the 4x but minus 2.*

Teacher: *So, the equation you were thinking about is the distance, y, is the rate of 3 meters per second times the number of seconds, but shift it down 2, or minus 2.*

Turn to your partner and discuss this strategy. How would you describe it? Why does it work?

Students turn and talk and the teacher listens in.

Teacher: *Let's come back together. Will you two share what you were talking about?*

Student: *We noticed that it seems like our mission every time was to find the y-intercept.*

Student: *Yeah, we had the rate, so we just needed to find the y-intercept and then we could write the equation.*

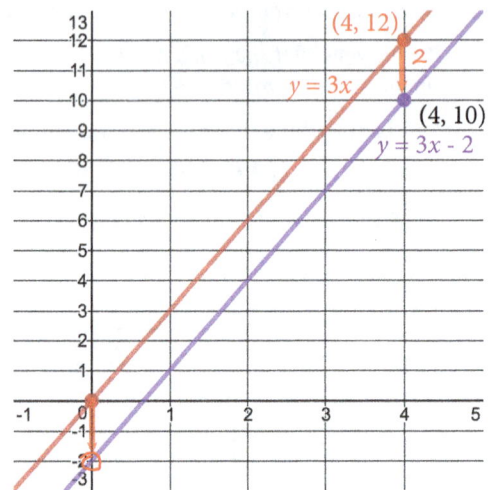

Teacher: *Let's put some words to this last strategy. Someone who did not use this thinking, please.*

Student: *It's like they know the rate so they pretend that the person started at the motion detector. Then they find out where the person would've been and shift the line to get there.*

Student: *It's like they know the y = bx part and they just have to find the a, so they shift it.*

Teacher: *How would you summarize some of the things that came up in this string today?*

Elicit the following:

- *Once you know the rate and the y-intercept, you can write an equation for a line.*

- *You can use the unit rate to walk back in time to find the y-intercept.*

- *You can use a non-unit rate, how far the walker went in a range of time, to walk back in time to find the y-intercept.*

- *You can use the rate b to write y = bx, and then the amount you need to shift the function up or down is the y-intercept.*

- *You can do your thinking with a graph or table, either one works.*

- *Some strategies help you come up with y = a + bx (b is the rate) and some help you get y = mx + b (b is the shift up or down).*

- *Using a non-unit rate is more efficient than using a unit rate.*

Algebra Problem Strings
©2017 Kendall Hunt Publishing

Sample Final Display

Your display could look like this at the end of the problem string:

rate of 2, start at 0 distance

rate of 2, (3, 7)

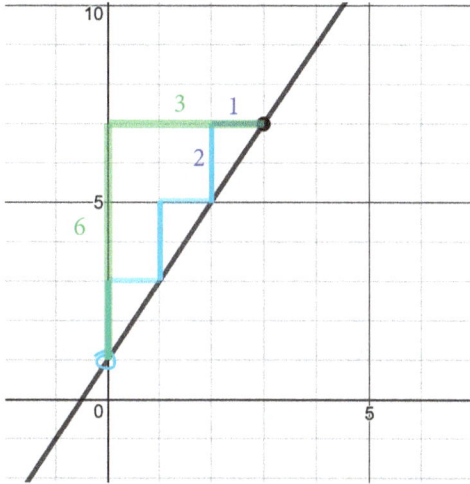

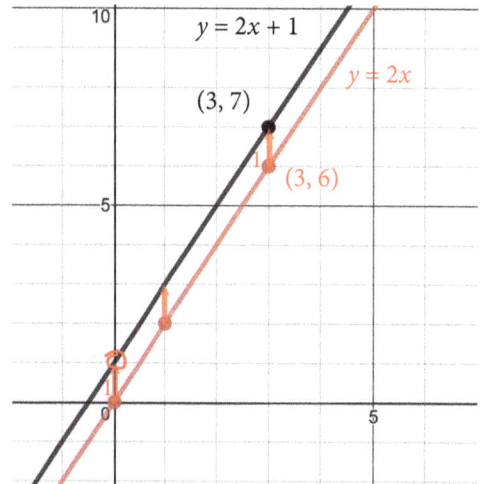

t	d
0	1
1	3
2	5
3	7

$y = 1 + 2x$

rate of 3, (4, 10)

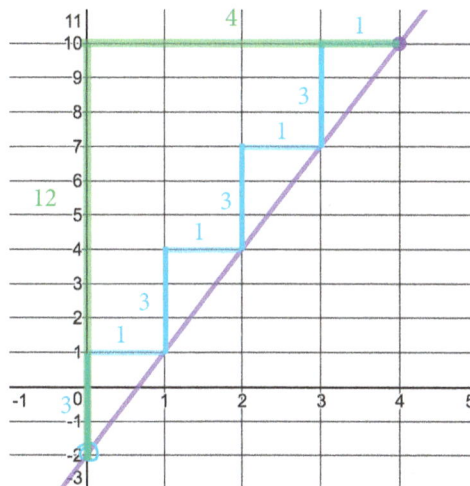

t	d
0	−2
1	1
2	4
3	7
4	10

$y = -2 + 3x$

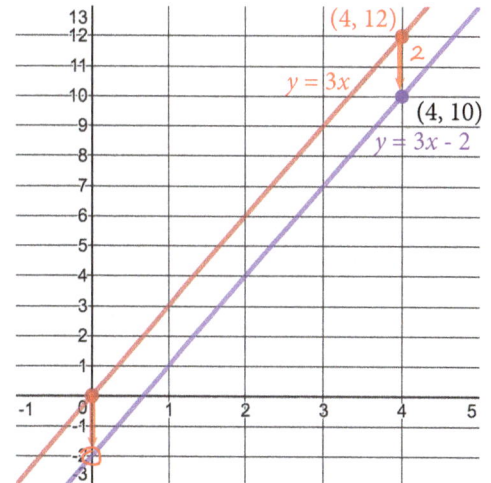

Mission: finding the *y*-intercept!

Backing up in time

using the unit rate (one second at a time)

using the non-unit rate (bigger span)

Use the rate to shift the line $y = 3x$ to find b in $y = 3x + b$.

(continued)

Facilitation Notes

This version of the problem string lists short notes for important teacher moves during the string. After you've done the string yourself and studied the relationships involved, you might make similar notes for the things you want a reminder of or deem important.

rate of 2, start at 0 distance	What if this represents a walk away from a motion detector? Quick. Plot points. Write y = 2x.
rate of 2, (3, 7)	Still walking. What is equation? Listen for unit rate, non-unit rate, transformation strategies. What do you know? What are you looking for? What's the equation of a line? Can you work backwards in time? Can you use the first problem to help? Linger. Repeat back. Explain. Use color. Note that unit rate and non-unit rate strategies think y = 1 + 2x, but transformation thinks y = 2x + 1
rate of 3, (4, 10)	Repeat. Can you step out of the problem, look at relationships, and challenge yourself to try a different strategy? Let's put some words to these strategies? Compare the unit rate and non-unit rate strategies. Discuss efficiency. How would you describe the transformation strategy? Connect "y-intercept first" to "rate first approach," y=-2+3x=3x-2.

Algebra Problem Strings
©2017 Kendall Hunt Publishing

4.5 | Writing Linear Equations 2

At a Glance

rate of 2 m/s, (30, 55)

rate of −5 m/s, (2, 4)

rate of −1.5 m/s, (4, −5)

Objectives
The goal of this problem string is to help students use more efficient strategies for finding the slope y-intercept form of the equation of a line.

Placement
This is the second in a series of six strings on writing linear equations. Use this problem string to strengthen students' linear equation writing of the slope y-intercept form of a line and prepare students for success with the point-slope form.

You can use this problem string to precede the work in textbook lesson 4.5 to continue to strengthen students' facility with the slope y-intercept form before students learn the point-slope form.

Guiding the Problem String
The three problems should each take about the same amount of time. As always, let students solve each problem any way they choose, but promote efficiency and sophistication by celebrating it. Help students see connections between the strategies, how each is using the unit rate. The first problem involves the point (30, 55) which should cause students to reconsider if they try to walk back to the y-intercept using the unit rate. Bring out the algebraic way of modeling the transformation strategy so that when students try the second problem with its negative rate, they have other options. Highlight that algebraic strategy and also the efficient non-unit rate strategy. For the last problem, with its negative rate and point in the fourth quadrant, let students try everything and compare.

About the Mathematics
In the first problem, the point (30, 55) is so far out from the y-axis that it is no longer efficient to use the unit rate strategy many students were using in the previous string. This may prompt students to try something more efficient.

In this problem string, we connect the transformation strategy (see problem string 4.4) to an algebraic solution using the equation $y = mx + b$. You might be inclined to make this procedural (take the equation, substitute stuff in, solve for b), but if it's connected to students' experience with the transformation strategy and the non-unit rate strategy, students will own more connections and relationships. They will also forget "how to do it" less and be more inclined to think and reason instead of reaching for rote memorized information and procedures.

Graph, tables, and equations are models. On these models, we can represent strategies—how students use relationships to find the y-intercept. We represent the transformation strategy in this string on two models: graphs and equations.

(continued)

Sample Interactions

Use the following as you plan how to elicit and model student strategies. This is not meant as a script, but as a view into the relationships involved and the intent of the problem string.

Teacher: *In our last problem string, we found the equations of lines to model motion walking away from a motion detector. Today's string is going to build on that. The first scenario is a walker going away at 2 meters per second. You know that at 30 seconds, she is 55 meters away. Write the equation of the line that models her walk.*	rate of 2 meters per second away, (30, 55)

Students work and the teacher circulates, noting students using a unit rate strategy, and looking for a non-unit rate strategy and a transformation strategy. To help some stumped students the teacher asks:

Teacher: *What do you know? How can you use that? Do you remember any ideas we used in the last string? How fast is she walking? For how long?*

Teacher: *Let's get started by noting what equations you found.*

Student: *I got $y = -5 + 2x$.*

Student: *The same, just written differently, $y = 2x - 5$.*

Teacher: *I saw some of you using the unit rate to count back one second at a time. Whew! That is a lot of work. Great perseverance. Some of you were using the rate in bigger jumps, by thinking about how far the walker went in all 20 seconds. That seemed to save some time. Please tell us about your thinking.*

Student: *I knew that since she was walking at 2 meters per second for 30 seconds, she would have walked 60 meters.*

Student: *She went a total of 3 times 20, or 60 meters, in 30 seconds. So she must have started at −5, like 5 meters behind the motion detector.*

Teacher: *I will model what you said on the graph and table like this. Go back all 30 seconds at once, and back all 60 meters at once. And there we are at the same starting point, the y-intercept.*

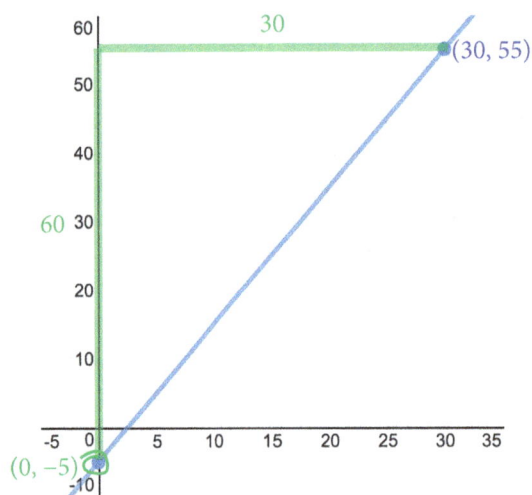

Student: *Wow. That could've saved me some time. I was going back one at a time.*

Teacher: *Yesterday's problems were just about as efficient using either the unit rate or the non-unit rate, but for this problem, I hear you are saying that the non-unit rate is more efficient. Nice thinking.*

Algebra Problem Strings
©2017 Kendall Hunt Publishing

Teacher: *But I saw some of you using that other strategy we worked with in the last string, where the line shifted. Who can tell us about that?*	
Student: *I was thinking about walking at 2 meters per second. If you did that, that's y = 2x. So then at 30 seconds, you would be at 60 meters. But at 30 seconds, you should be at 55 meters, so shift the whole line down 5.*	
Teacher: *So the y = 2x shifted down 5, that sounds like the other version of the equation, y = 2x − 5. Does anyone have any questions about that?*	
Teacher: *Turn and talk about these two strategies.*	
Students talk and then the teacher calls them back together.	
Student: *I am noticing that they both thought about the 30 and the 30 times 2 is 60.*	
Student: *I noticed that if you thought about the starting point first, you said the line is y = −5 + 2x, but if you thought about the rate first and then shifting, you said the line is y = 2x − 5.*	

Teacher: *I'd like to model this transformation strategy differently. Each time we've talked about that strategy, you start out thinking about the rate and the line that represents that rate if the walker started at 0, right? So, I've been modeling that on a graph. Let's look at it with equations. So, starting with that rate of 2 and starting at 0, you have the line y = 2x, right? But you know that at 30 seconds, it was supposed to be at 55 meters. So, you know the line will have a y-intercept. So, where is y = 2x at x = 30? At 60. So, you'll have to shift the 60 down 5 to get to the 55. So the shift is −5. What do you think? Can this algebraic stuff model your thinking too?*	$y = 2x$ — the line with rate 2 $(30, 55)$ — at 30 seconds, should be 55 meters $y = 2x + b$ — know the line will have y-intercept $55 = 2(30) + b$ — at 30, $y=2x$ is $y=2(30)=60$ $55 = 60 + b$ $b = -5$ — ah, must have to shift it down 5

Teacher: *The next problem in our string is a bit different. The walker is walking toward the motion detector, so a rate of −5, and at 2 seconds he is 4 meters from the motion detector.*	rate of −5 m/s, (2, 4)

(continued)

Students work, the teacher circulates, and then asks students to share a transformation strategy.

Teacher: *This one was a little weird, yes? A rate of −5 m/s? But some of you made that work. What was the equation?*

Student: *I got y = −5x + 14.*

Teacher: *Did anyone get anything different? No? Okay, tell us how you made that work.*

Student: *I thought about walking toward the motion detector at a rate of 5 m/s and that would look like y = −5x. So, at 2 seconds, that would be down at −10, but he was supposed to be at 4 meters, so shift the whole thing up 14, y = −5x + 14.*

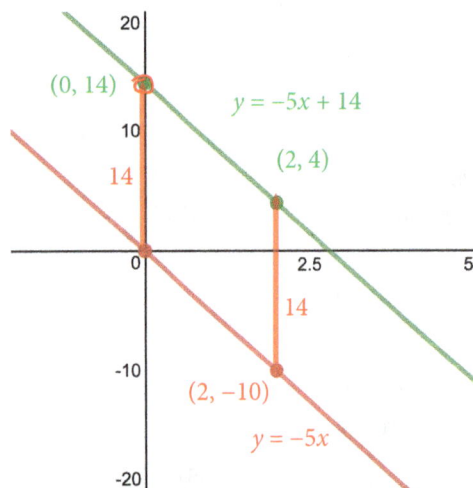

$$y = -5x$$
$$(2, 4)$$
$$y = -5x + b$$
$$4 = -5(2) + b$$
$$4 = -10 + b$$
$$b = 14$$

Teacher: *I'm a little curious. Did anyone try walking back?*

Student: *I did and for this problem, it was quite easy. You know (2, 4) and a rate of −5 m/s, so just back up in time 2 seconds to zero and move back 10 meters to 14 meters. So he was coming toward the motion detector.*

Teacher: *So, that's noteworthy. Mathematicians seek patterns and connections, but also efficiency. Let's try to keep all of our strategies in mind, all of the relationships we've been building to decide how to tackle this next problem.*

$$y = 14 - 5x$$

Teacher: *I wonder how you will think about the next problem in our string: a rate of −1.5 meters per second, and the point (4, −5). Does it make sense for me to talk about a walker with the point (4, −5)? It can? Does it need to? I wonder how you might find an equation of the line that has a rate of −1.5, and contains the point (4, −5)?*

Students work, the teacher circulates and then asks a student to explain the transformation strategy and a non-unit rate strategy. Model the transformation strategy algebraically.

Student: *I thought kind of like before. I need the equation of a line, y = mx + b. And we know the rate is −1.5 m/s, so we know y = −1.5x + b. But we need to shift the line to the point (4, −5). Rather than graphing anything I wrote an equation, −5 = −1.5(4) + b. So −1.5 times 4 is −3 times 2, so −6. And so b is 1.*

Student: *I thought about getting from 4 to 0 to find the y-intercept. Since the rate is −1.5, that's 4 times −1.5, or −6. So subtract −6 from −5 is 1. So the y-intercept is 1. I thought about 1.5 times 4 as one 4 and a half of 4, that's 4 plus 2, so 6.*

rate of −1.5 m/s, (4, −5)

$y = -1.5x$
$(4, -5)$
$y = -1.5x + b$
$-5 = -1.5(4) + b$
$-5 = -6 + b$
$b = 1$
$y = -1.5x + 1$

t	d
0	1
4	−5

$4 \langle \quad \rangle -6$

$y = -1.5x + 1$

Teacher: *Why didn't anyone use a unit rate strategy?*

Student: *I thought about it, but you'd go back by 1.5. It seemed easier to go back 6 all at once, because two 1.5s is 3, so four 1.5s is 6.*

Teacher: *Did anyone think about a transformation strategy on a graph but then solve it algebraically?*

Student: *Well, I thought about it that way to help me figure out the algebraic part.*

Student: *I didn't because thinking about the negative 5 was weird.*

Student: *You can think about the walker moving behind the motion detector, but since I had other ways to think about it, I chose not to.*

Teacher: *How would you summarize some of the things that came up in this string today?*

Elicit the following:

- *If you know the rate and the y-intercept, you can write the equation for a line.*
- *You can use the unit rate to walk back in time to find the y-intercept. This can be efficient if it's easy to get from the point to the y-intercept with the unit rate.*
- *You can use a non-unit rate, how far the walker went in a range of time, to jump back in time to find the y-intercept.*
- *You can use the rate m to write y = mx and then the amount you need to shift the function up or down is the y-intercept on a graph.*
- *You can use y = mx + b and the given rate and point to algebraically find b. This can be efficient when the numbers are big or unwieldy.*

(continued)

Sample Final Display

Your display could look like this at the end of the problem string:

rate of 2 m/s, (30, 55)

$y = 2x - 5$

$y = 2x$	the line with rate 2
$(30, 55)$	at 30 seconds, should be 55 meters
$y = 2x + b$	know the line will have y-intercept
$55 = 2(30) + b$	at 30, $y = 2x$ is $y = 2(30) = 60$
$55 = 60 + b$	
$b = -5$	ah, must have to shift it down 5

rate of −5 m/s, (2, 4)

$y = -5x + 14$

$y = -5x$
$(2, 4)$
$y = -5x + b$
$4 = -5(2) + b$
$4 = -10 + b$
$b = 14$

t	d
0	14
2	4

$2 \langle \quad \rangle 10$

$y = 14 - 5x$

rate of −1.5 m/s, (4, −5)

$y = -1.5x + 1$

$y = -1.5x$
$(4, -5)$
$y = -1.5x + b$
$-5 = -1.5(4) + b$
$-5 = -6 + b$
$b = 1$
$y = -1.5x + 1$

t	d
0	1
4	-5

$4 \langle \quad \rangle -6$

$y = -1.5x + 1$

Facilitation Notes

This version of the problem string lists short notes for important teacher moves during the string. After you've done the string yourself and studied the relationships involved, you might make similar notes for the things you want a reminder of or deem important.

rate of 2 m/s, (30, 55)	*Remember last string? Walking away at rate of 2 m/s. At 30 secs, 55 m away.* *Share non-unit, transformation. Why not use unit rate?* *Model transformation with y = mx + b, algebraically substituting in (30, 55).*
rate of −5 m/s, (2, 4)	*Walking toward. Listen for non-unit rate, transformation strategies: graph and algebraic models.* *Which is more efficient? Why?*
rate of −1.5 m/s, (4, −5)	*Can you use what you know when walking doesn't make as much sense?* *Find non-unit and algebraic transformation.* *Why not other strategies?*

Algebra Problem Strings
©2017 Kendall Hunt Publishing

4.6 Equivalent Expressions

At a Glance

12×8

12×18

$2(x+4)$

$(x+3)\cdot 7$

$-3(x-5)$

$-4(x+3)$

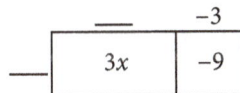

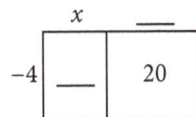

$-3(\underline{\quad}+4) = -3x + \underline{\quad}$

$2(x+\underline{\quad}) = \underline{\quad} -10$

Objectives
The goal of this problem string is for students to gain facility with using the distributive property of multiplication over addition or subtraction to find equivalent expressions.

Placement
You can use this problem string to get students ready to translate between the point-slope and the slope-intercept forms of the equation of a line.

Use this string to support textbook Lesson 4.6 Equivalent Algebraic Equations.

Guiding the Problem String
The first two problems are numerical and set the stage for using the distributive property of multiplication. The third and fourth problems introduce variables to represent parts of lengths of rectangles. The fifth and sixth problems transition to integers where length and area no longer make sense but rectangular diagrams can still be helpful models. Use the word *opposite* to help students reason about integer multiplication and later division. The next three problems are given as rectangular diagrams missing information where now students have to use the model. Record what happens in the rectangular diagrams with equations. Do not force students to draw the rectangular diagrams up to this point. Just use the model to represent the relationships. The last two problems are given as equations missing information. Have students draw the corresponding rectangular diagrams and fill in the missing parts.

About the Mathematics
The string begins where the teacher models multiplication with an open array, a model where the factors represent the length and width of the array and the product represents the area of the array. The rectangles should be roughly to scale when possible. The first factor represents the number of rows, the second the number of columns. Later, the string moves away from the array as an area model towards a rectangular diagram, meant to capture the general relationship between factors and their products. The factors (here monomial and binomial expressions) are analogous to the lengths and widths of the rectangle and the product to the area, but because length and area are not negative, we use the model as less a representation of physical relationships and more as a tool for computation.

Using the rectangular diagram as a model for the distributive property of multiplication can help prevent the error where students don't fully distribute, such as $2(x+4) \neq 2x+2$.

(continued)

Sample Interactions

Use the following as you plan how to elicit and model student strategies. This is not meant as a script, but as a view into the relationships involved and the intent of the problem string.

Teacher: *Today's problem string begins with a numerical problem. What is 12 times 8? You might know 12 times 8 or you might want to find a nice chunk to help you figure twelve 8s.* Brief think time.	12×8
Teacher: *What is 12 times 8?* **Students:** 96. **Teacher:** *So, a 12 by 8 rectangle has an area of 96? Yes?* As the teacher says the dimensions and area, the teacher draws and labels the rectangular open array.	$12 \times 8 = 96$ $12 \times 8 = 96$ *(open array: 8 across top, 12 down left, 96 inside)*
Teacher: *Did anyone think about ten 8s to help them?* **Student:** *Yeah, 10 times 8 is 80, two more 8s is 16. So 80 plus 16 is 96.* **Teacher:** *I'll model your thinking on this open array. A 10 by 8 has an area of 80 and a 2 by 8 has an area of 16 so put those together and get an area of 96. That's a nice way of figuring 12 times 8 if you didn't already know it.*	$12 \times 8 = 96$ *(open arrays: 8 top, 12 and 96; split into 10/80 and 2/16, bracket to 96)*
Teacher: *The next problem is 12 times 18. I wonder if you know anything that might be helpful?* Brief think time. If students are writing, the teacher circulates quickly, looking for students who are using the first problem, 12×8 and students who are thinking about ten 18s and two 18s. If students are not writing, the teacher waits until most students are ready and plans to ask who used the first problem.	12×18
Teacher: *What is 12 times 18?* **Student:** 216. **Teacher:** *So the area of a 12 by 18 is 216.* As the teacher says the dimensions and area, the teacher draws and labels the rectangular open array.	$12 \times 18 = 216$ *(open array: 18 across top, 12 down left, 216 inside)*
Teacher: *Did anyone use the first problem to help?* **Student:** *I did. We already know that 12 times 8 is 96 so just add 120.* **Teacher:** *Why 120?* **Student:** *Because that's 12 times 10.* **Teacher:** *So you can split the area into nice chunks you know and then add the chunks? Nice.*	$12 \times 18 = 216$ *(open array: 18 top split 8/10, 12 left, 96 and 120 inside, bracket to 216)*

Teacher: *Did anyone split the array into different chunks? Like ten 18s?*

Student: *You can find ten 18s, 180 and add that to two 18s, 36.*

Teacher: *Turn to your partner and discuss these two strategies and how using friendly chunks can help to find the product in a multiplication problem.*

As partners talk briefly, the teacher listens in.

$12 \times 18 = 216$

	18	
	8	10
12	96	120

216

	18	
12	10 180	
	2 36	

216

Teacher: *The next problem is also finding a product, but this time it's a little different. What is the area of something that has one dimension of 2 and the other dimension we only know is some value, x, plus 4.*

As the teacher says the problem, the teacher writes the expression and draws and labels the rectangle diagram. Brief think time.

$2(x+4)$

	x	4
2		

Teacher: *What is the area of a rectangle that has one dimension 2 and all we know about the other is that it is some value, x, plus 4?*

Student: *It's 2x plus 8.*

The teacher records this both as an equation and in the rectangle diagram.

$2(x+4) = 2x+8$

	x	4
2	$2x$	8

Teacher: *How do you know?*

Student: *2 times x is 2x.*

Teacher: *Could we think about this dimension of 2 as a 1 and a 1? So what is this piece?*

Student: *A one by x. And another one by x. So that's a two by x, 2x.*

Teacher: *So we can talk about equivalent expressions, two times the quantity x plus 4 is equivalent to 2x + 8. And we can talk about dimensions and area, if the dimensions of a rectangle are 2 and x + 4, then the area can be represented by 2x + 8.*

$2(x+4) = 2x+8$

	x	4
2	$2x$	8

	x	4
1	$1x$	8
1	$1x$	

(labeled 2)

The teacher repeats a similar conversation for the next problem, $(x+3) \cdot 7$.

$(x+3) \cdot 7 = 7x+21$

	7
x	$7x$
3	21

	7
x	x (1)
3	21

Seven rectangles that are 1 by x = 7x

Teacher: *The next problem is to find an equivalent expression for the opposite of 3 times x minus 5, $-3(x-5)$ without parentheses. It's not quite right to talk about dimensions and area anymore because negative length doesn't have meaning in this scenario, but we can still use a rectangular diagram to help us think about the multiplicative relationships.*

Brief think time.

$-3(x-5)$

	x	-5
-3		

(continued)

Lesson 4.6 • Equivalent Expressions (continued)

Teacher: *What's an equivalent expression?* **Student:** *–3x plus 15.* **Teacher:** *One way to think about that is with the word opposite instead of negative. Does anyone think of it that way?* **Student:** *We know that 3 times x is 3x. So the opposite of 3 times x is the opposite of 3x, or negative 3x.* **Teacher:** *What about the negative 3 times negative 5?* **Student:** *Since 3 times 5 is 15, the opposite of that is –15 and the opposite of negative 15 is positive 15.*	$-3(x-5) = -3x+15$ $\begin{array}{c|c|c} & x & -5 \\ \hline -3 & -3x & 15 \end{array}$
The teacher repeats a similar conversation for the next problem, $-4(x+3)$, using the word "opposite" to describe –4. Since 4 times 3 is 12, the opposite of 4 times 3 is the opposite of 12. Another way to think of that is that 3 times the negative 4 is negative 12. Both of these meanings can be helpful.	$-4(x+3) = -4x-12$ $\begin{array}{c|c|c} & x & 3 \\ \hline -4 & -4x & -12 \end{array}$
Teacher: *The next problem in the string is this rectangular diagram with some missing information. What's missing? What makes sense in these blanks?* The teacher draws the problem with the blanks on the board.	$\begin{array}{c|c|c} & \underline{} & \underline{} \\ \hline 2 & 2x & -8 \end{array}$
Student: *You just think about what times 2 is 2x, that's x. And what times 2 is –8, that's –4.* **Teacher:** *I'll record that in the rectangular diagram. What equation can we write? Since we started with the information inside the rectangular diagram, I'll write that first. Then what?* **Student:** *That's equal to 2 times parentheses x minus 4.* **Teacher:** *So, 2x – 8 is equivalent to 2 times the quantity x minus 4.*	$\begin{array}{c|c|c} & x & -4 \\ \hline 2 & 2x & -8 \end{array}$ $2x-8=2(x-4)$
The teacher repeats a similar conversation for the next two problems, giving students a bit more think time for each and asking students to explain where they started and why. Since you can start anywhere, the point is not to decide where one should start, but to bring out the flexibility in solving such problems.	$\begin{array}{c|c|c} & x & -3 \\ \hline 3 & 3x & -9 \end{array}$ $3x-9=3(x-3)$ $\begin{array}{c|c|c} & x & \underline{-5} \\ \hline -4 & \underline{-4x} & 20 \end{array}$ $-4(x-5)=-4x+20$
Teacher: *The next problem is slightly different. This time there are missing parts in the equation. What makes sense in the blanks and how would this equation look in a rectangular diagram?* Students work while the teacher circulates, looking for students who are using the rectangular diagram to help them.	$-3(\underline{}+4)=-3x+\underline{}$

184

Algebra Problem Strings
©2017 Kendall Hunt Publishing

Teacher: *First of all, tell us what goes in each blank and what the rectangular diagram looks like.* The teacher records students' answers as they explain. **Teacher:** *I noticed that some of you were filling in the rectangular diagram first. Tell us about that.* **Student:** *It helped me figure out what was missing. When I could see the −3 and the −3x, I knew that the top had to be x. Then when I filled in the equation, I could see how that makes sense too.* **Student:** *I started with the equation, but I asked myself the same question, what times −3 is −3x and also got x.*	$-3(\underline{\;x\;}+4)=-3x+\underline{-12}$ $\begin{array}{c c c}& x & 4 \\ -3 & \boxed{-3x} & \boxed{-12}\end{array}$
Teacher: *Alright my friends, let's end with this one. What belongs where and why?* After students explain, the teacher wraps up the string. **Teacher:** *When finding equivalent expressions without parentheses in this way, we are using the distributive property. In this problem, we were distributing the 2 over the expression x minus 5. Nice work everyone.*	$2(x+\underline{-5})=\underline{2x}-10$ $\begin{array}{c c c}& x & -5 \\ 2 & \boxed{2x} & -10\end{array}$

Teacher: *How would you summarize some of the things that came up in today's string?*

Elicit the following:

- *We can represent multiplication with equations and rectangular diagrams. If the factors are positive, the factors are the dimensions and the product is the area. If the factors are integers, we can use the rectangular diagram to represent the factors and products even though they no longer represent dimensions and area.*

- *The distributive property of multiplication over addition or subtraction can be represented with both equations and rectangular diagrams.*

- *The word opposite can be helpful when reasoning about multiplication and division of integers.*

- *When finding equivalent expressions with and without parentheses using the distributive property, don't forget to distribute to both terms (don't forget part of the rectangular diagram).*

(continued)

Sample Final Display

Your display could look like this at the end of the problem string:

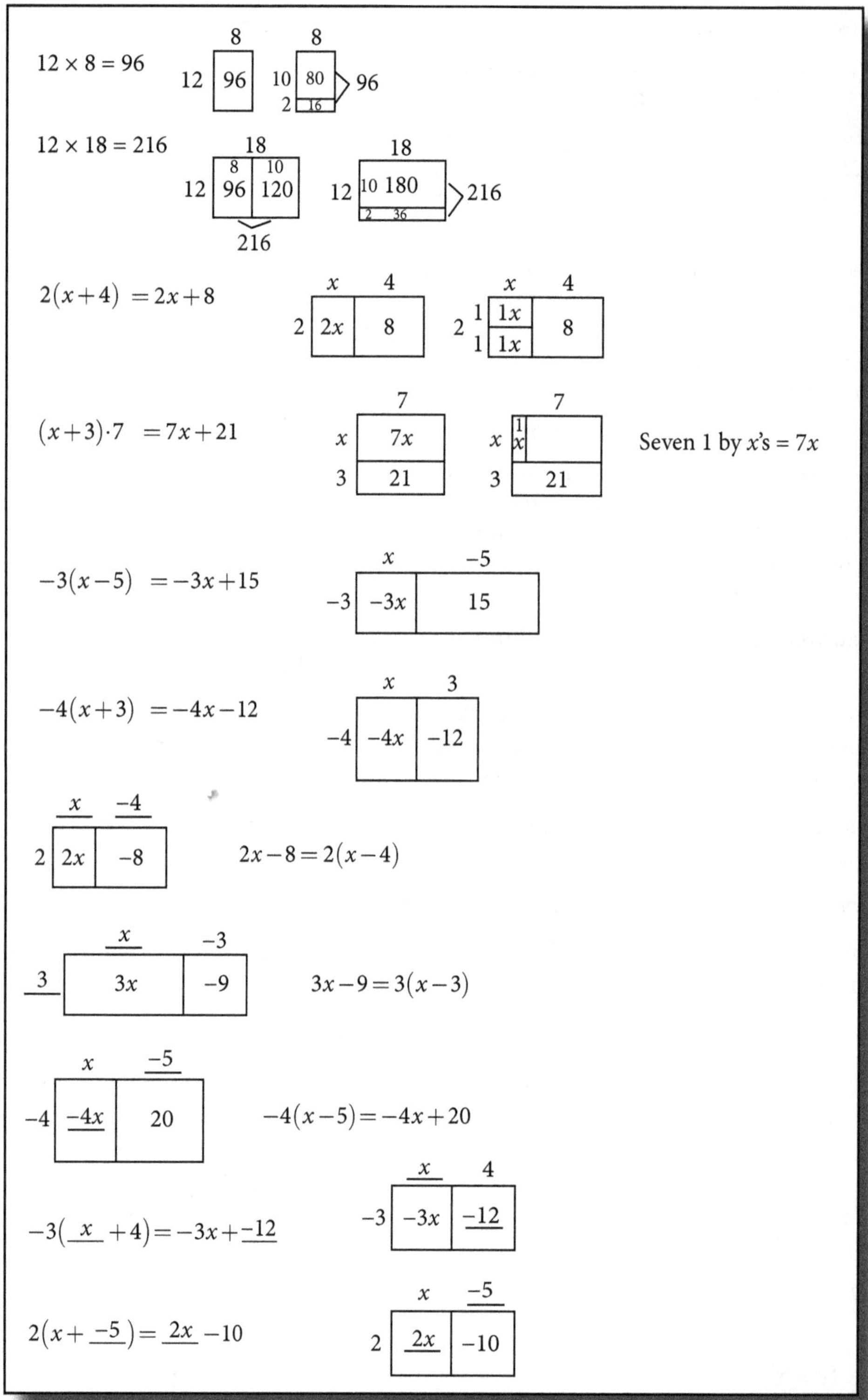

$12 \times 8 = 96$

$12 \times 18 = 216$

$2(x+4) = 2x+8$

$(x+3) \cdot 7 = 7x+21$ Seven 1 by x's $= 7x$

$-3(x-5) = -3x+15$

$-4(x+3) = -4x-12$

$2x-8 = 2(x-4)$

$3x-9 = 3(x-3)$

$-4(x-5) = -4x+20$

$-3(\underline{x}+4) = -3x+\underline{-12}$

$2(x+\underline{-5}) = \underline{2x}-10$

Facilitation Notes

This version of the problem string lists short notes for important teacher moves during the string. After you've done the string yourself and studied the relationships involved, you might make similar notes for the things you want a reminder of or deem important.

12×8	Do you know a nice chunk that could help you figure 12 8s? Draw and label an open array while saying dimensions and area. Keep it proportional. Did anyone think about 10 8s? Elicit (10+2)×8.
12×18	I wonder if you know anything that might be helpful? Draw, label, verbalize an open array. Keep it proportional. Did anyone use the first problem? Elicit 12×(8+10) and (10+2)×18.
$2(x+4)$	Still finding a product, but what if all we know is 2 by some length plus 4? Draw, label a rectangular diagram, keep the 2 and 4 proportional. Record equation and fill in rectangular diagram.
$(x+3)\cdot 7$	Repeat.
$-3(x-5)$	No longer area, but rectangular diagram can still help find an equivalent expression without parentheses. What is an equivalent expression? Use "opposite" to help reason about multiplying with integers.
$-4(x+3)$	Repeat.
	New format. What is the missing information? Fill in rectangular diagram and write equation.
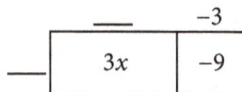	Repeat.
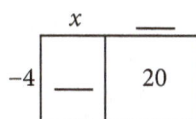	Repeat.
$-3(\underline{}+4)=-3x+\underline{}$	New format. What is the missing information? Draw and fill in rectangular diagram and write equation. Linger if needed.
$2(x+\underline{})=\underline{}-10$	Repeat. Name the distributive property if you haven't already.

4.7 Writing Linear Equations 3

At a Glance

Write the equation of the line that contains:

(5, 10) (3, 0)

(−3, −5) (0, 4)

(−4, −1) (−6, −½)

(−2, 10) (10, −2)

Objectives

The goal of this problem string is for students to gain facility with the point-slope form of the equation of a line by deciding if there are advantages to considering which point to use. The string also aims to help students parse out when using the slope y-intercept form or point-slope form is most efficient.

Placement

This is the third in a series of six strings on writing linear equations. Use this problem string after students have been introduced to the point-slope form of the equation of a line to strengthen student facility with finding the equation of a line when given two points.

You can use this problem string to support the work in textbook Lesson 4.7 Using Linear Equations after students have been introduced to the point-slope form of the equation of a line.

Guiding the Problem String

As you present each problem, asking students to find the equation of the line that contains the points, listen for students who are looking to the numbers before choosing a strategy.

The textbook develops the intercept form in lesson 3.5 as $y = a + bx$. The unit rate and non-unit rate strategies yield equations in this form. Textbook lesson 4.4 introduces the related form of slope-intercept which fits with the transformation strategy. Some students may still be using the rate strategies, regardless of the numbers involved. Model the non-unit rate strategy with a table for the first problem since a given point is close to the y-axis. Encourage students to choose more efficient and sophisticated strategies in the rest of the problems by modeling only those strategies.

The first problem should go quicker than the rest. The next three problems should each take about the same amount of time. As always, let students solve each problem any way they choose, but promote efficiency and sophistication by celebrating it. Let students try everything and then compare their strategies as a class. You might encourage students who finish early to look back at their work and consider other strategies to see if they can be more efficient.

About the Mathematics

The two main forms of the equation of a line that we use are:

Slope-intercept form $y = mx + b$

Point-slope form: $y = y_1 + b(x - x_1)$

The strategies to bring out in this problem string are about finding the slope between the points and after finding the slope, writing the equation using either form. When finding the slope, highlight both the slope formula and using distances between the coordinates with the direction of the line. When writing the equation of the line, highlight using the point-slope form with each of the points and help students develop a sense of using the easier, less unwieldy point if there is one. Also highlight using the slope-intercept form. Then ask students to compare, helping students develop a sense of when to use which form.

Important Questions

Use the following as you plan how to elicit and model student strategies.

- *What choices do you have when given two points to write the equation of a line containing the points?*

- *When might you use the slope formula? When might you use a distance and direction strategy to find the slope between two points? What about the points might nudge you toward either strategy?*

- *When might it be efficient to use the intercept form $y = a + bx$ and find a. What is true about the particular points that makes that form easy?*

- *When might it make the most sense to use the slope y-intercept form and solve for b? What is true about the particular points that makes that form easy?*

- *When might it be most efficient to use the point-slope form of the equation of a line? What is true about the particular points that makes that form easy?*

- *When you use the point-slope form, how are you choosing which point to use? The first point listed? What are some things to look for when choosing which point to use?*

How would you summarize some of the things that came up in this string today?

- *If one of the points is the y-intercept then it makes sense to use the slope y-intercept form.*

- *If one of the points is really close to the y-axis, you might use one of the rate strategies and the intercept form.*

- *If there are many negatives in the coordinates of the two points, consider finding the slope by finding distances and direction.*

- *If one of the points is the x-intercept, use that point in the point-slope form.*

- *If the slope is a fraction, look to use a point for which it is easier to find the fraction of its x-value.*

(continued)

Lesson 4.7 • Writing Linear Equations 3 (continued)

Sample Final Display

Your display could look like this at the end of the problem string:

$(5, 10)\ (3, 0)$ $m = \dfrac{10-0}{5-3} = \dfrac{10}{2} = 5$

$y = 5x - 15$

$y = 10 + 5(x - 5)$ $y = 0 + 5(x - 3)$ $y = 5x + b$
$y = 10 + 5x - 25$ $y = 5x - 15$ $0 = 5(3) + b$
$y = -15 + 5x$ $b = -15$
$y = 5x - 15$

x	y
0	−15
3	0

$3(\xrightarrow{\hspace{1cm}})15$ $y = -15 + 5x$

$(-3, -5)\ (0, 4)$ $m = \dfrac{-5-4}{-3-0} = \dfrac{-9}{-3} = 3$

$y = 3x + 4$

$\dfrac{\text{distance from } -5 \text{ and } 4 \text{ is } 9}{\text{distance from } -3 \text{ and } 0 \text{ is } 3} = 3$

line is increasing so positive 3 slope

$y = 4 + 3(x - 0)$ $y = 3x + 4$
$y = 4 + 3x$

$(-4, -1)\ (-6, -\frac12)$ $m = \dfrac{-1-(-\frac12)}{-4-(-6)} = \dfrac{-0.5}{2} = -0.25$

$y = -2 - \frac14 x$

$\dfrac{\text{distance from } -1 \text{ and } -\frac12 \text{ is } \frac12}{\text{distance from } -4 \text{ and } -6 \text{ is } 2}, \dfrac{\frac12}{2} = \dfrac{1}{4}$ $\times 2$

line is decreasing so $-\frac14$ slope

$y = -1 + -\frac14(x - (-4))$ $y = -\frac12 + -\frac14(x - (-6))$ $y = -\frac14 x + b$
$y = -1 - \frac14(x + 4)$ $y = -\frac12 - \frac14(x + 6)$ $-1 = -\frac14(-4) + b$
$y = -1 - \frac14 x - 1$ $y = -\frac12 - \frac14 x - 1\frac12$ $-1 = 1 + b$
$y = -2 - \frac14 x$ $y = -2 - \frac14 x$ $b = -2$
$y = -\frac14 x - 2$

$(-2, 10)\ (10, -2)$

$y = -x + 8$

$\dfrac{\text{distance from } -2 \text{ and } 10 \text{ is } 8}{\text{distance from } 10 \text{ and } -2 \text{ is } 8} = 1$

line is decreasing so -1 slope

$y = 10 + -1(x - (-2))$ $y = -2 + -1(x - 10)$ $y = -x + b$
$y = 10 - 1(x + 2)$ $y = -2 - 1x + 10$ $10 = -(-2) + b$
$y = 10 - x - 2$ $y = -x + 8$ $10 = 2 + b$
$y = -x + 8$ $b = 8$
$y = -x + 8$

Algebra Problem Strings
©2017 Kendall Hunt Publishing

Facilitation Notes

This version of the problem string lists short notes for important teacher moves during the string. After you've done the string yourself and studied the relationships involved, you might make similar notes for the things you want a reminder of or deem important.

(5, 10) (3, 0)	Find the equation of the line. What is the slope? Share point-slope with (5, 10), then (3, 0), then non-unit and transformation. Which do you want your brain to be inclined to next time? Why? Maybe the strategies are all about as efficient no matter which you use?
(−3, −5) (0, 4)	Repeat. How did you find this slope? Share formula and using distance with direction. Share using point-slope with (0, 4) then using slope y-intercept. Which is more efficient? Why?
(−4, −1) (−6, −½)	Repeat. How did you find this slope? Share formula and using distance with direction. Which strategy for finding slope works better for you with these numbers? Why? Share using point-slope with both points. Why might (−4, −1) be easier? 1/4 of 1. Share using slope y-intercept. Which do you want your brain to be inclined to next time? Why?
(−2, 10) (10, −2)	Repeat. How did you find this slope? Share using distance with direction. Share using point-slope with both points. Looking back, which do you like? Share using slope y-intercept. Which do you want your brain to be inclined to next time? Why?

Writing Linear Equations 4

At a Glance

x	y
−5	10
3	2
6	−1

x	y
4	2
10	8
15	13

x	y
−15	−15
−8	−8
20	20

x	y
−15	12
−7	10
5	−2

Objectives

The goal of this problem string is for students to practice choosing which form of the equation of a line to use when finding an equation to fit exact data. An accompanying goal is for students to begin to consider the standard form of the equation of a line.

Placement

This is the fourth in a series of six strings on writing linear equations. Use this problem string as students are considering the various forms of the equation of a line to use to fit a linear model to data. The data in this string is exactly linear, so the lines will be best fit lines. As students choose which points and which form to use, this can help students solidify the reasoning they use about data that is not exactly linear.

You can use this problem string to support the work in textbook Lesson 4.8 A Standard Linear Model.

Guiding the Problem String

This problem string is a series of four data sets for which students are to find the equation of the line. For each problem, elicit different strategies and then discuss which one(s) feel efficient or well suited for the given points. Focus your conversations not on what they did but what they want to be inclined to do the next time they work with similar data.

After the last problem, if it has not already come up, wonder aloud about the relationships between the x's and y's for each data set. Nudge students to notice the standard form of the equation of the line that can be found be either adding or subtracting the x's and y's because in these specific cases each data set has a particular sum or difference. This has the potential to help students realize that considering the data first can help them make a smart choice for finding the equation. The third problem is of particular note because it is the line $y = x$ and ideally students will now begin to recognize it without needing to do any other work.

About the Mathematics

The standard form of the equation of a line is $ax + by = c$. In this form, it can be easy to solve for either the x- or y-intercept by substituting zero for either x or y. When the line that contains the data points has a slope of 1 or −1, the standard form of the equation of the line can be apparent by looking at the sum or differences of the x's and y's. When the slope is not 1 or −1, the patterns are much harder to notice. While this is a relatively rare occurrence, slopes of exactly 1 or −1, using such data to help students realize the standard form can be a helpful first step to owning that form of the equation of a line.

Sample Interactions

Use the following as you plan how to elicit and model student strategies. This is not meant as a script, but as a view into the relationships involved and the intent of the problem string.

Teacher: *Let's warm up today with a problem string. The first problem is to find a line that fits this data. All of the data sets today are linear. Take some time to look at the data and make a plan. Use whatever strategy you think is best. If you get done and are satisfied, I want to encourage you to look back at the data and your line and wonder if you can now see anything even more efficient.* Students work while the teacher circulates, helping students get started by asking nudging questions, and looking for students who are solving for *b* using the slope-intercept form, using the point-slope form, and taking note of other strategies, especially anyone noticing that the sum of the *x*'s and *y*'s is 5.	$\begin{array}{c\|c} x & y \\ \hline -5 & 10 \\ \hline 3 & 2 \\ \hline 6 & -1 \end{array}$
Teacher: *First, what equation did you find?* **Student:** *y equals −x plus five.* **Student:** *I got basically the same thing, just the 5 first.* **Teacher:** *And we just did work yesterday about equivalent expressions. Do we agree those are equivalent?*	$y = -x + 5$
Teacher: *No matter what you did to find the line, I think I saw everyone finding the rate of change, or the slope, at some point. Who used the first two points? And who used the last two points? And did it matter?* **Student:** *You had already told us that the data was linear, so we could choose either. And it didn't matter because either set gives you the rate of −1 to 1. It might have been easier to find the short difference of 3 rather than the slightly bigger differences of 8.* **Student:** *Both times you had one negative number.* **Student:** *It might depend on whether you prefer to find the difference between −5 and 3 or between 2 and −1.*	$\dfrac{-8}{8} = \dfrac{-3}{3} = -1$
Teacher: *Nice analysis. Let's keep going. Someone who used the slope-intercept form, please share that with us.* As the student describes, the teacher records. **Teacher:** *Any questions about this? No? Did anyone choose a different point?* **Student:** *I did, but I think the work is about the same.*	$y = mx + b$ $10 = -1(-5) + b$ $10 = 5 + b$ $b = 5$ $y = -x + 5$
Teacher: *And someone who used the point-slope form? Tell us about that please.* As the student describes, the teacher records. **Student:** *Why did you choose that point and not just the first one?* **Student:** *It didn't have any negatives.*	$y = 2 - 1(x - 3)$ $y = 2 - x + 3$ $y = 5 - x$

(continued)

Teacher: *Looking back at these two strategies, do either of them strike you now as more efficient, like you want your brain to be inclined to choose that one the next time you run into data like this?*

Student: *I know I didn't really look at which point to use when. That's something I'd like to think about next time.*

Student: *Yeah, if there's a point that has no negatives, that seems smart.*

Student: *When you use a point that has a positive x-value in the point-slope form, then you don't have to mess with subtracting a negative. Like up there, the x minus 3 is easy to deal with. If you had chosen the point (–5, 10), then you'd have to deal with x minus –5. That's not hard, it's just one more thing to deal with.*

Teacher: *Those sound like helpful things to keep in mind.*

x	y
–5	10
3	2
6	–1

8, 3 (left brackets); –8, –3 (right brackets)

$y = -x + 5$

$$\frac{-8}{8} = \frac{-3}{3} = -1$$

Didn't matter which points, but might want to use smaller differences.

$y = mx + b$
$10 = -1(-5) + b$
$10 = 5 + b$
$b = 5$
$y = -x + 5$

$y = 2 - 1(x - 3)$
$y = 2 - x + 3$
$y = 5 - x$

Might want to use point with positive x-value.

Teacher: *Here's the next set of data. Find the line!*

Students work while the teacher circulates, looking for students who are solving for b using the slope-intercept form, using the point-slope form, and taking note of other strategies, especially anyone noticing that the difference of the x's and y's is 2.

This problem should go quicker than the first.

x	y
4	2
10	8
15	13

The teacher asks students to share and crafts a similar brief conversation about efficient or clever choices. Since all of the coordinates are positives, students mention that it might be smart to use the small numbers.

x	y
4	2
10	8
15	13

$y = x - 2$

$$\frac{6}{6} = \frac{5}{5} = 1$$

All of the coordinates are positive, use the small numbers.

$y = mx + b$
$2 = 1(4) + b$
$b = -2$
$y = x - 2$

$y = 2 + 1(x - 4)$
$y = 2 + x - 4$
$y = -2 + x$

The teacher continues in a similar manner with the third problem. In the discussion, the teacher elicits the relationship that each of the x-values are equal to the corresponding y-values and records that relationship as an equation and sentence.

x	y
–15	–15
–8	–8
20	20

$y = x$

$$\frac{7}{7} = \frac{28}{28} = 1$$

$y = mx + b$
$20 = 1(20) + b$
$b = 0$
$y = x + 0$

$y = 20 + 1(x - 20)$
$y = 20 + x - 20$
$y = x$

$y = x$ All of the y's are the same as the x's.

Algebra Problem Strings
©2017 Kendall Hunt Publishing

The teacher continues in a similar manner with the last problem. In the discussion, the teacher asks students to consider why it might be more efficient to use either the slope-intercept or the point-slope form.

Student: *Either way, you will have to find the slope.*

Student: *If you use slope-intercept, once you find b, you are practically done, just stick it and the slope in.*

Student: *Yes, but if you use point-slope, then you could be done at the first step. It's not simplified, but you could type it into a grapher in that form.*

Teacher: *So there might be some pros and cons to each. And maybe what you are about to do with the line might influence which strategy you choose. It sounds important to own both of these strategies so that you can choose.*

x	y
−15	18
−7	10
5	−2

$y = -x + 3$

$\dfrac{-8}{8} = \dfrac{-12}{12} = -1$

You have to find b and then write an equation you can type.

$y = mx + b$
$18 = -1(-15) + b$
$18 = 15 + b$
$b = 3$
$y = -x + 3$

You can type in and graph right now, without doing any simplifying.

$y = 18 - 1(x - (-15))$
$y = 18 - x - 15$
$y = 3 - x$

If you want to use the positive y-value of 18.

Teacher: *Okay, great work. Um, I'm looking at the board and I'm noticing a pattern with some of these x's and y's. Did anyone else notice any patterns happening when you look from an x to a y in a table? Something about adding or subtracting the x's and y's?*

If no students respond after think time, the teacher suggests adding the x-values and the y-values of the first table.

Teacher: *For example, I noticed that if I add the −5 and 10, that's 5.*

Student: *Hmmm. That's weird. You can add the 3 and 2 and get 5.*

Student: *And the 6 and −1 and get 5. What does that mean?*

Student: *They all add to be 5.*

x	y
−5 +	10 → = 5
3	2
6	−1

8 ... −8
3 ... −3

Teacher: *You mean that every x plus every y is 5? I think I could record that as x + y = 5. What do you think?*

Student: *Yes, that works.*

Student: *And it's the same as the other equation that we wrote, y = −x + 5.*

Student: *It is!*

Teacher: *This form of the equation of a line, where x and y are on the same side of the equation and the constant is on the other side is called the standard form.*

x	y
−5 +	10 → = 5
3	2
6	−1

8 ... −8
3 ... −3

$x + y = 5$

Every x + every y is 5.

$y = -x + 5$

$\dfrac{-8}{8} = \dfrac{-3}{3} = -1$

Didn't matter which points, but might want to use smaller differences.

$y = mx + b$
$10 = -1(-5) + b$
$10 = 5 + b$
$b = 5$
$y = -x + 5$

$y = 2 - 1(x - 3)$
$y = 2 - x + 3$
$y = 5 - x$

Might want to use point with positive x-value.

(continued)

Teacher: *Take some time and look over all of our problems today. I wonder if there are any other relationships like this? After you've looked and thought, turn and talk to your partner about what you are thinking.*

After students have thought and talked with a partner, the teacher pulls students together and records the standard form of the equation of the lines for each problem. See the Sample Final Display for how that might look.

Teacher: *Interesting stuff. What are you wondering about?*

Student: *It has me thinking that we could save a bunch of work if we recognize those patterns.*

Teacher: *Do you think all sets of colinear points have patterns that are so easy to see? Is there something special about these lines that might make it so that the pattern morphs right into the equation of the line? What do all of these lines have in common? We'll do some more work with finding equations of lines and you can look for these patterns.*

Teacher: *How would you summarize some of the things that came up in this string today?*

Elicit the following:

- *Look at the points first and make a smart choice about which points to use when you find the slope and when you find the equation of the line.*

- *No matter what strategy you use, you'll have to find the slope.*

- *Look for patterns. Let the data help you decide which strategy to choose. When you're done, now that you own the relationships, look back and decide if there was a better choice.*

Algebra Problem Strings
©2017 Kendall Hunt Publishing

Sample Final Display

Your display could look like this at the end of the problem string:

x	y
$-5 +$	10
3	2
6	-1

$= 5$
-8
-3

8
3

$x + y = 5$

Every x + every y is 5.

$y = -x + 5$

$\dfrac{-8}{8} = \dfrac{-3}{3} = -1$

Didn't matter which points, but might want to use smaller differences.

$y = mx + b$
$10 = -1(-5) + b$
$10 = 5 + b$
$b = 5$
$y = -x + 5$

$y = 2 - 1(x - 3)$
$y = 2 - x + 3$
$y = 5 - x$

Might want to use point with positive x-value.

x	y
4	2
$10 -$	8
15	13

$= 2$

$x - y = 2$ Every x – every y is 2.

$y = x - 2$

$\dfrac{6}{6} = \dfrac{5}{5} = 1$

All of the coordinates are positive, use the small numbers.

$y = mx + b$
$2 = 1(4) + b$
$b = -2$
$y = x - 2$

$y = 2 + 1(x - 4)$
$y = 2 + x - 4$
$y = -2 + x$

x	y
-15	-15
$-8 =$	-8
20	20

$x - y = 0$ Every x – every y is 0. $y = x$ All of the y's are the same as the x's.

$y = x$

$\dfrac{7}{7} = \dfrac{28}{28} = 1$

$y = mx + b$
$20 = 1(20) + b$
$b = 0$
$y = x + 0$

$y = 20 + 1(x - 20)$
$y = 20 + x - 20$
$y = x$

You have to find b and then write an equation you can type.

You can type in and graph right now, without doing any simplifying.

x	y
-15	18
-7	10
$5 +$	-2

$= 3$

$x + y = 3$ Every x + every y is 3.

$y = -x + 3$

$\dfrac{-8}{8} = \dfrac{-12}{12} = -1$

$y = mx + b$
$18 = -1(-15) + b$
$18 = 15 + b$
$b = 3$
$y = -x + 3$

$y = 18 - 1(x - (-15))$
$y = 18 - x - 15$
$y = 3 - x$

If you want to use the positive y-value of 18.

(continued)

Facilitation Notes

This version of the problem string lists short notes for important teacher moves during the string. After you've done the string yourself and studied the relationships involved, you might make similar notes for the things you want a reminder of or deem important.

x	y
−5	10
3	2
6	−1

Find the line that fits the data. All of the data sets today are colinear.
Seek for efficiency.
Elicit slope with 2 sets of points.
Elicit finding b with slope-intercept and using point-slope.
Looking back, does anything strike you as efficient, like you want your brain to be inclinded to choose the next time?

x	y
4	2
10	8
15	13

Repeat.
Quicker.
Small numbers might be nice.

x	y
−15	−15
−8	−8
20	20

Repeat.
Quicker.
Each x is the same as its corresponding y. That's y = x!

x	y
−15	12
−7	10
5	−2

Repeat.
What are some reasons to find b using the slope-intercept form?
Why use the point-slope form?
Look back. See any patterns?
I am noticing that the x plus the y in problem #1 all add to 5.
Think then partner share.
Record equations and sentences to describe the standard forms.
Why does the standard form fall out? Anything these lines have in common?

At a Glance

x	y
48	52
52	48
99	1

x	y
−5	−8
5	4
11	11.2

x	y
3	−97
99	−1
105	5

x	y
−13	13
−7	7
21	−21

Objectives

The goal of this problem string is to build students' ability to let the points influence their choice of strategy when finding the equation of the line between points.

Placement

This is the fifth in a series of six strings on writing linear equations. Use this problem string to continue to help students wisely choose a strategy to find the equation of the line between points. Students should have been working with the slope-intercept and point-slope forms and at least introduced to the standard form of the equation of a line.

You can use this problem string before textbook Lesson 4.9 Correlation and Causation.

Guiding the Problem String

Facilitate this string as detailed in problem string 4.8, by having students find the equation of the line that fits each data set using any strategy they choose. The first problem can readily be modeled with the standard form. The second problem has a slope of 1.2. Compare strategies and ask students to consider which point to choose to work well with the slope. The third problem can be modeled with the standard form but it might be hard for students to find. If so, model more strategies and then compare with the standard form. The last problem is the line $y = -x$. If students do not recognize it, finding the equation using other strategies and then comparing those equations to realize that each x is equal to the opposite of its corresponding y-value; this should help that become more automatic.

When you ask students to share for problems 1, 3, and 4, decide if students could use the reinforcement of seeing the different strategies (find their own errors, compare for efficiency) or if you can skip to having a student share the standard form strategy. This is not about discounting student work or shaming students into using the "correct" strategy, it is about celebrating efficiency and looking to the problem before deciding on a strategy.

About the Mathematics

When looking for data that allows the standard form of a line to obviously stand out, it can be helpful to realize that it seems to work best when a and b are 1 or −1. Other coefficients complicate the relationship and it is harder to see the connection just from the data. So, we have a limited number of a and b combinations, but we are unlimited on the c-values. In this problem string, we use the patterns: $x + y = c$ and $x - y = c$. Sometimes students can also pick up on $-x + y = c$, so you may want to try those in future work with students.

(continued)

Sample Interactions

Use the following as you plan how to elicit and model student strategies. This is not meant as a script, but as a view into the relationships involved and the intent of the problem string.

Teacher:	*Today's problem string involves data that is linear. For each of these problems, use what you've been learning. Take some time to look at the data and make a plan. Use whatever strategy you think is best. If you get done and are satisfied, I want you to look back at the data and your equation and wonder if you can now see anything even more efficient. Look for patterns.*			

x	y
48	52
52	48
99	1

Teacher:	*What equation did you find that includes these points? Everyone agree?*
	Did anyone find any patterns that were helpful? How? What did you notice? How did that help?
	Each x plus each y is 100? What equation is that?
	If you used a different strategy, is your equation equivalent?
	If you used a different strategy, can you see the pattern now? What might you look for in the future?

x	y	
48 +	52	= 100
52	48	
99	1	

$x + y = 100$ Each x + each y is 100.

Teacher:	*Find an equation for these points. Then look back at your work and the points and reconsider your choices. Could you do anything more efficient?*

x	y
−5	−8
5	4
11	11.2

Teacher:	*What equation did you find that includes these points? Anyone disagree?*
	No one found any patterns this time? We'll look at that in a bit.
	What did you find for the rate of change, the slope? Did you then use the fraction form or the decimal form? Why?
	Who used the slope-intercept form to find b? Tell us about that. What point did you choose? Why? How did the slope of 1.2 influence your choice? How could it have?
	Who used the point-slope form? Tell us about that. What point did you choose? Why? How did the slope of 1.2 influence your choice? How could it have?

x	y
−5	−8
5	4
11	11.2

$m = \dfrac{12}{10} = \dfrac{6}{5} = 1\frac{1}{5} = 1.2$

Choose the fraction or decimal, whichever works better with the numbers!

$y = mx + b$
$4 = \frac{6}{5}(5) + b$
$4 = 6 + b$
$b = -2$
$y = \frac{6}{5}x - 2$

$y = 4 + \frac{6}{5}(x - 5)$
$y = 4 + \frac{6}{5}x - 6$
$y = -2 + \frac{6}{5}x$

Choose a point that works nicely with the fraction!

No way we could've seen this pattern!

$y = \frac{6}{5}x - 2$
$5y = 6x - 10$
$6x - 5y = 10$

Algebra Problem Strings
©2017 Kendall Hunt Publishing

		x	y
Teacher: *Here is the next problem. Find the equation for these points? What are the points suggesting you do? Again, when you've got something, look back and seek efficiency.*		3	−97
		99	−1
		105	5

Teacher: *What equation did you find that includes these points? Anyone disagree?*

What did you find for the rate of change, the slope?

Who used the slope-intercept to find b? Tell us about that. What points did you choose? Why?

Who used the point-slope form? Tell us about that. What point did you choose? Why?

Did the slope of 1 make anyone think of a different strategy?

Did anyone see any patterns in the x and y columns? Each x subtract each y is 100? Is that equivalent to the other forms of the equations?

Could the slope of 1 have helped remind you to look to see if you could see a pattern? Might it in the future?

x	y
3	−97
99	−1
105	5 = 100

$$m = \frac{6}{6} = \frac{96}{96} = 1$$

$x - y = 100$ Each x − each y is 100.

$y = mx + b$
$5 = 1(105) + b$
$5 = 105 + b$
$b = -100$
$y = x - 100$

$y = 5 + 1(x - 105)$
$y = 5 + x - 105$
$y = -100 + x$

$y = x - 100$
$x - y = 100$

		x	y
Teacher: *Okay, last problem. Find the equation of the line. What might you be looking for to help you decide on a strategy?*		−13	13
		−7	7
		21	−21

Teacher: *What equation did you find that includes these points? Everyone agree?*

What did you find for the rate of change, the slope?

Did anyone find a pattern? How did it help you?

x	y
−13	13
−7 +	7 = 0
21	−21

Look for and use patterns when you can!

$x + y = 0$ Each x + each y is 0.

$x = -y$ Each x = opposite of each y.

Teacher: *How would you summarize some of the things that came up in this string today?*

Elicit the following:

- *Look at the points first and make a smart choice about which points to use when you find the slope and when you find the equation of the line.*

- *If the slope is a fraction, look to use points that are easier to use with that fraction.*

- *Look for patterns. Let the data help you decide which strategy to choose.*

(continued)

Sample Final Display

Your display could look like this at the end of the problem string:

x	y	
48 +	52	= 100
52	48	
99	1	

$x + y = 100$ Each x + each y is 100.

x	y
−5	−8
5	4
11	11.2

$m = \dfrac{12}{10} = \dfrac{6}{5} = 1\frac{1}{5} = 1.2$

Choose the fraction or decimal, whichever works better with the numbers!

$y = mx + b$
$4 = \frac{6}{5}(5) + b$
$4 = 6 + b$
$b = -2$
$y = \frac{6}{5}x - 2$

$y = 4 + \frac{6}{5}(x - 5)$
$y = 4 + \frac{6}{5}x - 6$
$y = -2 + \frac{6}{5}x$

Choose a point that works nicely with the fraction!

No way we could've seen this pattern!

$y = \frac{6}{5}x - 2$
$5y = 6x - 10$
$6x - 5y = 10$

x	y	
3	−97	
99	−1	
105 −	5	= 100

$m = \dfrac{6}{6} = \dfrac{96}{96} = 1$

$y = mx + b$
$5 = 1(105) + b$
$5 = 105 + b$
$b = -100$
$y = x - 100$

$y = 5 + 1(x - 105)$
$y = 5 + x - 105$
$y = -100 + x$

$y = x - 100$
$x - y = 100$

$x - y = 100$ Each x − each y is 100.

x	y	
−13	13	
−7 +	7	= 0
21	−21	

Look for and use patterns when you can!

$x + y = 0$ Each x + each y is 0.
$x = -y$ Each x = opposite of each y.

Algebra Problem Strings
©2017 Kendall Hunt Publishing

Facilitation Notes

This version of the problem string lists short notes for important teacher moves during the string. After you've done the string yourself and studied the relationships involved, you might make similar notes for the things you want a reminder of or deem important.

x	y
48	52
52	48
99	1

Find the line that fits the data. All of the data sets today are colinear.
Seek for efficiency.
Elicit the pattern of each x plus each y is 100, x + y = 100.
If you didn't see the pattern, is your equation equivalent?
Could the slope of 1 help remind you to look for patterns in the future?

x	y
−5	−8
5	4
11	11.2

What equation fits these points?
How did you use the slope, as fraction or decimal? Why?
Elicit using slope-intercept and finding b. Which point did you use? Why?
Elicit point-slope. Which point did you use? Why?
Compare.
Find the equivalent standard form. That would have been hard to just "see"!

x	y
3	−97
99	−1
105	5

What do these points suggest?
Seek for efficiency.
Elicit using slope-intercept and finding b.
Elicit point-slope.
Elicit the pattern of each x subtract each y is 100, x − y = 100.
If you didn't see the pattern, is your equation equivalent?
Could the slope of −1 help remind you to look for patterns in the future?

x	y
−13	13
−7	7
21	−21

Elicit the pattern that each x is the opposite of each y, y = −x
Elicit the pattern that each x plus each corresponding y is 0, x + y = 0.
If you didn't see the pattern, is your equation equivalent?
Looking for patterns can really save a lot of work!

4.10 Writing Linear Equations 6

At a Glance	Objectives

Objectives
The goal of this problem string is for students to determine an efficient strategy for finding the equation of a line between two points.

At a Glance

x	y
−3	10
3	4
7	0

x	y
−15	27
0	22
8	−18

x	y
−8	28
8	−12
12	−22

Placement
This is the sixth in a series of six strings on writing linear equations. Use this problem string to continue to help students analyze the data to judiciously choose a strategy to find the equation of the line. As students are strategizing about how to most efficiently find the equation, they are also getting a lot of practice with feedback.

You can use this problem string to reinforce the learning in textbook Chapter 4 Functions and Linear Modeling.

Guiding the Problem String
As always, let students find the equations as they will, but encourage them to look to the data to find patterns and choose wisely by highlighting efficient strategies and helping students generalize their ideas.

The first problem has the potential for students to notice that the sum of each x and its corresponding y is 7. Highlight this pattern to standard form strategy. The second problem provides students with the y-intercept. Highlight the strategy of using the y-intercept in the slope-intercept form and have students generalize the idea of using the y-intercept when it's given. The third problem is messier and will probably require more work. Highlight the idea that you can look for patterns and relationships to help, but when you don't see any, you can fall back to other tried and true strategies that work for any linear data.

About the Mathematics
The problem strings in this series of writing linear equations all deal with perfectly linear data. This is purposeful as problem strings are mini-lessons. You will want to make sure that your students also have experience finding lines that fit data that is not perfectly linear.

Important Questions

Use the following as you plan how to elicit and model student strategies.

- *What did you notice when you first looked at the points that influenced your choice of strategy?*
- *Which points did you use to find the slope? Why?*
- *What strategy did you use to find the equation of the line? Why?*
- *What might you look for when you are finding the equation of a line? If you find what you are looking for, how will you use it?*

How would you summarize some of the things that came up in this string today?

- *When you see a pattern, use it.*
- *When you are given the y-intercept, use it.*
- *Look to the points to help you determine what strategy to use and which points to use.*
- *Try to use points that make the computation easier if possible. Think about the direction of the line (increasing, decreasing) to help make sense of the slope.*

Sample Final Display

Your display could look like this at the end of the problem string:

(continued)

Facilitation Notes

This version of the problem string lists short notes for important teacher moves during the string. After you've done the string yourself and studied the relationships involved, you might make similar notes for the things you want a reminder of or deem important.

x	y
−3	10
3	4
7	0

Find the line that fits the data. All of the data sets today are colinear.
Seek for efficiency.
Elicit the pattern of each x plus each y is 7, x + y = 7.
If you didn't see the pattern, is your equation equivalent?
Generalize: if you can see a pattern, use it!

x	y
−15	27
0	22
8	−18

What about this data?
What is the rate of change? How did you find it?
Did anyone notice and use the y-intercept?
Generalize: if you are given the y-intercept, use it!

x	y
−8	28
8	−12
12	−22

Find the line that fits the data.
What is the rate of change? How did you find it?
Elicit slope-intercept to find b.
Elicit point-slope.
Compare.
Generalize: if you can't see a pattern, use what you know and choose the points carefully!

Algebra Problem Strings
©2017 Kendall Hunt Publishing

5.0 | Division of a Sum

At a Glance	Objectives
	The goal of this problem string is for students to simplify expressions involving division that they may encounter when solving by substitution.

At a Glance

$$816 \div 8 = \frac{816}{8}$$

$$792 \div 8 = \frac{792}{8}$$

$$\frac{4x}{4}$$

$$\frac{4x+8}{4}$$

$$\frac{3x}{3}$$

$$\frac{3x-9}{3}$$

$$\frac{5x}{2}$$

$$\frac{5x+3}{2}$$

$$\frac{4x-3}{8}$$

$$\frac{10-2x}{-5}$$

Objectives
The goal of this problem string is for students to simplify expressions involving division that they may encounter when solving by substitution.

Placement
You can use this problem string before students solve systems of linear equations by substitution where they are asked to solve a linear equation in standard form for one of the variables.

Use this problem string as you begin textbook Chapter 5: Systems of Equations and Inequalities.

Guiding the Problem String
The first two problems are numerical examples where splitting up the numerator with addition or subtraction can help solve the division problem. These problems set the stage for the three partner-problem sets, where the first problem is part of solving the second problem. Spend time as needed parsing out the fraction result in the last set where students reason about $5 \div 2$ and $3 \div 2$. Begin the last two problems by asking students to create their own helper problems.

Throughout the string, use the words *divided by* not *over* as you describe the division problems. The word *over* is a positional word that does not describe a mathematical relationship. Talk about dividing out common factors and multiplying by the resulting one rather than using the word *cancel*. Cancel often becomes a catch-all phrase that students apply without reasoning and thereby misapply.

About the Mathematics
Solving for y in an equation like $2x - 5y = 10$, can look like:

$$2x - 5y = 10$$
$$-5y = -2x + 10$$
$$y = \frac{10 - 2x}{-5}$$

This problem string can set students up for success when they reach this point in such solutions.

(continued)

Sample Interactions

Use the following as you plan how to elicit and model student strategies. This is not meant as a script, but as a view into the relationships involved and the intent of the problem string.

Teacher: *Let's have a little fun with a short problem string today. The first question is to find 816 divided by 8. I can record that with this division symbol, ÷, and like this. What is 816 divided by 8?* If students immediately start writing, the teacher might encourage them to try to think about the relationships, but just like all problem strings, the teacher allows students to solve the problem in their own way.	$816 \div 8 = \dfrac{816}{8}$
Teacher: *Did anyone think about 800 divided by 8? Some of you? Could you? Could you split up 816 into 800 and 16 and think first about 800 divided by 8? What is 800 divided by 8?* **Student:** 100	$816 \div 8 = \dfrac{816}{8} = \dfrac{800+16}{8} = \dfrac{800}{8} + \dfrac{16}{8}$ $= 100 +$
Teacher: *And we have this 16 leftover? We already know 800 divided by 8 is 100, but we need 816 divided by 8. So 16 divided by 8 is?* **Student:** *Two.* **Teacher:** *So 816 divided by 8 is this 100 and this 2, so 102. Turn to your partner and talk about this strategy.* Students briefly turn and talk while the teacher listens in.	$816 \div 8 = \dfrac{816}{8} = \dfrac{800+16}{8} = \dfrac{800}{8} + \dfrac{16}{8}$ $= 100 + 2 = 102$
Teacher: *I wonder if that might influence how you find this next problem, 792 divided by 8.*	$792 \div 8 = \dfrac{792}{8}$
The teacher repeats as with the first problem, letting students solve the problem and then eliciting an *over* partial quotients strategy as modeled. **Teacher:** *So, when dividing, you can split up the numerator by addition or subtraction and solve each resulting division problem. It seems like that might come in handy.*	$792 \div 8 = \dfrac{792}{8} = \dfrac{800-8}{8} = \dfrac{800}{8} - \dfrac{8}{8}$ $= 100 - 1 = 99$
Teacher: *Next problem. What is 4x divided by 4 equivalent to? And how do you know?*	$\dfrac{4x}{4}$
Teacher: *What do you think?* **Student:** *It's x.* **Teacher:** *Does anyone agree? Other answers? No? How do you know?*	$\dfrac{4x}{4} = x$
Student: *You just cancel the 4s.* **Teacher:** *I've heard a few of you talking about canceling. What's really going on here? What does 4x divided by 4 mean?*	$\dfrac{\cancel{4}x}{\cancel{4}}$

Algebra Problem Strings
©2017 Kendall Hunt Publishing

Student: *I think about it like you're dividing the top and bottom by 4, so that's one times x or just x.* **Teacher:** *I can record your thinking this way.*	$$\dfrac{\overset{1}{\cancel{4}}x}{\underset{1}{\cancel{4}}} = \dfrac{1 \cdot x}{1} = x$$
Student: *I think about 4 divided by 4 all times x. So that's one times x.* **Teacher:** *I'll model your thinking this way.*	$$\dfrac{4x}{4} = \dfrac{4}{4} \cdot x = 1 \cdot x = x$$
Teacher: *Did anyone think about what 4x means? What does 4x mean?* **Student:** *It means 4 times x. So 4 times something divided by 4 is just that thing.* **Teacher:** *I'll model your thinking on an open number line. If we have 4 times something, x in this case, here and let's say 0 is here, where is 4x divided by 4?* **Student:** *It's a fourth of the way and it's x.* **Teacher:** *And what if 0 was over here? Does that change what 4x divided by 4 is?* **Student:** *No, now it's to the right, but it's still x at one-fourth of 4x.*	
Teacher: *Okay, the next problem is to find an equivalent expression to 4x plus 8 all divided by 4.* Students work. The teacher circulates, looking for students who are breaking up the numerator.	$$\dfrac{4x + 8}{4}$$
Teacher: *I saw some of you split up the numerator. Tell us about that please.* **Student:** *I thought about 4x divided by 4 and 8 divided by 4.* **Teacher:** *And we already know what 4x divided by 4 is. What is 8 divided by 4?* **Student:** *Two.* **Teacher:** *So again we can split up the numerator with addition and solve each resulting problem.*	$$= \dfrac{4x}{4} + \dfrac{8}{4}$$ $$= x + 2$$
Teacher: *The next problem is a quick one. What is 3x divided by 3?* **Student:** *It's x.* **Teacher:** *Why?* **Student:** *Three times something divided by three is that thing.* **Student:** *Three divided by three is 1, so it one x.*	$$\dfrac{3x}{3}$$ $$\dfrac{3x}{3} = x$$ $$\dfrac{3}{3} \cdot x = 1x = x$$

(continued)

Teacher: *The next problem, completely unrelated, is 3x minus 9 all divided by 3?* Brief think time.	$$\dfrac{3x-9}{3}$$
Teacher: *What are you thinking? What is 3x minus 9 all divided by 3?* **Student:** *It's just x minus 3.* **Teacher:** *Why?* **Student:** *It's like 3x divided by 3 and 9 divided by 3. And you keep the subtraction.* **Teacher:** *What do we think? Does it make sense to break up the numerator like this? Turn to your partner and talk about this strategy.*	$$\dfrac{3x-9}{3}=\dfrac{3x}{3}-\dfrac{9}{3}$$ $$=x-3$$
Teacher: *Okay, what's 5x divided by 2?* Brief think time.	$$\dfrac{5x}{2}$$
Teacher: *What is 5x divided by 2?* **Student:** *Well, five divided by 2 is 2.5, but I'm not sure what to do with that.* **Teacher:** *Who agrees that 5 divided by 2 is 2.5? Yes? What are we going to do with it?* **Student:** *It's just still times x. So it's ⁵⁄₂x.* **Student:** *Or 2.5x.*	$$\dfrac{5x}{2}=\tfrac{5}{2}x=2.5x$$
Teacher: *Does that help with the next problem, 5x plus 3 all divided by 2?*	$$\dfrac{5x+3}{2}$$
Student: *Sure! It's just 2.5x plus, um, 1.5?* **Student:** *Yeah, or ⁵⁄₂x plus ³⁄₂.*	$$\dfrac{5x+3}{2}=\dfrac{5x}{2}+\dfrac{3}{2}$$ $$=2.5x+1.5=\tfrac{5}{2}x+\tfrac{3}{2}$$
Teacher: *Let's look back at this string. Notice that for many of these problems, we had a helper problem that helped solve the next problem. That was handy. For this next problem, we don't have a helper. What helper might you write?* **Student:** *4x divided by 8.* Brief think time.	$$\dfrac{4x-3}{8}$$
Teacher: *What is 4x minus 3 all divided by 8?* As the students answer, the teacher models the strategy.	$$=\dfrac{4x}{8}-\dfrac{3}{8}$$ $$=\tfrac{1}{2}x-\tfrac{3}{8}=0.5x-0.375$$

The teacher repeats with the last problem.	$\dfrac{10-2x}{-5} = \dfrac{10}{-5} - \dfrac{2x}{-5}$ $= -2 + \frac{2}{5}x = -2 + 0.4x$

Teacher: *How would you summarize some of the things that came up in this string today?*

Elicit the following:

- *The fraction bar can mean division.*

- *We can split up the numerator with addition or subtraction and solve each resulting problem (expression).*

- *It can be helpful to find a simpler problem to solve.*

- *Dividing can be like simplifying a fraction.*

Sample Final Display

Your display could look like this at the end of the problem string:

$$816 \div 8 = \frac{816}{8} = \frac{800+16}{8} = \frac{800}{8} + \frac{16}{8} \qquad\qquad 792 \div 8 = \frac{792}{8} = \frac{800-8}{8} = \frac{800}{8} - \frac{8}{8}$$
$$= 100 + 2 = 102 \qquad\qquad\qquad = 100 - 1 = 99$$

$$\frac{4x}{4} = x \qquad \frac{\cancel{4}^{1}x}{\cancel{4}_{1}} = \frac{1 \cdot x}{1} = x \qquad \frac{4x}{4} = \frac{4}{4} \cdot x = 1 \cdot x = x$$

$$\frac{4x+8}{4} = \frac{4x}{4} + \frac{8}{4}$$
$$= x + 2$$

$$\frac{3x}{3} = x \qquad \frac{3}{3} \cdot x = 1x = x$$

$$\frac{3x-9}{3} = \frac{3x}{3} - \frac{9}{3}$$
$$= x - 3$$

$$\frac{5x}{2} = \frac{5}{2}x = 2.5x$$

$$\frac{5x+3}{2} = \frac{5x}{2} + \frac{3}{2}$$
$$= 2.5x + 1.5 = \frac{5}{2}x + \frac{3}{2}$$

$$\frac{4x-3}{8} = \frac{4x}{8} - \frac{3}{8}$$
$$= \frac{1}{2}x - \frac{3}{8} = 0.5x - 0.375$$

$$\frac{10-2x}{-5} = \frac{10}{-5} - \frac{2x}{-5}$$
$$= -2 + \frac{2}{5}x = -2 + 0.4x$$

(continued)

Facilitation Notes

This version of the problem string lists short notes for important teacher moves during the string. After you've done the string yourself and studied the relationships involved, you might make similar notes for the things you want a reminder of or deem important.

$816 \div 8 = \dfrac{816}{8}$	What is 816 divided by 8? We can write it both ways. Did anyone think about 800 divided by 8?
$792 \div 8 = \dfrac{792}{8}$	What is 792 divided by 8? We can write it both ways. Did anyone think about 800 divided by 8? When dividing, we can break up the numerator by addition or subtraction and solve each resulting division problem. Handy.
$\dfrac{4x}{4}$	What is 4x divided by x? How do you know? If students use "cancel", talk about what is actually happening.
$\dfrac{4x+8}{4}$	What is 4x plus 8 all divided by 4? I wonder if we know anything that might be helpful. How do you know?
$\dfrac{3x}{3}$	Quick.
$\dfrac{3x-9}{3}$	Who split up the numerator? Does this strategy make sense? Turn and talk to a partner.
$\dfrac{5x}{2}$	Elicit fractions and decimals.
$\dfrac{5x+3}{2}$	Can the previous problem help?
$\dfrac{4x-3}{8}$	Look up. The string has helper problems. What could be a helper problem for this one? How does that helper problem help?
$\dfrac{10-2x}{-5}$	What could be a helper problem for this one? How does that helper problem help?

Algebra Problem Strings
©2017 Kendall Hunt Publishing

5.1 Systems of Linear Equations

At a Glance

5 sandwiches & 4 cookies cost $62

4 sandwiches & 5 cookies cost $55

6 sandwiches & 3 cookies cost $69

3 sandwiches & 6 cookies cost $48

10 sandwiches & 8 cookies cost $124

8 sandwiches & 10 cookies cost $110

2 sandwiches & 7 cookies cost $41

1 sandwich & 8 cookies cost $34

How much is a cookie?

How much is a sandwich?

Objectives

The purpose of this problem string is for students to associate a combination of quantities with a system of linear equations, to have students articulate the important differences between a single linear equation and a system, and to develop students' emergent ideas about how to solve such a system.

Placement

Use this problem string to help introduce systems of linear equations.

You can use this problem string to introduce and support textbook Lesson 5.1 Solving Systems of Equations.

Guiding the Problem String

The first problem sets the context and starts the conversation about a line representing a infinite set of points. The second problem creates the system of linear equations and students begin to realize that the intersection point represents the data point the two scenarios have in common. The third and fourth problems represent an exchange of a sandwich for a cookie from the first and second problems respectively. The fifth and sixth problems are doubles of the first two problems, representing double orders. The rest of the problems represent further exchanges of a sandwich for a cookie. As students graph each equation, they begin to realize the importance and meaning of the intersection point and that equivalent equations have the same graphs.

As you facilitate the string, highlight students' reasoning. Always encourage students to justify a claim they are making. It can be tempting when students provide their thinking to evaluate it yourself ("Good" or "I agree" or even "That makes sense to me"). Instead, take a more neutral stance here, pivoting toward your entire class—take the reasoning and claims to the class to evaluate as a community.

About the Mathematics

This system of linear equations represent combinations of changing quantities of sandwiches and cookies. Not all linear systems can be represented easily as combinations of changing quantities, particularly those with rational number or negative coefficients, but students can use the intuition they develop here to help make sense of the more complicated equations later. You and your students can refer back to this scenario as students acquire more formal, symbolic strategies for solving systems such as elimination and substitution.

Too often the sole goal of linear systems is for students to solve them. But this narrow view can overshadow all the reasoning students can do with a system, on students' eventual pathway to solving. For example, in this string students can be invited to think about the relationship between the price of a sandwich and the price of a cookie. This can be captured as an exchange—if you trade your sandwich for a cookie the price decreases by $7—and this swapping of two items of varying prices can be symbolically represented as:

$$C + 7 = S \text{ and } S - 7 = C$$

where C is the price of a cookie and S is the price of a sandwich. This type of reasoning about exchanges and equivalence can play out powerfully as students acquire formal strategies for solving.

(continued)

Sample Interactions

Use the following as you plan how to elicit and model student strategies. This is not meant as a script, but as a view into the relationships involved and the intent of the problem string.

Teacher: *Today I brought a situation for us to think about together. And as I reveal more and more of the story, I'd like you to keep asking yourself a few questions: What do I know? Does this make sense? What makes me think this? How could I convince everyone of my thinking?*

I'll write those up here so you can keep coming back to them.

Okay, so the other day I picked up lunch for the math department. They sent me with some cash and asked me to go get the order. I bought 5 sandwiches and 4 cookies for a total of $62.

Take a moment with your partner to discuss: What can you say for sure about the sandwiches or cookies from this café? What are you not so sure about?

Partners talk while the teacher listens in and, if necessary, nudges students to find some combinations that work, especially the intercepts, where either the cookies are free or the sandwiches are free.

Questions to consider:

- What do I know?
- Does it make sense?
- What makes me think this?
- How could I convince everyone of my thinking?

5 sandwiches & 4 cookies cost $62

Teacher: *I heard some interesting ideas. Will you share what you said to your partner with all of us?*

Student: *Sure. We were saying that for sure you got 5 sandwiches and 4 cookies and it cost you $62. But beyond that, we tried guessing the prices and we weren't totally sure.*

Teacher: *Does anyone think they know for sure, based on this image of what happened, which costs more, a sandwich or a cookie?*

Student: *Sandwiches usually cost more than cookies, but maybe they are those giant gourmet cookies. Who knows?*

Teacher: *Did anyone find a combination that would work?*

If no one suggests, the teacher could ask students to consider if the cookies were free.

Student: *Well, it's possible that you got the cookies for free. So then the 5 sandwiches cost the total $62.*

Teacher: *I'm going to model that as 5 times the cost of the sandwiches plus 4 times the cost of the cookies which is zero and that's $62. So, how much would the sandwiches cost?*

Student: *That means that 5 times the cost of a sandwich is $62. So, what is 62 divided by 5? I know 60 divided by 5 is 12.*

Student: *And 2 divided by 5 is two-fifths or four-tenths, 0.4. So a sandwich costs $12.40.*

Teacher: *What if the sandwiches were free? How much would each cookie cost?*

Student: *That means that 4 times the cost of a cookie is $62? So, $15.50.*

Teacher: *How did you find that?*

Student: *$62 divided by 4 is like 60 divided by 4 and 2 divided by 4, so 15 and ½, $15.50.*

($?, $0)

$$5S + 4(0) = \$62$$

$$\frac{62}{5} = \frac{60}{5} + \frac{2}{5} = 12 + 0.4 = 12.4$$

$S = 12.4$, so ($12.40, $0)

$$5(0) + 4C = \$62$$

$$\frac{62}{4} = \frac{60}{4} + \frac{2}{4} = 15 + 0.5 = 15.5$$

$C = 15.5$, so ($0, $15.50)

Algebra Problem Strings
©2017 Kendall Hunt Publishing

Teacher: *So, now we have two possibilities. I'm going to list these in a table.*

Number of Sandwiches	Cost($) of Sandwiches	Number of Cookies	Cost($) of Cookies	Total Cost($)
5	12.40	4	0	62
5	0	4	15.50	62

Teacher: *What other possibilities are there?*

Student: *If the sandwiches are $10, then that's $50, so $12 left which means that each cookie is $3.*

Student: *We've got another one, $12 for the sandwiches and $0.50 for the cookies.*

Teacher: *How are you finding these?*

Student: *We picked nice numbers for sandwiches and then the rest, whatever was left over, was for cookies.*

Teacher: *Is that what you were all doing? Yes? I wonder if there might be a way to find all of the possibilities?*

Number of Sandwiches	Cost($) of Sandwiches	Number of Cookies	Cost($) of Cookies	Total Cost($)
5	12.40	4	0	62
5	0	4	15.50	62
5	10	4	3	62
5	12	4	0.50	62

Teacher: *Looking at the table, what relationships do you see?*

Student: *It's always the sandwiches plus the cookies to get $62.*

Student: *It's 5 times the sandwiches and 4 times the cookies.*

Teacher: *It's the number of sandwiches and the number of cookies equals $62?*

Student: *No, it's the cost of the cookies and the cost of the sandwiches together is $62.*

Teacher: *So it looks like we've got 5 times the cost of the sandwiches plus 4 times the cost of the cookies equals $62.*

Number of Sandwiches	Cost($) of Sandwiches	Number of Cookies	Cost($) of Cookies	Total Cost($)
5	12.40	4	0	62
5	0	4	15.50	62
5	10	4	3	62
5	12	4	0.50	62
5	S +	4	C =	62

Teacher: *Let's graph this equation.*

I wonder if this graph can help us find some other possibilities? What do you think?

If the grapher allows, enter the equation as shown. If not, have students solve for C.

Student: *We can look at those points on the line.*

Teacher: *Turn to your partner and talk about this graph and the situation.*

The students turn and talk. Then the teacher leads a short conversation about the points in the first quadrant having meaning in the context, but the points in the other quadrants do not. The teacher then traces on the line in the first quadrant and leads a short conversation about other possible combinations of prices for a total of $62.

(0, 15.5)

$5S + 4C = 62$

(12.4, 0)

(continued)

Teacher: *How about now? What if I went back to the same café the next day and picked up a different order?* *Let's all study this order for 30 seconds first. Here I want you to think about:* *What's changing? What's staying the same? What does that tell you?* Think time. *Turn and tell your partner what you are thinking.*	4 sandwiches & 5 cookies cost $55
Teacher: *What are you thinking?* **Student:** *For sure the sandwiches are more expensive.* **Teacher:** *Who agrees with this idea?* *Okay, many but not all of us. Could you convince the skeptics of this?* **Student:** *I'll try. From the first order to the second there are a few important differences: you went from 5 sandwiches to 4, and you went from 4 cookies to 5. This means sandwiches decreased by 1 and cookies increased by 1. And while this happened the price went down $7. So we think this means that by taking out a sandwich and replacing it with a cookie your price went down. So, sandwiches are definitely more expensive.* **Teacher:** *Anyone not yet convinced by this argument?*	5 sandwiches & 4 cookies cost $62 down one sandwich — up one cookie — down $7 4 sandwiches & 5 cookies cost $55
Teacher: *Okay, what else?* **Student:** *I think we could say that the sandwiches are $7.* **Teacher:** *What makes you say that?* **Student:** *Like Toby said, the price went down $7 when you took out a sandwich.* **Teacher:** *What do we think?* **Student:** *I don't think we know that for sure, but I get this reasoning. We also added a cookie. I think the difference between the sandwich and the cookie is $7. Because it's like a trade—get rid of a sandwich and put a cookie in its place save $7.* **Student:** *Oh, yeah, I want to change my thinking on that. I'm convinced.*	$S - C = \$7$
Teacher: *So, this second day, what is an equation to represent the cost of sandwiches, the cost of cookies, and the total cost?* **Student:** *It's like the last one, but this time it's 4S plus 5C is $55.*	$4S + 5C = \$55$

Algebra Problem Strings
©2017 Kendall Hunt Publishing

Teacher: *What would this graph look like? Can you find some nice points?*	$4(0) + 5C = 55, C = 11$
Student: *We can do the zero thing. If the sandwiches are free, then it's 5 times the cost of the cookies is $55, so a cookie costs $11.*	$(0, 11)$ $4S + 5 (0) = 55, S = 13.75$
Student: *The other one is not as nice, but 4 times the cost of a sandwich is 55. Four times 13 is 52 and then there's the left over 3 divided by 4, that's 0.75, so $13.75.*	$\dfrac{55}{4} = \dfrac{52}{4} + \dfrac{3}{4} = 13 + 0.75 = 13.75$ ($13.75, 0)

Teacher: *I'll plot these points. What are these points called?*

Student: *Intercepts. It's where the line intersects the axes.*

Teacher: *What are your thoughts looking at this graph?*

Student: *The lines intersect.*

Teacher: *What do you think that means? Anything? Why?*

Student: *I wonder if that's the cost.*

Student: *For each line, there are many different possibilities, but if you look at them together, I bet that's what the stuff actually costs.*

Student: *Because it's where the first day and the second day have the same combinations of costs.*

Student: *What is that point where they cross? Is it (10, 3)?*

Teacher: *I'll scroll over there while you think about (10, 3). What does the point (10, 3) mean in this scenario?*

Student: *It's $10 for a sandwich and $3 for a cookie. On both days.*

Teacher: *It looks like $10 for a sandwich and $3 for a cookie works for both days. We already had that as one of our possibilities for the first day. Does it also work for the second day?*

Student: *Yes, because $40 and $15 is $55.*

Teacher: *So, now I can plan because I know how much the sandwiches and the cookies cost at that place.*

(continued)

Teacher: *Okay, for the next problem, what can you tell me about this order? I bought 6 sandwiches and 3 cookies and it cost me $69? Was I at the same café or somewhere else? How do you know?*

The teacher leads a short conversation, eliciting the equation for the situation and the difference of one more sandwich and one less cookie being $7 more.

Teacher: *What do you think the graph will look like? How will it compare? Predict please.*

After short think time, the teacher displays the graph.

Teacher: *What do you notice?*

Student: *It has the same intersection point. All three lines intersect there.*

Student: *That means you are at the same place. The costs are the same per sandwich and per cookie.*

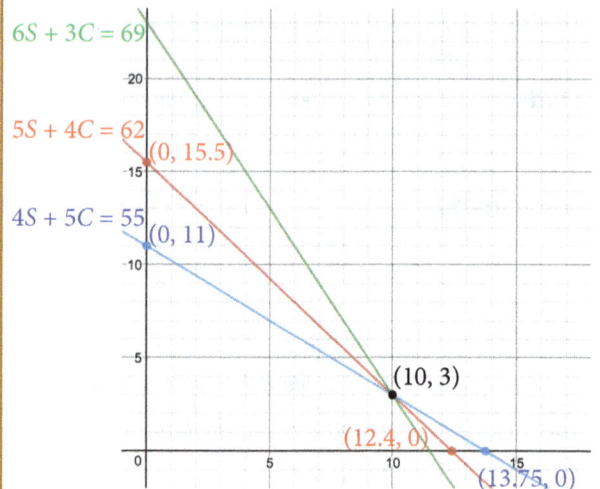

[Graph showing three lines: $6S + 3C = 69$, $5S + 4C = 62$, $4S + 5C = 55$. Points labeled: $(0, 15.5)$, $(0, 11)$, $(10, 3)$, $(12.4, 0)$, $(13.75, 0)$. All lines intersect at $(10, 3)$.]

Teacher: *For the next problem, what can you tell me about this order? I bought 3 sandwiches and 6 cookies and it cost me $48? Was I at the same café again? How do you know?*

The teacher leads a short conversation, eliciting the equation for the situation, the difference of one less sandwich and one more cookie being $7 less, the graph, and the same intersection point.

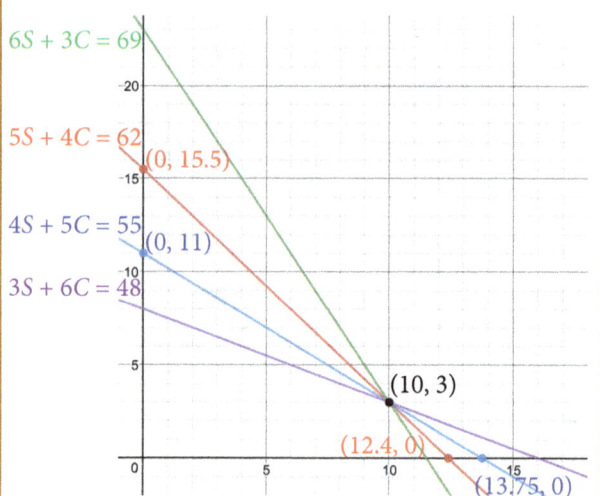

[Graph showing four lines: $6S + 3C = 69$, $5S + 4C = 62$, $4S + 5C = 55$, $3S + 6C = 48$. Points labeled: $(0, 15.5)$, $(0, 11)$, $(10, 3)$, $(12.4, 0)$, $(13.75, 0)$. All lines intersect at $(10, 3)$.]

Algebra Problem Strings
©2017 Kendall Hunt Publishing

Teacher: *What does this next problem mean in terms of ordering? I bought 10 sandwiches and 8 cookies and it cost me $124? Was I at the same café? How do you know?*

Student: *It's double the red order. From 5 sandwiches to 10, and 4 cookies to 8. And the money doubled as it should. Makes sense.*

Teacher: *How do you think the graph of this line will look?*

Student: *I don't know.*

Student: *It seems to me like they are equivalent. The red one and this one. It's just double.*

Student: *Shouldn't that be bigger somehow, double?*

Student: *I think they will still have the same intercepts.*

Teacher: *Let's look. Here is the graph.*

The teacher displays the graph.

Teacher: *What's going on?*

Student: *They are the same! That's interesting.*

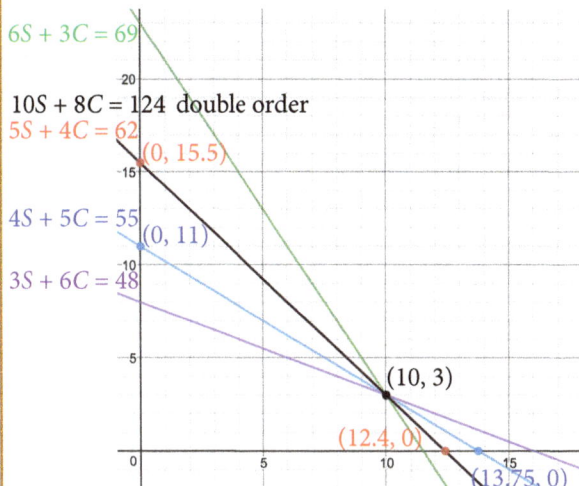

$6S + 3C = 69$

$10S + 8C = 124$ double order
$5S + 4C = 62$
$(0, 15.5)$

$4S + 5C = 55$
$(0, 11)$

$3S + 6C = 48$

$(10, 3)$

$(12.4, 0)$
$(13.75, 0)$

Teacher: *How about this next order, 8 sandwiches and 10 cookies for $110? Does it relate to any of the others?*

Student: *It's another double. I bet it will look like it's double.*

Teacher: *What does everyone think? Predict what you think the graph will look like.*

Think time.

Teacher: *What do you think?*

Student: *The graphs will look the same as the blue one. If you divide the whole equation by 2, it's the same as the $4S + 5C = 55$. They are equivalent.*

The teacher displays the graph.

Student: *Yep, the same graph. Same intersection point. If you double the order, you double the price.*

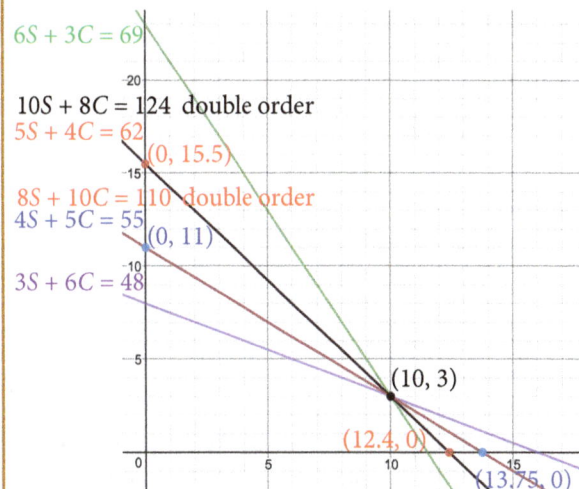

$6S + 3C = 69$

$10S + 8C = 124$ double order
$5S + 4C = 62$
$(0, 15.5)$

$8S + 10C = 110$ double order
$4S + 5C = 55$
$(0, 11)$

$3S + 6C = 48$

$(10, 3)$

$(12.4, 0)$
$(13.75, 0)$

(continued)

Teacher: *The next order for us to think about is 2 sandwiches and 7 cookies totaling $41. Same café? How do you know?*

Student: *Yes, look at the purple graph. It's 1 sandwich less and 1 cookie more and $7 difference.*

Student: *Also, 2 times $10 is $20 and 7 times $3 is $21. $20 plus $21 is $41. So that checks out.*

The teacher displays and labels the graph.

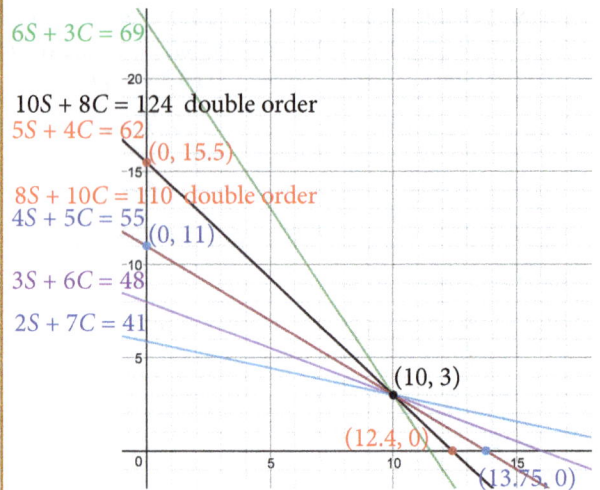

$6S + 3C = 69$
$10S + 8C = 124$ double order
$5S + 4C = 62$
$(0, 15.5)$
$8S + 10C = 110$ double order
$4S + 5C = 55$
$(0, 11)$
$3S + 6C = 48$
$2S + 7C = 41$
$(10, 3)$
$(12.4, 0)$
$(13.75, 0)$

The teacher repeats for the next problem, one sandwich and eight cookies cost $34.

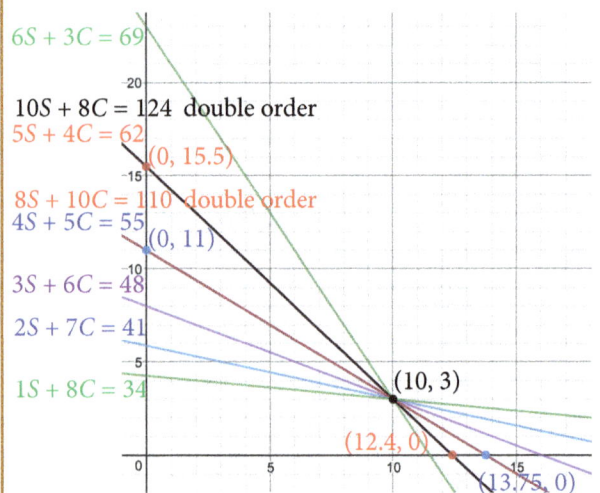

$6S + 3C = 69$
$10S + 8C = 124$ double order
$5S + 4C = 62$
$(0, 15.5)$
$8S + 10C = 110$ double order
$4S + 5C = 55$
$(0, 11)$
$3S + 6C = 48$
$2S + 7C = 41$
$1S + 8C = 34$
$(10, 3)$
$(12.4, 0)$
$(13.75, 0)$

Teacher: *Last question for today: What if I just bought 9 cookies? If we follow the pattern of one less sandwich and one more cookie?*

Student: *That is $27.*

Teacher: *Earlier you said that having one order isn't enough to know the price of the items, but here's an example of how that can be done. So, what's going on here?*

Student: *Ahhh, if we find an equation that only has one variable, then we can solve. So like $5S + 4C = 62$ has all of these possibilities, but $9C = 27$ is no problem. One variable.*

Teacher: *Sounds like getting a combination where one of the variables "goes away" seems to be helpful.*

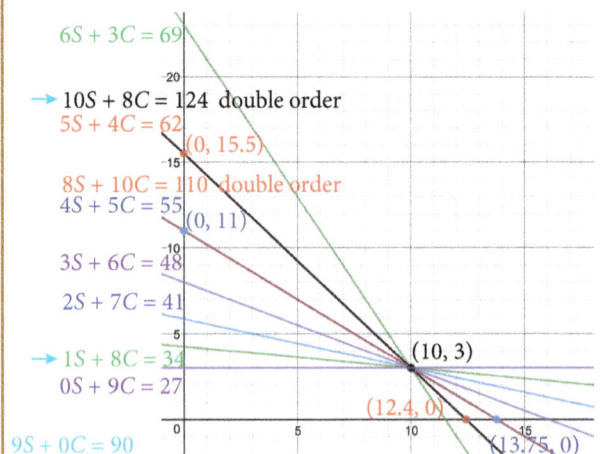

$6S + 3C = 69$
$\rightarrow 10S + 8C = 124$ double order
$5S + 4C = 62$
$(0, 15.5)$
$8S + 10C = 110$ double order
$4S + 5C = 55$
$(0, 11)$
$3S + 6C = 48$
$2S + 7C = 41$
$\rightarrow 1S + 8C = 34$
$0S + 9C = 27$
$(10, 3)$
$(12.4, 0)$
$(13.75, 0)$
$9S + 0C = 90$

Teacher: *I wonder… is that the only way? Any way to find the price of a sandwich using the orders we have here? In other words, if we didn't already know the $10 per sandwich and $3 per cookies from the graph, could we have used the equations to find the price of the sandwich? Do you see any relationships that could help?*

Students turn and talk while the teacher listens in.

Teacher: *What are you thinking?*

Student: *My partner and I tried looking for ones that were somehow similar so we compared the two orders where there were 8 cookies.*

Teacher: *These two?*

Student: *Yep.*

Teacher: *How did these help you?*

Student: *You can see that the cookies stayed the same, but you removed 9 sandwiches and the price dropped $90.*

Teacher: *Let me pause you. Who can finish this line of reasoning?*

Student: *Got it. We just made another combination, in a way, where 9 sandwiches is the same as $90. Sandwiches are $10 each.*

Teacher: *Nice work everyone.*

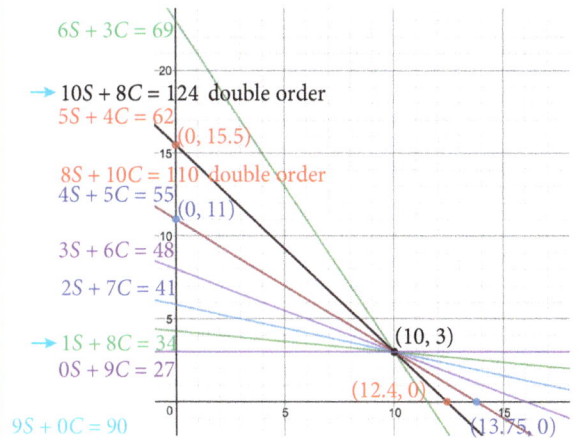

Graph equations:

$6S + 3C = 69$

→ $10S + 8C = 124$ double order
$5S + 4C = 62$
(0, 15.5)

$8S + 10C = 110$ double order
$4S + 5C = 55$
(0, 11)

$3S + 6C = 48$

$2S + 7C = 41$

→ $1S + 8C = 34$
$0S + 9C = 27$

$9S + 0C = 90$

(10, 3)

(12.4, 0)

(13.75, 0)

Teacher: *How would you summarize some of the things that came up in this string today?*

Elicit the following:

- *One equation with two variables represents many different combinations.*
- *You can find the intercepts of an equation by solving for x when y = 0 and vice versa.*
- *The intersection of two equations represents the combination that they share.*
- *Doubles of equations are equivalent and graph the same line.*
- *If you can find an equivalent equation with only one variable, you can find the cost of that variable.*

(continued)

Sample Final Display

Your display could look like this at the end of the problem string:

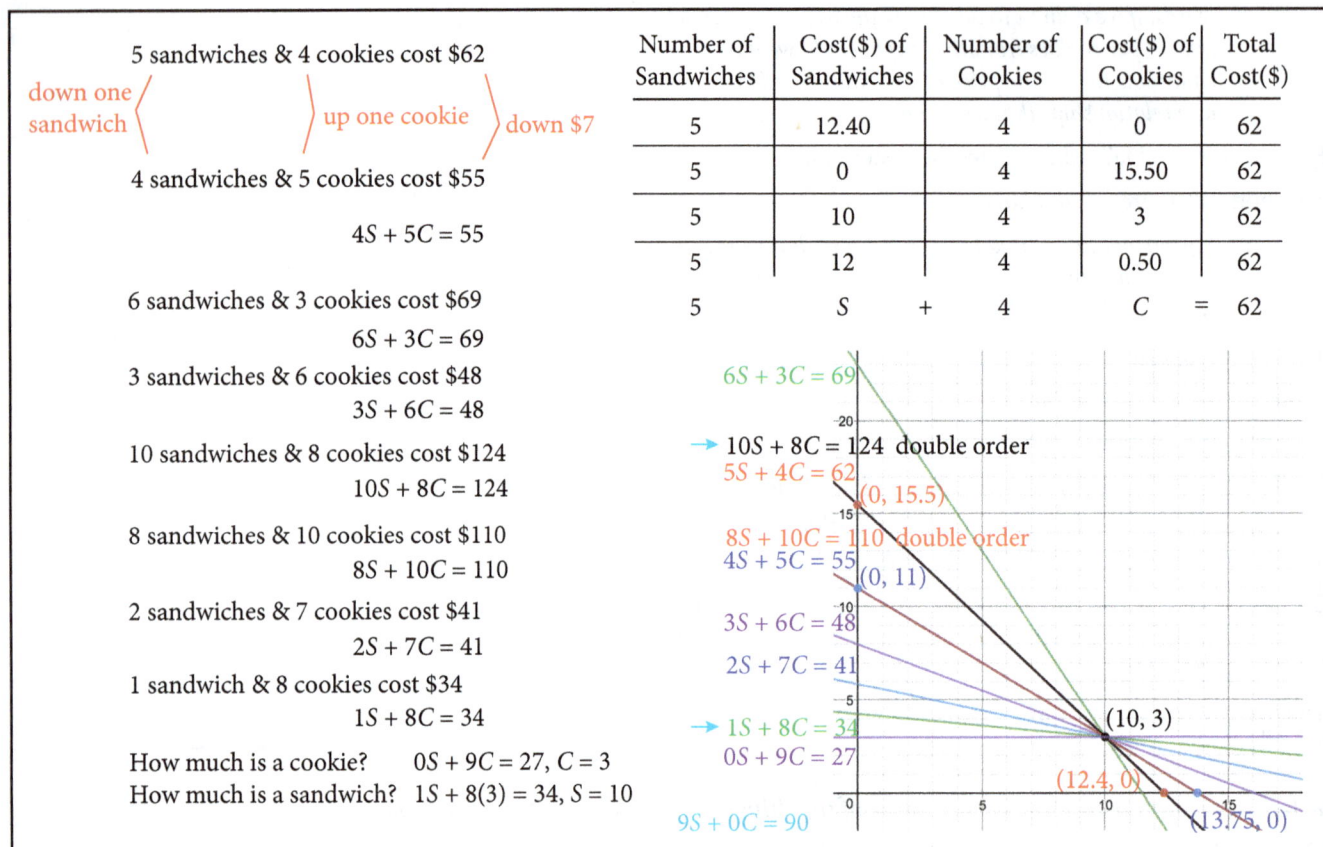

5 sandwiches & 4 cookies cost $62

down one sandwich / up one cookie \ down $7

4 sandwiches & 5 cookies cost $55

$$4S + 5C = 55$$

6 sandwiches & 3 cookies cost $69

$$6S + 3C = 69$$

3 sandwiches & 6 cookies cost $48

$$3S + 6C = 48$$

10 sandwiches & 8 cookies cost $124

$$10S + 8C = 124$$

8 sandwiches & 10 cookies cost $110

$$8S + 10C = 110$$

2 sandwiches & 7 cookies cost $41

$$2S + 7C = 41$$

1 sandwich & 8 cookies cost $34

$$1S + 8C = 34$$

How much is a cookie? $0S + 9C = 27, C = 3$

How much is a sandwich? $1S + 8(3) = 34, S = 10$

Number of Sandwiches	Cost($) of Sandwiches	Number of Cookies	Cost($) of Cookies	Total Cost($)
5	12.40	4	0	62
5	0	4	15.50	62
5	10	4	3	62
5	12	4	0.50	62
5	S +	4	C =	62

$6S + 3C = 69$

$\rightarrow 10S + 8C = 124$ double order

$5S + 4C = 62$

(0, 15.5)

$8S + 10C = 110$ double order

$4S + 5C = 55$

(0, 11)

$3S + 6C = 48$

$2S + 7C = 41$

$\rightarrow 1S + 8C = 34$

(10, 3)

$0S + 9C = 27$

(12.4, 0)

$9S + 0C = 90$

(13.75, 0)

Algebra Problem Strings
©2017 Kendall Hunt Publishing

Facilitation Notes

This version of the problem string lists short notes for important teacher moves during the string. After you've done the string yourself and studied the relationships involved, you might make similar notes for the things you want a reminder of or deem important.

5 sandwiches & 4 cookies cost $62	*Ask yourself:* *What do I know?* *Does it make sense?* *What makes me think this?* *How could I convince everyone of my thinking?* *Partner talk.* *Did anyone find a combination that would work?* *Find intercepts. Make a table. List a few points.* *Write an equation based on student discussion.* *Graph, label equation and intercepts.* *Trace two more points. What do the points mean in the situation?*
4 sandwiches & 5 cookies cost $55	*What if I went to the same café and got a different order?* *What's changing? What's staying the same? What does that tell you?* *Elicit the noticing of down one sandwich, up one cookie is a change of $7.* *Graph, label. What does the intersection point mean?*
6 sandwiches & 3 cookies cost $69	*What about this new order? Am I at the same café?* *Did it follow the same pattern?* *Graph, label.*
3 sandwiches & 6 cookies cost $48	*Repeat. Quick.*
10 sandwiches & 8 cookies cost $124	*How does this order relate to the other orders?* *Double the order, double the price. Predict the graph.* *Graph, label.*
8 sandwiches & 10 cookies cost $110	*Repeat. Quick.*
2 sandwiches & 7 cookies cost $41	*How does this order relate to the other orders?* *Graph, label.*
1 sandwich & 8 cookies cost $34	*Repeat. Quick.*
How much is a cookie? How much is a sandwich?	*What if I just bought 9 cookies?* *What if we follow the pattern of one less sandwich and one more cookie?* *Can one equation help us find the cost? In the case of just one variable?* *Any way to find the price of the sandwich using the orders we have here?*

5.2 | Parallel and Perpendicular Slopes

At a Glance

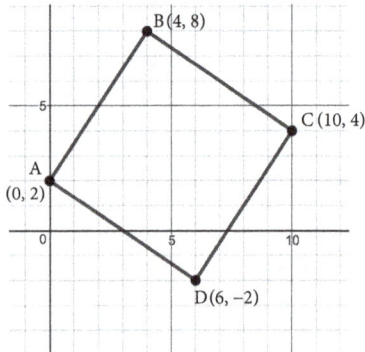

Find the equation of $\overline{AD}$.

Find the equation of $\overline{BC}$.

Find the equation of $\overline{AB}$.

Find the equation of $\overline{CD}$.

Objectives
The goal of this problem string is for students to use the properties of squares to gain intuition about the slopes of parallel and perpendicular lines.

Placement
You can use this problem string to introduce the relationships of the slopes of parallel and perpendicular lines. Students should have prior experience finding the equation of a line given a point and slope.

Use this problem string as you begin textbook Lesson 5.2 Parallel and Perpendicular Lines to help students gain intuition for the slope of parallel and perpendicular lines.

Guiding the Problem String
This short problem string is intended to have students use the context of sides of a square to review writing the equations of lines given a point and a slope, while planting notions about the slopes of parallel and perpendicular lines.

Show students the sketched square with the vertices labeled. The first question asks students to find the equation of side AD, which contains the *y*-intercept and another point. Nudge students to choose the slope-intercept form as the more efficient strategy in the future by having students share both it and the point-slope form and compare. The second problem is to find the equation containing the side parallel to the first. Notice if students use that they are parallel from the definition of a square. Elicit the idea in the discussion. The third problem is to find the line of one of the sides that is perpendicular to the first two. Watch to see if anyone notices the relationship between the slopes of the perpendicular lines. You might wonder about it aloud. The fourth problem is to find the equation of the last side using relationships.

If your students will be doing the Investigation in textbook lesson 5.2, facilitate the problem string quickly, getting students warmed up to find the equations of the sides of rectangles and make conjectures in the investigation. If not, spend more time eliciting, naming, and discussing the patterns between the slopes of parallel and perpendicular lines.

About the Mathematics
Parallel lines have equivalent slopes. The slopes of perpendicular lines are opposite reciprocals. Use *opposite reciprocals* instead of *negative reciprocals* to help students make sense of the relationship. If students are thinking about *negative reciprocals*, students might see a line with a slope like $-\frac{1}{2}$ and when asked to find the slope of a perpendicular line, think, "But that slope is already negative!" If they are thinking about *opposite reciprocals*, they have the opportunity to consider that the opposite of a negative number is the positive number.

Sample Interactions

Use the following as you plan how to elicit and model student strategies. This is not meant as a script, but as a view into the relationships involved and the intent of the problem string.

The teacher projects a grid on the board and draws and labels the square as shown.

Teacher: *For today's problem string, we are going to review some things that we have done before. Here is a square. The first problem is to find the equation of side AD. Just note it on your paper as you find the equation.*

As students work briefly, the teacher circulates, looking for students who use the point-slope or the slope-intercept form of a line.

Teacher: *What equation did you find?*

Student: *I got $y = -\frac{2}{3}x + 2$.*

Teacher: *Did anyone get anything different?*

Student: *I found $y = -\frac{2}{3}(x-6) - 2$.*

Teacher: *Are those different? Equivalent? How do you know?*

Student: *The first one is just using the point (0, 2) and the second one is using the point (6, −2).*

Student: *Yeah, and they use the same slope so they are equivalent.*

Teacher: *Are we convinced?*

Student: *I multiplied the second one out and they're the same.*

$$y = -\frac{2}{3}x + 2$$

$$y = -\frac{2}{3}(x-6) - 2$$

Teacher: *So, if we graphed both of them, they should be the same line and that line should include this side of the square? Let's look.*

The teacher displays the graphs of the equations using the class grapher.

Student: *It looks like they are the same.*

Teacher: *Since they are equivalent, I'll just label the line with one of the equations.*

The teacher labels the line with the slope-intercept form of the line.

(continued)

Teacher: *For our second problem today, find the equation of side BC.*

As students work, the teacher circulates, looking for students who are finding the slope between the two points and students who are using the notion that the sides of a square are parallel and therefore the slopes are equivalent.

Teacher: *What equation did you find?*

Student: *I got $y = -\dfrac{2}{3}(x-4)+8$.*

Student: *And I found $y = -\dfrac{2}{3}(x-10)+4$.*

Teacher: *What do we think?*

Student: *The equations are the same. They just used the other point.*

Teacher: *Let's graph both of them. Yep, the same. Will you please tell us how you found the slope?*

Student: *I found the ratio of the difference in the y's and the x's, like normal.*

Student: *Actually, I just assumed that the slope would be the same as the slope for the first problem, side AD. Because they are parallel.*

Student: *Nice. That would have saved me time.*

Student: *Sure enough. The slopes are the same. I hadn't even noticed. So the lines are parallel. Yes, sides of a square are parallel. Okay, I get it.*

Teacher: *Do you think that's always true, that the opposite sides of a square are parallel? And does that always mean that the slopes are the same? Let's keep thinking about that.*

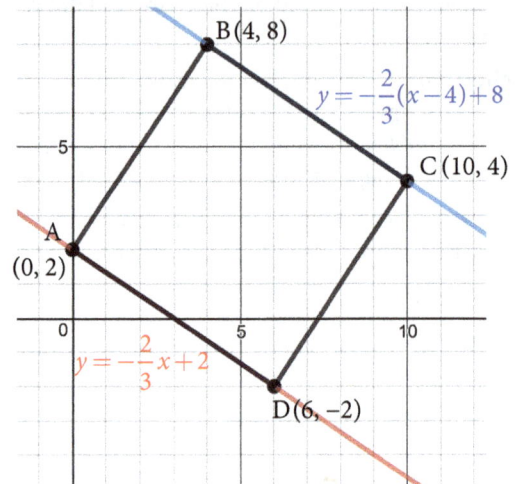

Teacher: *For our third problem today, find the equation of the line that contains the side AB.*

As students work briefly, the teacher circulates, looking for students who use the point-slope or the slope-intercept form of a line. If students are using the point-slope form, the teacher wonders aloud about which point students are using, nudging students to notice the given *y*-intercept if it's within the students' zone of proximal development. The teacher also looks for a student who has written the slope as a fraction, ⅔, to maximize the chance that students will wonder about the relationship between this slope and the slopes of the previous two sides.

Algebra Problem Strings
©2017 Kendall Hunt Publishing

Teacher: *What equation did you two find?*

Student: *We got* $y = \frac{3}{2}x + 2$.

Teacher: *Everyone else, what point did they use?*

Student: *They used the (0, 2), the y-intercept.*

Teacher: *That's efficient, isn't it? Does anyone notice any potential patterns on the board right now?*

Student: *Well, we already said we think that the parallel sides have the same slope.*

Student: *We could use that to find the equation of the last side.*

Teacher: *Any other insights or wonderings?*

Student: *Two of the sides have negative slopes and the other two have positive slopes.*

Student: *All of the slopes have 2s and 3s in them. That seems to be right for the parallel lines, but I'm not sure about the 90 degree lines.*

Teacher: *You're not sure about the perpendicular lines?*

Student: *Right.*

Teacher: *Let's keep looking for patterns that might be helpful.*

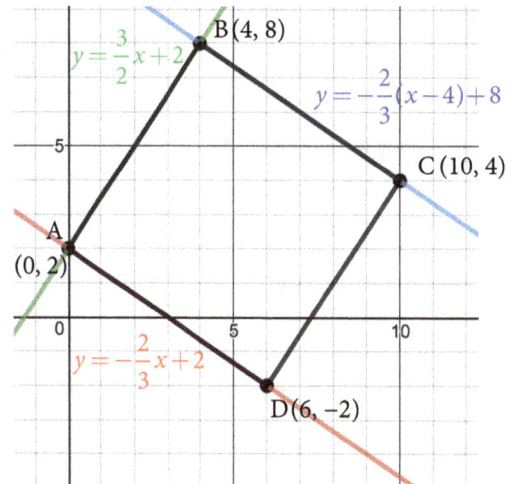

Teacher: *For our next problem today, let's predict first. Predicting, thinking, envisioning, what do you think the equation is for side CD? What do we know?*

Brief think time.

Teacher: *Turn to your partner. What do we know?*

Students briefly turn and talk while the teacher listens in.

Teacher: *What do we know?*

Student: *DC is parallel to AB. So maybe it has the same slope?*

Teacher: *The lines DC and AB are parallel? Agree/disagree? Everyone agrees? So, can you just find an equation?*

Student: *Yeah, we can just use the slope of 1.5 and either of the points.*

Teacher: *Choose one of the points and let's graph it.*

Student: *Let's use (10, 4) since neither coordinate is negative. So* $y = \frac{3}{2}(x - 10) + 4$.

Teacher: *Here is the graph. What do you think?*

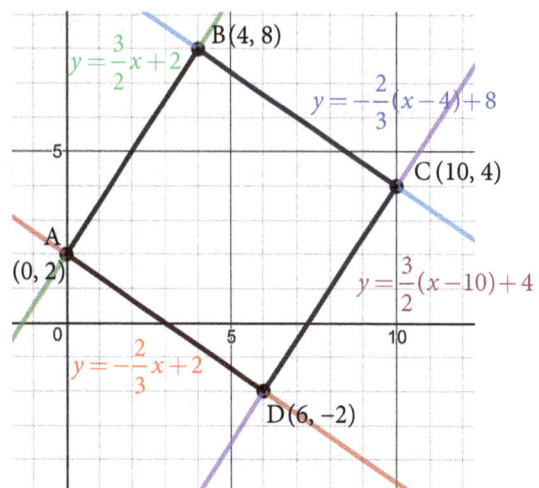

(continued)

Student: *It looks like parallel lines have the same slope.*

Student: *And maybe perpendicular lines have slopes with the same numbers in them.*

Teacher: *Any other way to describe the relationship between −⅔ and 3/2?*

Student: *They are flipped.*

Student: *Upside down.*

Teacher: *We call that a reciprocal.*

Student: *And one is positive and one is negative.*

Teacher: *So they are the opposite sign of each other. Opposite reciprocals. Anyone wondering if this is true just for these lines or for all perpendicular lines?*

Teacher: *How would you summarize some of the things that came up in this string today?*

Elicit the following:

- *Squares have opposite sides that are parallel and adjacent sides that are perpendicular.*

- *It looks like parallel lines have the same slopes.*

- *It looks like perpendicular lines have opposite reciprocal slopes.*

Sample Final Display

Your display could look like this at the end of the problem string:

Find the equation of $\overline{AD}$.

$$y = -\frac{2}{3}x + 2 \qquad y = -\frac{2}{3}(x-6)-2$$

Find the equation of $\overline{BC}$.

$$y = -\frac{2}{3}(x-4)+8 \qquad y = -\frac{2}{3}(x-10)+4$$

Find the equation of $\overline{AB}$.

$$y = \frac{3}{2}x + 2$$

Find the equation of $\overline{CD}$.

$$y = \frac{3}{2}(x-10)+4$$

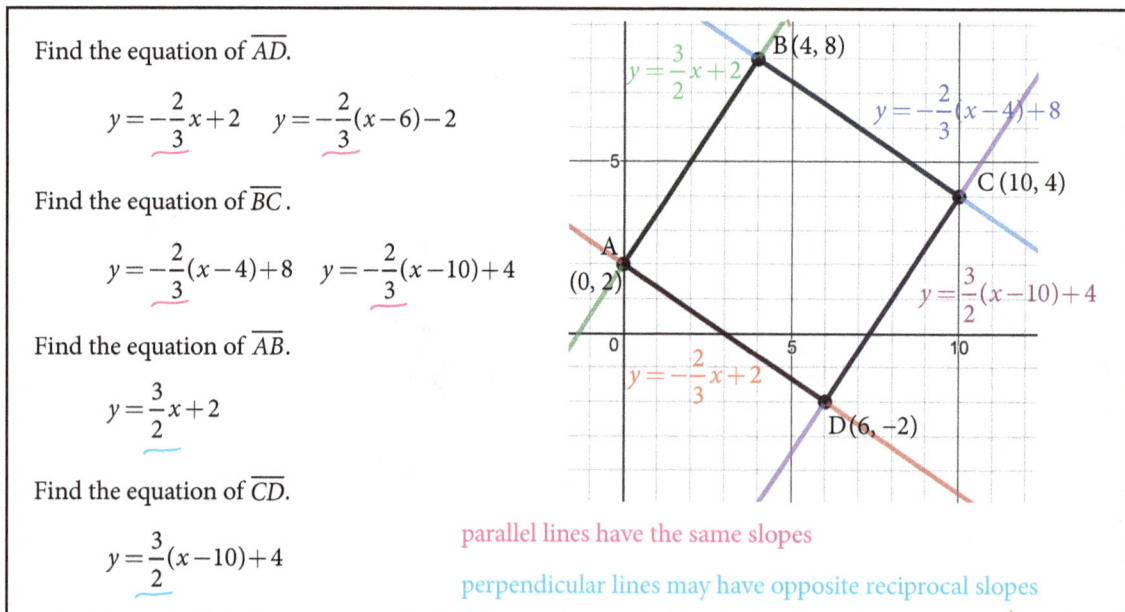

$B(4, 8)$

$y = \frac{3}{2}x + 2$

$y = -\frac{2}{3}(x-4)+8$

$C(10, 4)$

$A(0, 2)$

$y = \frac{3}{2}(x-10)+4$

$y = -\frac{2}{3}x + 2$

$D(6, -2)$

parallel lines have the same slopes

perpendicular lines may have opposite reciprocal slopes

Algebra Problem Strings
©2017 Kendall Hunt Publishing

Facilitation Notes

This version of the problem string lists short notes for important teacher moves during the string. After you've done the string yourself and studied the relationships involved, you might make similar notes for the things you want a reminder of or deem important.

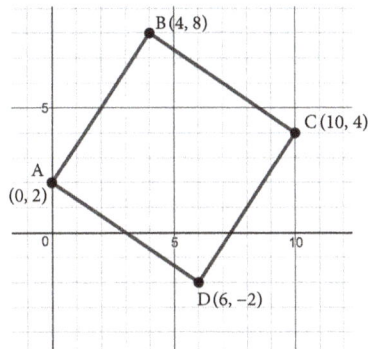

Find the equation of $\overline{AD}$.	Let's review. This is a square. Elicit point-slope and slope-intercept. Label the line.
Find the equation of $\overline{BC}$.	Elicit point-slope and using the fact that the sides are parallel. Label the line.
Find the equation of $\overline{AB}$.	Elicit slope-intercept using the slope written as a fraction, 3/2. Label the line. Any noticings or wonderings?
Find the equation of $\overline{CD}$.	Predict. What do we know? Elicit point-slope. Label the line. What are you thinking about the slopes of parallel lines? Perpendicular lines?

5.3 | Solving for One Variable

At a Glance	Objectives

Objectives
The goal of this problem string is for students to use relationships to solve for one variable given a two variable linear equation. These equations are such that one might use substitution when solving a system of equations.

Solve for x:

$x + 3y = 8$

$x - 3y = 2$

Placement
You can use this short problem string to help students reason using linear relationships to solve a two variable linear equation for one of the variables.

Use this problem string before you begin textbook Lesson 5.3 Solving Systems of Equations Using Substitution.

Solve for y:

$2x + y = 3$

$3x - y = 9$

Guiding the Problem String
The crux of this problem string is to ask students where a variable is given the beginning information modeled as locations on an open double number line. Facilitate each problem by writing the equation and modeling it as locations on an open double number line. Then ask where the specified variable is. As students describe the relationships, "If $x + 3y$ is there, then x has to be to the left," mark a distance of $3y$ to the left.

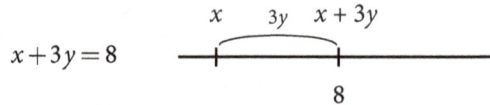

$$x + 3y = 8$$

Then the bottom of the open double number line follows, "If x is $3y$ to the left of $x + 3y$, then what is $3y$ to the left of 8?"

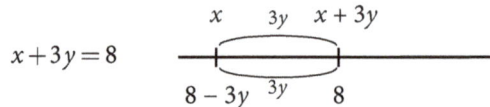

$$x + 3y = 8$$

The first two problems ask students to solve for x and the last two for y. On the last problem, model the strategy of finding the opposite of y first and also the strategy of getting y positive first. Throughout the string, model the relationships on the open double number line first, then say the same relationships as you model those relationships using horizontal and vertical equations.

About the Mathematics
Using the open double number line shifts the question from "what is x?" to "where is x?" See problem string lessons 2.7, 2.8, and 3.6 for more information about the open double number line as a tool for solving equations.

Sample Interactions

Use the following as you plan how to elicit and model student strategies. This is not meant as a script, but as a view into the relationships involved and the intent of the problem string.

Teacher: *Let's start today with a quick problem string. The first problem is this equation: $x + 3y = 8$. I'm going to model it on a double open number line. And my question is, if this equation is true, if $x + 3y$ is in the same location as 8, where is x?*	$x + 3y = 8$ number line labeled $x + 3y$ above and 8 below
Teacher: *Where is x? It's at $8 - 3y$? Why?* *So, I hear you saying that if $x + 3y$ is here, then x has to be back $3y$. Because $x + 3y$ is to the right, x has to be to the left. And if we subtract $3y$ on the top of the number line, then we need to subtract $3y$ on the bottom as well. What is 8 subtract $3y$? It's just $8 - 3y$.*	$x + 3y = 8$ $x = 8 - 3y$ number line labeled x, $3y$, $x + 3y$ above and $8 - 3y$, $3y$, 8 below
Teacher: *We can model that thinking the way we just did on the open double number line. We can also model that thinking using equations horizontally.*	$x + 3y - 3y = 8 - 3y$ $x = 8 - 3y$
Teacher: *We can also model that thinking using equations vertically.*	$\begin{aligned} x + 3y &= 8 \\ -3y & \quad -3y \\ \hline x &= 8 - 3y \end{aligned}$
Teacher: *So, what if I told you that in our next problem, $x - 3y$ is in the same location as 2? Again my question is where is x?*	$x - 3y = 2$ number line labeled $x - 3y$ above and 2 below

The teacher elicits the solution and student thinking and models the reasoning on the open double number line and horizontal and vertical equations.

$x - 3y = 2$ number line labeled $x - 3y$, $3y$, x above and 2, $3y$, $2 + 3y$ below
 $x = 2 + 3y$

$x - 3y + 3y = 2 + 3y$
 $x = 2 + 3y$

$\begin{aligned} x - 3y &= 2 \\ +3y & \quad +3y \\ \hline x &= 2 + 3y \end{aligned}$

Teacher: *The next problem is a little different. We start with $2x + y$ being in the same location as 3, but this time we want to know where y is.*	$2x + y = 3$ number line labeled $2x + y$ above and 3 below

The teacher elicits the solution and student thinking and models the reasoning on the open double number line and horizontal and vertical equations.

$2x + y = 3$ number line labeled y, $2x$, $2x + y$ above and $3 - 2x$, $2x$, 3 below
 $y = 3 - 2x$

$2x + y - 2x = 3 - 2x$
 $y = 3 - 2x$

$\begin{aligned} 2x + y &= 3 \\ -2x & \quad -2x \\ \hline y &= 3 - 2x \end{aligned}$

(continued)

Teacher: *The last problem today is to find where y is if we know that 3x − y is in the same location as 9. Where is y?*	$3x - y = 9$
Teacher: *So, several of you have found where the opposite of y is. We write that −y. Let's model that on the board.* **Student:** *I thought about moving left because y plus 3x would be there to the right, so y has to be back to the left.* **Teacher:** *Thoughts on that?* **Student:** *I think it has to be negative y. You can think about 3x − y as 3x plus negative y.* **Student:** *Oh right. Okay, so negative y is back 3x.*	$3x - y = 9$
Teacher: *Then what is on the bottom?* **Student:** *9 minus 3x.* **Teacher:** *So what we have now is that the opposite of y is in the same location as 9 − 3x. So where is y?*	$3x - y = 9$
Student: *If the opposite of y is there, then y is going to be on the other side of zero.* **Teacher:** *Where is zero?* **Student:** *I don't think we know. Wait, we do know that it's to the left of 9, but I don't think we know how far.* **Teacher:** *I'm going to scoot everything to the right a bit so we can put zero somewhere to the left.*	
Teacher: *Where should I put zero?* **Student:** *To the left.* **Student:** *And so now y has to be on the other side of zero.* **Teacher:** *Turn and talk to your partner. Why is the variable y there?*	
Teacher: *If the opposite of negative y is on the other side of zero on the top of the number line, what is at the same location? How do you know?* *What is the opposite of 9 − 3x?*	

Algebra Problem Strings
©2017 Kendall Hunt Publishing

Teacher: *So where is y?*

And that same thinking can be shown using equations.

$$3x - y = 9$$

$$y = -9 + 3x$$

$$3x - y - 3x = 9 - 3x$$
$$-y = 9 - 3x$$
$$y = -9 + 3x$$

$$3x - y = 9$$
$$\underline{\quad -3x \qquad\quad -3x}$$
$$-y = 9 - 3x$$
$$y = -9 + 3x$$

Teacher: *Not all of you found where the opposite of y is first. Some of you found where 3x is first. Tell us about that.*

$$3x - y + y = 9 + y$$
$$3x = 9 + y$$

$$3x - y = 9$$
$$\underline{\quad +y \qquad\quad +y}$$
$$3x = 9 + y$$

Teacher: *So, once you knew where 3x was, how did that help you find y?*

You know that y is the left of 9 + y? I'm going to erase the first part, showing 3x − y is in the same place as 9.

So, if y is to the left of 9 + y by 9, what is on the top of the number line? In other words, what is 9 to the left of 3x? Yes, 3x − 9.

So 3x − 9 = y or using the commutative property, y = 3x −9.

$$3x - y + y = 9 + y$$
$$3x = 9 + y$$

$$3x - 9 = 9 + y - 9$$
$$3x - 9 = y$$
$$y = 3x - 9$$

$$3x - y = 9$$
$$\underline{\quad +y \qquad\quad +y}$$
$$3x = 9 + y$$
$$\underline{\quad -9 \qquad\quad -9}$$
$$3x - 9 = y$$

Teacher: *Let's compare these two strategies. Turn to your partner, what do you think? How would you describe each? Which are you more inclined to do?*

Good work. Nice thinking using relationships.

Teacher: *How would you summarize some of the things that came up in this string today?*

Elicit the following:

- *We can think about equations as each side of the equation being in the same location.*

- *If x is on one side of zero, the opposite of x is on the other side of zero.*

- *If we have a linear equation in two variables, we can solve for one of the variables using relationships.*

(continued)

Sample Final Display

Your display could look like this at the end of the problem string:

$$x + 3y = 8$$
$$x = 8 - 3y$$

$$x + 3y - 3y = 8 - 3y$$
$$x = 8 - 3y$$

$$x + 3y = 8$$
$$\underline{-3y-3y}$$
$$x = 8 - 3y$$

$$x - 3y = 2$$
$$x = 2 + 3y$$

$$x - 3y + 3y = 2 + 3y$$
$$x = 2 + 3y$$

$$x - 3y = 2$$
$$\underline{+3y+3y}$$
$$x = 2 + 3y$$

$$2x + y = 3$$
$$y = 3 - 2x$$

$$2x + y - 2x = 3 - 2x$$
$$y = 3 - 2x$$

$$2x + y = 3$$
$$\underline{-2x-2x}$$
$$y = 3 - 2x$$

$$3x - y = 9$$
$$y = -9 + 3x$$

$$3x - y - 3x = 9 - 3x$$
$$-y = 9 - 3x$$
$$y = -9 + 3x$$

$$3x - y = 9$$
$$\underline{-3x-3x}$$
$$-y = 9 - 3x$$
$$y = -9 + 3x$$

$$3x - y + y = 9 + y$$
$$3x = 9 + y$$

$$3x - 9 = 9 + y - 9$$
$$3x - 9 = y$$
$$y = 3x - 9$$

$$3x - y = 9$$
$$\underline{+y+y}$$
$$3x = 9 + y$$
$$\underline{-9-9}$$
$$3x - 9 = y$$

Algebra Problem Strings
©2017 Kendall Hunt Publishing

Facilitation Notes

This version of the problem string lists short notes for important teacher moves during the string. After you've done the string yourself and studied the relationships involved, you might make similar notes for the things you want a reminder of or deem important.

$x + 3y = 8$	Write the equation and sketch a number line with x + 3y in the same location as 8. Where is x? How do you know? We can also model this thinking using equations. Model on horizontal and vertical equations using the same relational language.
$x - 3y = 2$	Repeat.
$2x + y = 3$	This time we want to know where y is. Repeat modeling on an open double number line and with equations.
$3x - y = 9$	Repeat. Where is y? Elicit finding -y first. Elicit getting y positive first.

5.4 | Choose a Strategy

At a Glance

Choose a strategy to solve each system.

$$8x + 2y = 60$$
$$-8x - 5y = 72$$

$$-4x + 3y = -20$$
$$y = 5x + 8$$

$$2x + 4y = 72$$
$$3x + 5y = 80$$

$$y = 2x - 12$$
$$y = 3x - 12$$

Objectives

The purpose of this string is to deepen students' intuition for choosing an efficient strategy when solving a system of linear equations.

Placement

After students have had some experience solving systems using substitution and elimination, they can explore which strategy makes more sense for a given system and why.

Use this problem string after students have solved systems of linear equations using substitution and elimination in textbook lessons 5.3 and 5.4.

Guiding the Problem String

This problem string takes the form of a strategy share, a structure that differs considerably from most of the string structures students have experienced. Consider it a conversation about choosing strategy and not a chance to develop relationships from one problem to another. You can choose whether to simply have students defend their choice of strategy to the class, or encourage them to also solve the system using those chosen strategies.

The first system suggests elimination because of the $8x$ and $-8x$ terms. The second system suggests solving using substitution because one equation has a variable already isolated with a coefficient of one. The third system might be treated as a combination and solved by making new combinations until one variable disappears (has a zero coefficient). Finally, the last system could prompt students to envision the graph before trying any other strategy. Students could describe two intersecting lines that cross at $(0, -12)$ and have different slopes. With that idea alone they have solved the system.

Be focused throughout on the idea that the form and features of a system give us clues about which strategy will be smoothest. This includes the kinds of values that are present and their "friendliness" to one another, the form of each of the linear equations, and whether it's possible to envision the graphs.

This discussion and the reasoning that you are making public for the entire class is designed to help students reinforce the habit of pausing to analyze a system before choosing a strategy. In this way the conversation is designed to help them understand what exactly to look for and how to interpret what they notice as they decide on a strategy. Be sure to ask students why they are not choosing certain strategies as you want to make that thinking public too. Be careful to avoid making hard and fast rules about when each strategy should be used. Students should be encouraged to use the relationships they own and continue to construct new relationships. The point of this conversation is to make useful relationships more visible.

About the Mathematics

Students may lock in a preferred strategy for solving systems instead of thinking flexibly from system to system. Students can get consumed by the mechanics of steps of a procedure, losing sight of what it means to solve a system and the meaning of their solution. You can begin to bring all of these ideas up during this problem string, being mindful of the fact that much of their work with systems may be new, and therefore, still in development.

Algebra Problem Strings
©2017 Kendall Hunt Publishing

Sample Interactions

Use the following as you plan how to elicit and model student strategies. This is not meant as a script, but as a view into the relationships involved and the intent of the problem string.

Teacher: *Before we start, let's think about what it means to solve a system:* *When we are thinking about systems, what will the "answer" look like and how do you know?* *What does it tell us?* *How could this be helpful?* *Can you give a real life example of how a solution to a system tells us something important?* *When we get a solution, what exactly is that? What does it mean?* *Who can remind us of the strategies we have so far?*	
Teacher: *Take a look at this system of equations and ask yourself what might be an efficient strategy to solve this system? You don't actually have to solve it, but get ready to convince all of us of your strategy choice.*	$8x + 2y = 60$ $-8x - 5y = 72$

Some responses might include:

- *You can solve anything by graphing. If your grapher makes you solve for y, then graphing may not be the most efficient.*
- *Solving by substitution would require solving for one of the variables because none of them are already isolated. If we had to solve for a variable, I would choose solving for y in the first equation because at least there are common factors.*
- *Choosing elimination makes sense because the 8x and −8x terms will easily add to 0 and then you can just solve for y.*

Teacher: *How about this one?* *What are you noticing here that helps you choose a strategy?* *For those of you who are pretty convinced there is a very clear strategy here, what are you seeing that makes you say this?* *Let's get some other ideas out for us to consider.*	$-4x + 3y = -20$ $y = 5x + 8$

Some responses could include:

- *Substitution jumps out as a wise choice because the second equation is already solved for y. We can now just substitute 5x + 8 for y in the first equation.*
- *Elimination does not make as much sense because none of the variables will add to zero with first scaling one of the equations. If we had to use elimination, it would make sense to scale the y = 5x + 8 by −3. If you tried to eliminate the x terms, you would have to scale both equations.*
- *It would be easy to solve this by graphing if you had access to the internet, but if not, you would probably have to solve for y in the first equation if your graphing calculator only graphs in the form y =.*

Teacher: *Okay, what about this system of equations?* *Have you seen a system set up like this before?* *What's going to make this system easy for you to solve?* *Is anyone thinking about it a different way—you didn't envision a combination here?*	$2x + 4y = 72$ $3x + 5y = 80$

(continued)

Some responses could include:

- *We could graph it to find the intersection point, if there is one and if we have a device connected to the internet. But if we have to solve for y to graph, then graphing may not be a great choice.*

- *Substitution does not make sense because we would have to solve for one of the variables. If we had to solve for a variable, it might be easiest to solve for x in 2x + 4y = 72 by first dividing the whole equation by 2 and then x = 36 − 2y.*

- *But elimination seems like a good strategy. You could scale the first equation by 3 and the second by −2. That would keep the constants from getting as big as they would if you scaled the first by 5 and the second equation by −4. Another option is to divide the first equation by 2, and then scale it by −3.*

- *Another option is to recognize the pattern that 2x + 4y = 72 is down an x and a y from 3x + 5y = 80. We could continue that pattern until we get to no x's and some y's.*

$$3x + 5y = 80$$
$$2x + 4y = 72$$
$$1x + 3y = 64$$
$$0x + 2y = 56$$
$$y = 28$$

Teacher: *Last one to think about today.*	$y = 2x - 12$
Take a look at this system and before you do anything else, just notice how it's related to the other systems and how it's different. Turn and tell your partner what you notice.	$y = 3x - 12$
Given the ways that this system is structured, what strategies make sense here?	
Who feels really confident in their strategy choice and wants to try to convince us all?	
So let's now ask everyone—were you convinced by the strategy or do you think there might be an easier way to solve this one?	

Some responses could include:

- *We can picture this one and we don't even need to graph it. Because they share the same y-intercept but have different slopes, we can already picture the solution in our minds. It's just the y-intercept, (0, −12).*

- *We could use substitution or elimination, but thinking about the graph is so easy let's not.*

Teacher: *Today I put four really different systems in front of us. And I did that so that we would have to pause and really think about the best possible strategy—the one that would make our "solving lives" easier.*	• *substitution*
	• *elimination*
I want us to be able to make smart choices about a strategy so I'd like to put together a chart that we can leave in the room to remind us of this conversation. Will you have a conversation now with your partner in which you decide together how you know when to use each of these strategies?	• *graphing*
	• *envisioning the graph*
	• *combinations*
The teacher lists these and eventually makes a public anchor chart, codifying this shared knowledge.	

Teacher: *How would you summarize some of the things that came up in this string today?*

Elicit the following:

- *Look at the structure of the problem before you decide on a strategy.*

- *When you decide on a strategy to try, still think about making efficient choices.*

- *Look for relationships between the coefficients.*

Sample Final Display

Your display could look like this at the end of the problem string:

$$8x + 2y = 60$$
$$-8x - 5y = 72$$

will add to zero

$$-4x + 3y = -20$$ already solved for y
$$y = 5x + 8$$

$$3x + 5y = 80$$

$$2x + 4y = 72$$
$$3x + 5y = 80$$

$$2x + 4y = 72$$
$$1x + 3y = 64$$
$$0x + 2y = 56$$

can reason using combinations

$$y = 28$$

$$y = 2x - 12$$
$$y = 3x - 12$$

same y-intercept $(0, -12)$

(continued)

Facilitation Notes

This version of the problem string lists short notes for important teacher moves during the string. After you've done the string yourself and studied the relationships involved, you might make similar notes for the things you want a reminder of or deem important.

	When we are thinking about systems, what will the "answer" look like and how do you know?
	What does it tell us?
	How could this be helpful?
	Could you give an example of how a solution to a system would tell us something important to know?
	When we get a solution, what exactly is that? What does it mean?
	Who can remind us of the strategies we have so far?
$8x + 2y = 60$	What might be an efficient strategy to solve this system?
$-8x - 5y = 72$	You don't have to solve, but be ready to convince us of your strategy choice.
	Why or why not graphing? Substitution? Elimination? Thinking about combinations?
$-4x + 3y = -20$	What do you notice that helps you choose a strategy?
$y = 5x + 8$	Why or why not other strategies?
$2x + 4y = 72$	Have you seen a system set up like this before?
$3x + 5y = 80$	Why combinations?
	Why or why not other strategies?
$y = 2x - 12$	How is this system related to the other systems?
$y = 3x - 12$	What structure do you see that might be helpful?
	Who feels confident in their strategy choice and wants to try to convince us?
	Let's make an anchor chart of some of our ideas.

Sample Anchor Chart

$8x + 2y = 60$	**Elimination**
$-8x - 5y = 72$	• When you have "opposite terms" like 9x and –9x
	• When you can add or subtract the equations to get one of the variables to be zero
$-4x + 3y = -20$	**Substitution**
$y = 5x + 8$	• When one equation is in slope-intercept form and the other is not
	• When a variable is isolated with a coefficient of one
	• When substituting one variable for the other will be friendly
$2x + 4y = 72$	**Combinations**
$3x + 5y = 80$	• When you can tell the story of the combinations
	• When both of the variables are on one side and there's a value or quantity on the other side
	• If you can make new equations by combining and those new equations make it easy to isolate a variable.
$y = 2x - 12$	**Envisioning the graph**
$y = 3x - 12$	• Same slope—lines are parallel
	• Same y-intercept—it's the solution

Algebra Problem Strings
©2017 Kendall Hunt Publishing

5.5 | Inequalities in One Variable

At a Glance

$x = 3$

$x \geq 3$

$-x = 3$

$-x \geq 3$

$-x < -5$

$3x = -6$

$3x \leq -6$

$-3x \leq -6$

$-4x \geq 12$

Objectives

The purpose of this string is to help students learn to reason about solving inequalities in one variable by modeling relationships on an open double number line.

Placement

This problem string can be used to introduce or strengthen the notion that when solving inequalities in one variable, one must consider the effects of negatives on the variable and the inequality symbol.

Use this problem string after the investigation in textbook lesson 5.5 to support the learning from the investigation about the relationships involved in reversing the inequality symbol in certain one-variable inequalities.

Guiding the Problem String

This problem string is in the structure of helper-clunker problems. Model the initial inequality and then elicit student reasoning about the location of zero and opposites as appropriate. Be purposeful with your use of the words *opposite* and *negative* to help students draw on their intuition that the opposite of *x* must be on the other side of zero from *x*.

The first problem sets the stage for the next three problems. The fifth problem, $-x < -5$, is a clunker for you to see how students reason without a helper. The next three problems starting with $3x = -6$ are related. Ask students to sketch the relationships and to work with a partner to find equivalent expressions. The last problem is a second clunker for students to try their hand at without helpers.

About the Mathematics

A common rule for solving inequalities is that when multiplying or dividing by a negative number, you must reverse the inequality. This little understood rule can instead come naturally as students reason about opposites and their places on the number line. If *x* is on one side of zero, then the opposite of *x* must be on the other side of zero. This plays out with the sets of numbers represented by inequalities and helps students understand that the equivalent inequalities represent the same relationships. Some examples are shown:

$x = a$ if and only if $-x = -a$

$-x = b$ if and only if $x = -b$

$-x < -c$ if and only if $x > c$

$-x \geq c$ if and only if $x \leq -c$

(continued)

Sample Interactions

Use the following as you plan how to elicit and model student strategies. This is not meant as a script, but as a view into the relationships involved and the intent of the problem string.

Teacher: *We'll start today's problem string with a quick $x = 3$. I'll just sketch that here. If x is here at 3, where is 0?*	$x = 3$
Teacher: *What if we know that there are numbers, x, greater than or equal to 3? What does that look like?* *Where are all the numbers that are greater than 3?* *We can show that on a number line by putting a filled in circle on 3 because of the equality and we color in the part of the number line that represents all of those numbers greater than 3.* *Is x one value or a set of values? Why?*	$x \geq 3$
Teacher: *The next problem for us to consider is that the opposite of x equals 3. I'll put that on a number line so that the opposite of x and 3 are in the same location. Where is zero?* *Where is x?*	$-x = 3$ $x = -3$
Teacher: *What if the opposite of x is greater than or equal to 3?* *Where is zero?* *Where is x?* *What are the opposites of all of the numbers we represented by the arrow to the right? Does that make sense?* *Let's try a number that obviously works for $x \leq -3$, like -100. Is the opposite of negative 100 (positive 100) greater than 3? Yes?*	$-x \geq 3$ $x \leq -3$ $-100 \leq -3$ $-(-100) \geq 3$ $100 \geq 3$
Teacher: *The next problem is the opposite of x is less than negative 5. What would that look like on a number line?* *Open or closed circle? Why?* *Where is zero? Where is x?* *How can thinking about opposites help?*	$-x < -5$ $x > 5$
Teacher: *What about if $3x = -6$, where is zero?* *Where is x?* *How do you know?*	$3x = -6$ $x = -2$

Algebra Problem Strings
©2017 Kendall Hunt Publishing

Teacher: *What if $3x \leq -6$? Sketch that on your paper. Talk with a partner about how to find x.* *Did you use an open or closed circle? Why?* *How did you find x?* *Does x represent one value or a set of values? How can we represent that set using inequalities? A number line?*	$3x \leq -6$ $x \leq -2$ (number line: $3x$ at -6, x at -2, 0)
Teacher: *The next problem is $-3x \leq -6$. What does that look like on a number line? Work with a partner to find the set represented by x.* *Where is 3x? Did finding zero help?*	$-3x \leq -6$ (number line: $-3x$ at -6) $3x \geq 6$ (number line: $-3x$ at -6, 0, $3x$ at 6) $x \geq 2$ (number line: $-3x$ at -6, 0, x at 2, $3x$ at 6)
Teacher: *The last problem in our problem string is $-4x \leq 12$. Please represent that on a number line and find an inequality that represents x. After you have worked, compare your thinking with your partner and get ready to convince us of your reasoning.*	$-4x \leq 12$ (number line: $-4x$ at 12) $4x \geq -12$ (number line: $4x$ at -12, 0, $-4x$ at 12) $x \geq -3$ (number line: $4x$ at -12, x at -3, 0, $-4x$ at 12)
Teacher: *Look back at the board. Sometimes people refer to the change in the inequality symbol from less than in the problem to greater than in the solution as "reversing the inequality." Why might they do that? When does that happen? Why does that happen? What does that have to do with opposites?*	

Teacher: *How would you summarize some of the things that came up in this string today?*

Elicit the following:

- *It can be helpful to start with what you know by sketching relationships on a open double number line.*

- *If you know a location for x, the location of the opposite of x, −x, is on the other side of zero.*

- *If $x = a$, then the solution is the single value a. If $x \leq a$, $x \geq a$, $x > a$, or $x < a$, then the solution is a set of values.*

(continued)

Sample Final Display

Your display could look like this at the end of the problem string:

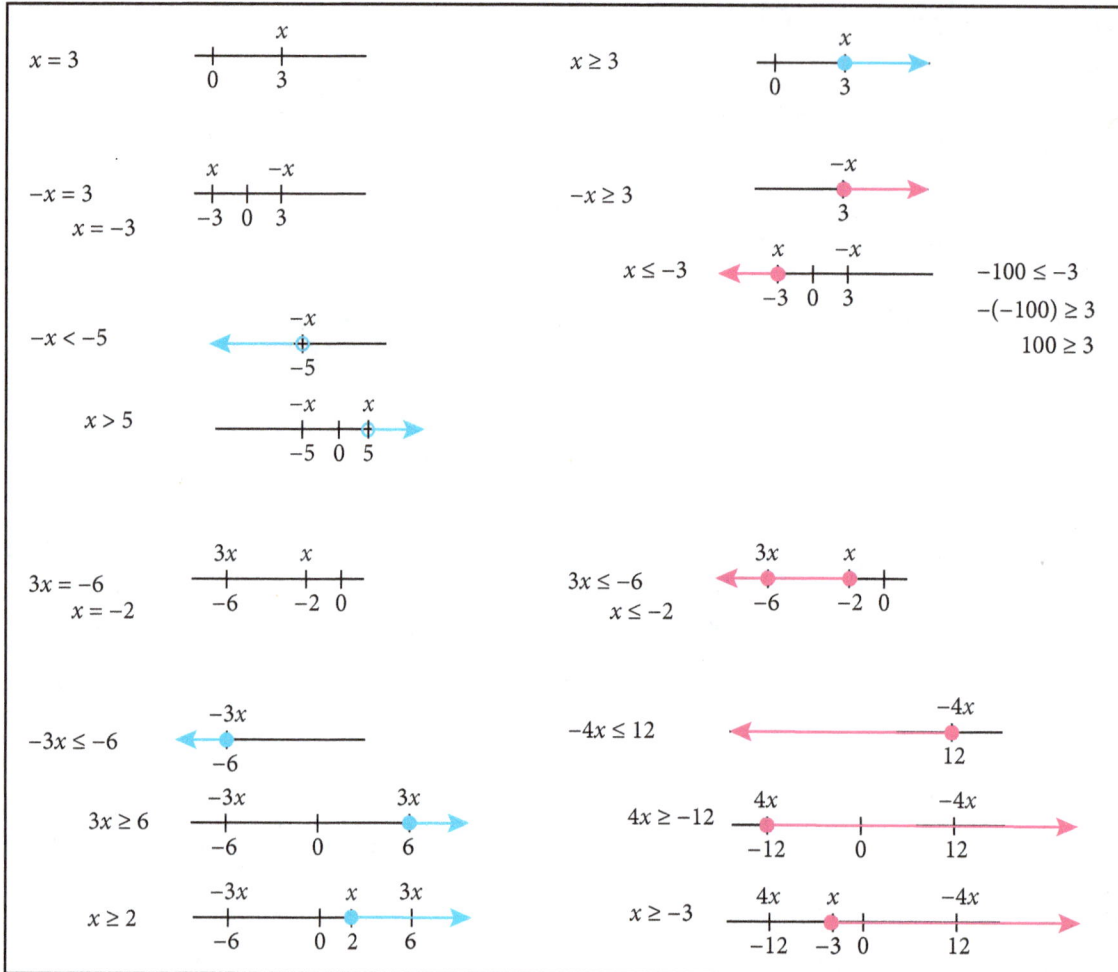

$x = 3$

$-x = 3$
$x = -3$

$-x < -5$

$x > 5$

$3x = -6$
$x = -2$

$-3x \le -6$

$3x \ge 6$

$x \ge 2$

$x \ge 3$

$-x \ge 3$

$x \le -3$ $-100 \le -3$
$-(-100) \ge 3$
$100 \ge 3$

$3x \le -6$
$x \le -2$

$-4x \le 12$

$4x \ge -12$

$x \ge -3$

Algebra Problem Strings
©2017 Kendall Hunt Publishing

Facilitation Notes

This version of the problem string lists short notes for important teacher moves during the string. After you've done the string yourself and studied the relationships involved, you might make similar notes for the things you want a reminder of or deem important.

$x = 3$	Quick. Sketch number line. Set the stage.
$x \geq 3$	Quick. Sketch number line. Briefly mention and use filled in circle.
$-x = 3$	Quick. Use the word "opposite" instead of "negative." If the opposite of x is at 3, where is x?
$-x \geq 3$	If the opposite of x is greater than or equal to 3, where is x?
$-x < -5$	If the opposite of x is less than negative 5, where is x? Briefly mention and use open circle.
$3x = -6$	Quick. Set the stage for the next 2 problems.
$3x \leq -6$	Sketch on your paper. Work with a partner. Did you use closed or open circle? How did you find x? Does x represent one value or a set of values?
$-3x \leq -6$	What does this one look like on a number line? Work with a partner to find the set represented by x. Where is 3x? Did finding zero help? Where is x?
$-4x \leq 12$	Clunker. What can students do without a helper? Look back. Sometimes referred to as "reversing the inequality." Why? When? What does that have to do with opposites?

5.6 | Graphing Standard Form

At a Glance

$$2x + y = 4$$

x	y
0	4
2	0

$$3x - 5y = -15$$

$$(0, \underline{}) \quad (\underline{}, 0)$$

$$4x + 3y = -12$$

x	y
0	
	0

$$5x - 6y = 30$$

$$2x + y = -2$$

Objectives

The purpose of this string is help students construct the strategy of using the x- and y-intercepts to graph the equation of a line when given in standard form.

Placement

This problem string can be used anytime after students have facility solving for y and graphing with the slope-intercept form. You may want to use this string before graphing linear inequalities in two variables.

Use this string before textbook Lesson 5.6 Graphing Inequalities in Two Variables.

Guiding the Problem String

This problem string uses the structure of equivalence to potentially cause students to wonder why the graphs are the same. As they analyze the problems, students may notice that the intercepts are readily available to use in sketching the graph of the line. The problems in this string come in related pairs.

The first problem is to graph a line given in standard form. The second problem is to find the line given two points. These two points just happen to be the intercepts for the first line and therefore the two lines are equivalent. Help students notice the equivalence by graphing on the display grapher. The third problem is to graph a new line given in standard form. The fourth problem is to find the intercepts of that line. The fifth and sixth problems are a similar pair except the intercepts are given in a table. Often students can more readily see the zeros trading places in one or the other format (ordered pair versus table) so we use both in the string. The last two problems are to sketch lines given in standard form and give you an opportunity to see if students are taking up the notion of finding and using intercepts. Each of the lines in the string have coefficients of x and y that are factors of the constant. The last problem is purposefully designed to be easy to use either the slope-intercept form or intercepts so that you can have a conversation about considering the numbers in the problem before choosing a strategy.

About the Mathematics

Using the intercepts is a friendly strategy to graph linear equations in two variables especially when the line is given in standard form and a and b are factors of c for $ax + by = c$.

Sample Interactions

Use the following as you plan how to elicit and model student strategies. This is not meant as a script, but as a view into the relationships involved and the intent of the problem string.

Teacher: *Today's problem string starts with some review. For the first problem, sketch a graph of this line, $2x + y = 4$.* Students work and the teacher circulates, looking for any students who found intercepts, solved for y, and noting other strategies. Because the teacher wants the strategy of using the intercepts to emerge later, the teacher does not choose to have that strategy shared yet.	$2x + y = 4$

Algebra Problem Strings
©2017 Kendall Hunt Publishing

Teacher: *I'm projecting a blank grid on the board. I'll use it to help us sketch the graph. Please tell us how you sketched the graph.* **Student:** *I got y on one side.* **Teacher:** *Why did you solve for y?* **Student:** *Then I could graph the y-intercept and the slope.* **Teacher:** *Tell us the equation you found.* **Student:** *I got $y = 4 - 2x$. So, I started at 4.* **Teacher:** *You have the y-intercept at 4?* **Student:** *Yes, and then over 1 and down 2 so the point (1, 2) and then I just drew the line.* **Teacher:** *Like this? Great. Does anyone disagree? No? So, sometimes we solve for y when we want to solve a system by substitution and sometimes because it can help us sketch a graph.*	
Teacher: *The next review type problem is to find the equation of the line that contains these two points in this table. When you've got the equation of this line, compare with your partner.* Students work and the teacher circulates, looking for students who notice that one of the points is the *y*-intercept.	
Teacher: *I noticed that you were talking about your partner's strategy. What were you noticing?* **Student:** *Well, I just saw two points and so I used the point-slope equation. But my partner used slope-intercept because you gave us the y-intercept. Wish I would have seen that.* **Teacher:** *That's pretty efficient. So, what equation did you both get?* **Student:** *The same one as the first problem, $y = 4 - 2x$.*	
Teacher: *Well, that's interesting. Same as the first problem? Hmmm. Thoughts on that?* **Student:** *They are the same line.* **Student:** *Yeah, look, if you go over 2 and down 1 again, you get to the other point, (2, 0).*	

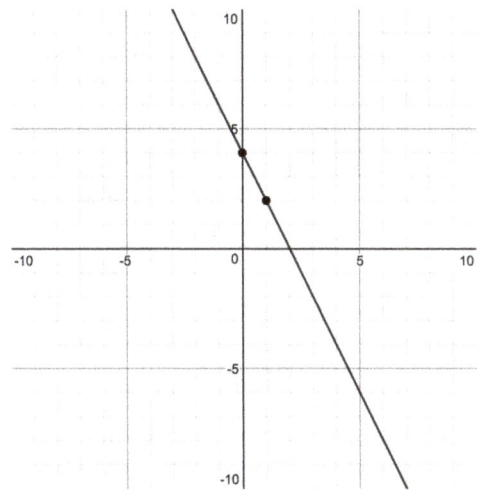

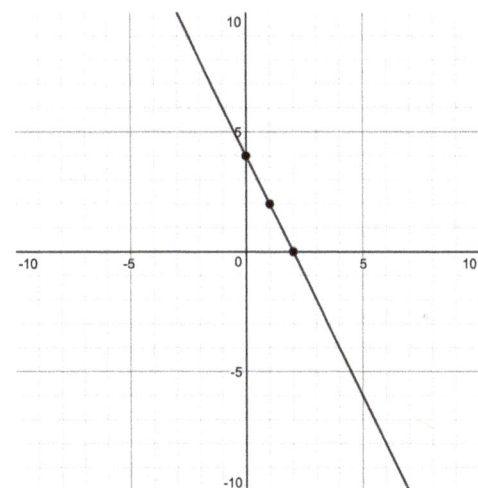

(continued)

Teacher: *And that's the same as the first problem, 2x + y = 4? They are equivalent ways of writing the same line?* **Student:** *Yes.* **Teacher:** *Just because we can, let's graph this using the technology display grapher. It used to be that we would have to solve for y in order to graph with technology, but with this grapher (Desmos), we can type in the equation in standard form. Here it is. Do the graphs look the same?* **Student:** *They are same. It is right on top of the sketch.* **Teacher:** *So we could label this line with both equations. What else do we know because of problem 2?* **Student:** *That the line has those points from problem 2.* **Teacher:** *Anything special about those points?* **Student:** *They are the intercepts.* **Teacher:** *That's a thing that mathematicians do. They notice relationships that may come in handy later.*	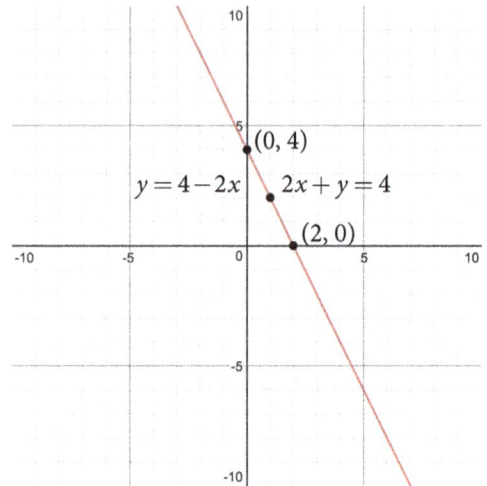
Teacher: *I'll quickly sketch a graph over here so that if we use the display grapher later, we'll have a record.* The teacher sketches a quick graph with the intercepts prominently shown so that students may refer to it later. The teacher also clears the display grapher to get ready for the next problem.	$2x + y = 4$ $y = 4 - 2x$ table: x / y : 0,4 (y-intercept); 2,0 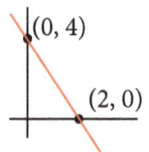
Teacher: *The next problem in our string is to sketch the graph of the line 3x − 5y = −15.* **Student:** *Couldn't we just type it in like we just did the problem before?* **Teacher:** *Yes, and we will. But your task is to sketch the graph first. You might think about what you know and see if you can find an efficient strategy.* While students work, the teacher circulates looking for students solving for y and noting other strategies.	$3x - 5y = -15$

Algebra Problem Strings
©2017 Kendall Hunt Publishing

Teacher: *Okay, I saw you two solve for y again. What equivalent equation did you find?*

Student: *We got* $y = 3 + \dfrac{3}{5}x.$

Teacher: *So, how did you graph that?*

Student: *The y-intercept is 3. Then from there, over 5 and up 3. That's the point (5, 6).*

The teacher sketches the points and uses the slope to sketch the line over the projection of the display grapher grid.

Teacher: *Like this?*

Student: *Yeah.*

Teacher: *Great review so far everyone.*

Teacher: *Let's change it up a bit. For the next problem, we'll refer to this previous equation,* $3x - 5y = -15$. *Here's what we have, two ordered pairs, two points, that are on that line. But there is missing information. Please find that information. Does everyone understand the question? Yes? Then would someone please restate the question?*

Student: *We have that line and we need to find the points on that line that have the zeros in the x and in the y.*

Students work briefly.

$(0, \underline{\quad})\ (\underline{\quad}, 0)$

Teacher: *How did you find the missing coordinates? Did anyone notice something familiar about the first point?*

Student: *Yes, it's just the y-intercept. We already knew that was (0, 3).*

Teacher: *So you didn't really do any work at all, just looked at what you already knew? Nice. But you didn't do that for the second point. What did you do?*

Student: *I knew that the x was 0 so I plugged that in.*

Teacher: *What did that look like?*

Student: *3x − 5(0) = −15. So 3x = −15 and x is −5. So, it's the point (−5, 0).*

$(0, \underline{3})\ (\underline{\quad}, 0)$

$3x - 5(0) = -15$

$3x = -15$

$x = -5$

$(0, \underline{3})\ (\underline{-5}, 0)$

(continued)

Teacher: *I'm going to type that standard form equation into the grapher. Here it is. Any thoughts?*

Student: *It graphed the same line as the one we sketched.*

Student: *And we can see that other point, (−5, 0), too.*

Teacher: *So, again, either equivalent equation can represent this line. And here are its intercepts.*

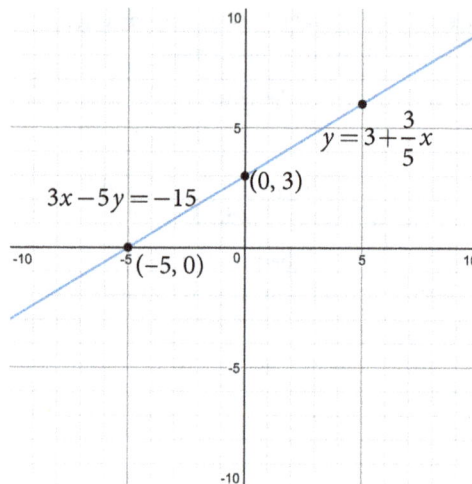

$3x - 5y = -15$

$y = 3 + \dfrac{3}{5}x$

$(0, 3)$

$(-5, 0)$

Teacher: *I have a question about this before we go on. You knew what the y-intercept was because we had already done the work to solve for y. What if we hadn't? Could we have found the y-intercept another way?*

Student: *Yeah, actually that's what we did. We didn't recognize that we already had the y-intercept, so we substituted x = 0 into the equation.*

As the student explains, the teacher records.

Teacher: *So either way we could find the y-intercept?*

$3(0) - 5y = -15$

$-5y = -15$

$y = 3$

Teacher: *Could someone please verbalize how to find the intercepts if we have the equation of the line?*

Student: *You plug in 0. If you want the y-intercept, you put in x is 0. If you want the x-intercept, you put in y is 0.*

Teacher: *So, you can evaluate the equation at x = 0 or y = 0? It seems like you just said those sort of switched, for x, you did y is 0? Who can chime in here?*

Student: *If you want the y-intercept, that's where x is zero and y is some number. If you want the x-intercept, that's where y is 0 and x is some number.*

Teacher: *Does everyone agree with that? Turn and talk to your partner about that.*

As partners talk, the teacher sketches the line and labels with both equations and intercepts so that students can refer to these later.

$3x - 5y = -15$ $y = 3 + \dfrac{3}{5}x$

$(0, \underline{\ 3\ })$ $(\underline{-5}, 0)$

$3x - 5(0) = -15$ $3(0) - 5y = -15$

$3x = -15$ $-5y = -15$

$x = -5$ $y = 3$

$(0, 3)$

$(-5, 0)$

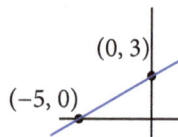

Teacher: *I wonder if any of that might influence how you solve this next problem, which is to sketch a graph of the line* $4x + 3y = -12$.

$4x + 3y = -12$

Algebra Problem Strings
©2017 Kendall Hunt Publishing

As the students work the teacher circulates, looking at and listening to student strategies. If many students seem to own the idea of looking for intercepts to sketch the graph, the teacher may choose to have those students share that intercepts strategy and then skip the next problem in the string. This problem, $4x + 3y = -12$, has a structure with coefficients for both x and y so that it may nudge students to not solve for y but instead find the intercepts, especially since 4 and 3 are nice factors of 12. This teacher decides to proceed by having students share solving for y and graphing the slope-intercept form and then finding intercepts in the next problem, giving more students the opportunity to consider intercepts before that strategy is highlighted.

Teacher: *Tell us about your strategy.*

Student: *We just solved for y and graphed the intercept and used the slope of negative four-thirds. So* $y = -\dfrac{4}{3}x - 4$, *the y-intercept is –4. Go over three and down 4 to (3, –8).*

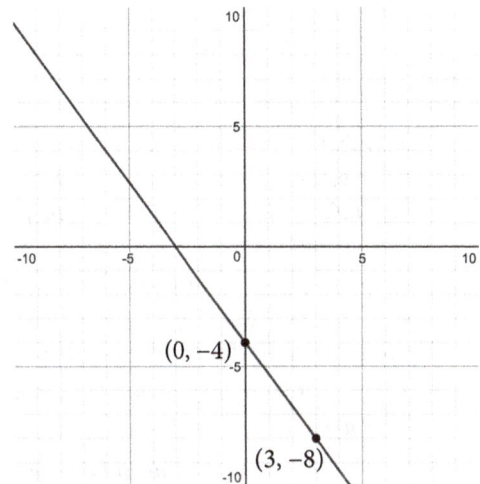

Teacher: *Great. And the next problem is to use that same equation,* $4x + 3y = -12$ *to find the missing coordinates of these points.*

Students work briefly.

$$4x + 3y = -12$$

x	y
0	
	0

Teacher: *What are the missing coordinates?*

Student: *When x is 0, y is –4. When y is 0, x is –3.*

Teacher: *How are you finding the other coordinate when one is 0?*

Student: *I'm noticing that one of the things goes away.*

Teacher: *One of the terms, like 4x or 3y?*

Student: *Yeah, because it's 0 times something so that's 0. So you end up with the other term equals –12. Then you just divide –12 by 3 or by 4, depending on which you're doing.*

Teacher: *Who understands what she's saying?*

Student: *When the line is in this form and you're looking for the intercepts, you can just substitute 0 which makes one of the terms 0. So really you end up just dividing the –12 by the coefficient of the variable you want.*

Student: *Yeah, you just have to keep track of which intercept.*

$$4x + 3y = -12$$

x	y
0	–4
–3	0

$$4x = -12 \qquad 3y = -12$$
$$x = -3 \qquad\; y = -4$$

Teacher: *Why do I keep asking you to find the intercepts? Could we use them in some way to graph the line?*

Student: *Do you mean that we could find the intercepts, graph them, and just connect them?*

Teacher: *What do you think? Is that a valid graphing strategy? Turn to your partner and discuss your thoughts on that.*

(continued)

Teacher: *The next problem in our string today is to graph the line* $5x - 6y = 30$. Students work and the teacher circulates, noting students who are solving for *y*, trying to graph with the intercepts, and any other strategies.	$5x - 6y = 30$

Teacher: *Let's hear from someone who solved for y first please.* **Student:** *I did. I got* $y = \frac{5}{6}x - 5$. *Start at* $(0, -5)$ *and go over 6 and up 5. That's* $(6, 0)$. *Connect them.* **Teacher:** *Now, will you please tell us what you were doing? I think I saw you finding intercepts.* **Student:** *Yes, I did not solve for y. I just found the intercepts like we were talking about. Funny thing is, they are the same points that you just got when you solved for y and used the slope.* **Teacher:** *Did anyone else find the intercepts by substituting zero in for x and for y? Some of you? What do you think about that strategy for this problem, for these numbers?* **Student:** *Well, it was pretty nice because both 5 and 6 are factors of 30 so it was easy to divide and get a nice number.* **Teacher:** *That's interesting. That might be something to look for.*	

Teacher: *The last problem today is to graph* $2x + y = -2$. *Before you start, think about what strategy might be efficient.* After students work briefly, the teacher has a student share the intercepts strategy. **Teacher:** *What do you think of using this intercepts strategy for this problem, for these numbers?* **Student:** *At first I wasn't sure, because there's only one y.* **Teacher:** *The coefficient of y is 1. That was confusing?* **Student:** *A little but then I realized it made it even easier. When x is 0, y is just −2. We didn't even have to divide for that one!* **Student:** *True, but it also made it pretty nice to just graph with the slope-intercept. That equation just pops out,* $y = -2x - 2$. **Teacher:** *So, it's important to keep thinking and considering what relationships to use. Nice!*	

Teacher: *How would you summarize some of the things that came up in this string today?*

Elicit the following:

- *To graph a line given in standard form, we can change it to the equivalent slope-intercept by solving for y.*

- *To find the x-intercept, substitute y = 0. To find the y-intercept, substitute x = 0.*

- *To graph a line given in standard form, we can find the intercepts and connect them.*

Algebra Problem Strings
 ©2017 Kendall Hunt Publishing

Sample Final Display

Your display could look like this at the end of the problem string:

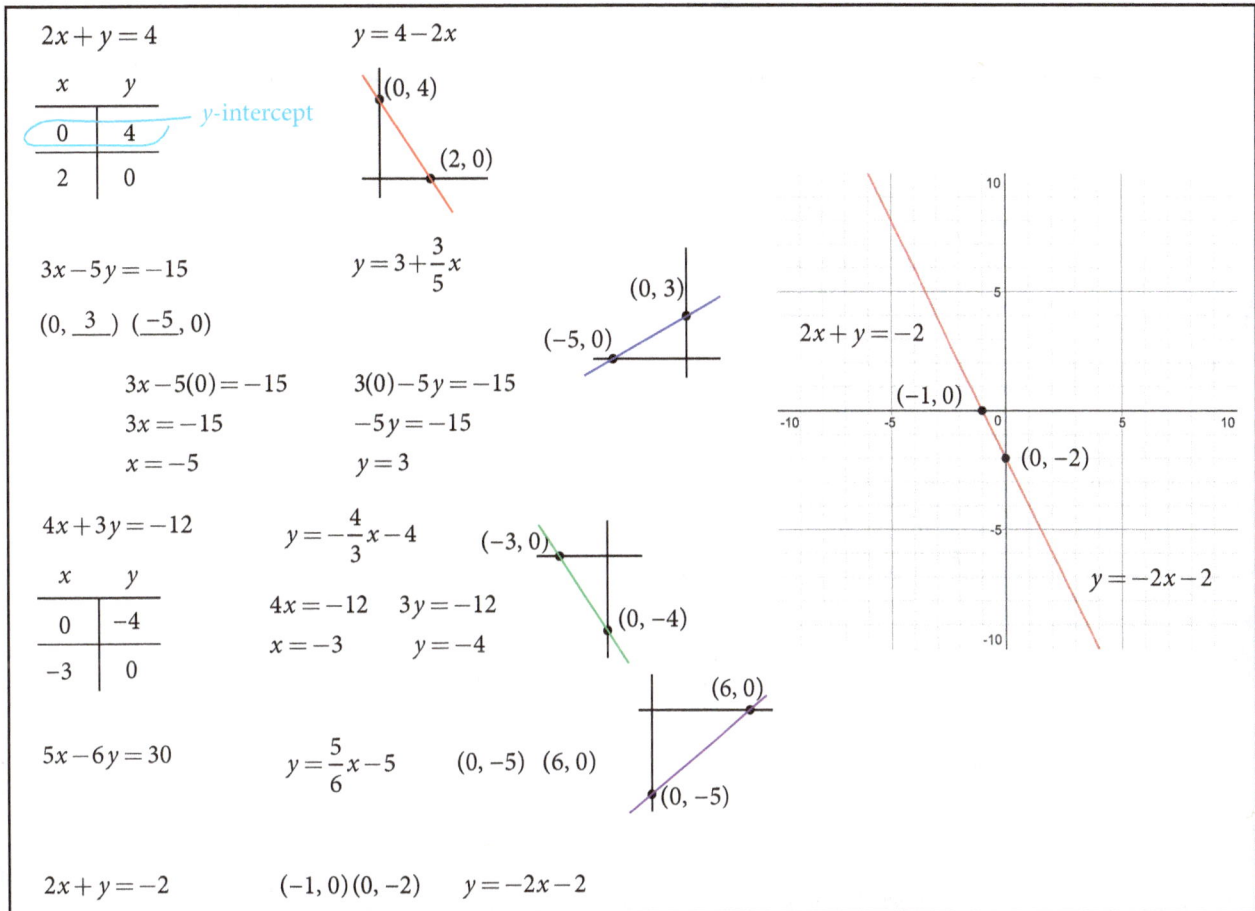

$2x + y = 4$ $y = 4 - 2x$

x	y
0	4
2	0

y-intercept $(0, 4)$ $(2, 0)$

$3x - 5y = -15$ $y = 3 + \dfrac{3}{5}x$

$(0, \underline{3})$ $(\underline{-5}, 0)$

$3x - 5(0) = -15$ $3(0) - 5y = -15$
$3x = -15$ $-5y = -15$
$x = -5$ $y = 3$

$(0, 3)$ $(-5, 0)$

$2x + y = -2$ $(-1, 0)$ $(0, -2)$ $y = -2x - 2$

$4x + 3y = -12$ $y = -\dfrac{4}{3}x - 4$ $(-3, 0)$

x	y
0	-4
-3	0

$4x = -12$ $3y = -12$
$x = -3$ $y = -4$ $(0, -4)$

$5x - 6y = 30$ $y = \dfrac{5}{6}x - 5$ $(0, -5)$ $(6, 0)$ $(6, 0)$ $(0, -5)$

$2x + y = -2$ $(-1, 0)(0, -2)$ $y = -2x - 2$

(continued)

Facilitation Notes

This version of the problem string lists short notes for important teacher moves during the string. After you've done the string yourself and studied the relationships involved, you might make similar notes for the things you want a reminder of or deem important.

$2x + y = 4$ $\begin{array}{c\|c} x & y \\ \hline 0 & 4 \\ \hline 2 & 0 \end{array}$	Let's review a bit. Sketch a graph. Elicit solving for y= and sketching using y-intercept and slope. Find the equation of the line between these two points. Compare with partner. Elicit slope-intercept because given y-intercept. Elicit that it is the same line as first problem, equivalent forms. Anything special about these points? Intercepts. Enter standard form into display grapher.
$3x - 5y = -15$	Sketch this line. Clear the display grapher and sketch small graph showing intercepts while students working. Elicit slope-intercept form.
$(0, \underline{\quad}) \ (\underline{\quad}, 0)$	Using problem 3, find the missing coordinates for these points. How did you find them? Enter standard form into display grapher. Any noticings? Elicit equivalent forms and intercepts. Ask students to verbalize finding the intercepts given the standard form of the line. Sketch small graph showing intercepts.
$4x + 3y = -12$	I wonder if that might influence your sketching a graph of this one. If students are finding intercepts, model that strategy and skip next question. If not, elicit slope-intercept.
$\begin{array}{c\|c} x & y \\ \hline 0 & \\ \hline & 0 \end{array}$	Using the equation above, find the missing coordinates for these points. How are you finding these intercepts? Why do I keep asking you to find intercepts? Could they be helpful when graphing?
$5x - 6y = 30$	Sketch a graph of this line. Elicit slope-intercept and intercepts strategies. What about this problem makes finding intercepts friendly?
$2x + y = -2$	Sketch a graph of this line. Before you start, consider what you might try. Elicit intercepts strategy and slope-intercept. Why are both friendly for this problem?

Algebra Problem Strings
 ©2017 Kendall Hunt Publishing

5.7 | What's Your Solution?

At a Glance	Objectives
	The goal of this problem string it to support students' understanding of the different solution possibilities of different kinds of equations and inequalities and systems of equations and inequalities.

$x = -3$

$x \geq -3$

$x < -3$

$2x - 6 = 0$

$y = 2x - 6$

$y > 2x - 6$

$y \leq 2x - 6$

$y = 2x - 6$
$y = \frac{1}{2}x$

$y > 2x - 6$
$y \leq \frac{1}{2}x$

Placement

This string could come after students have worked on graphing systems of linear equations and as students begin working with linear inequalities.

You could use this string before or during your work with the textbook Lesson 5.7 Systems of Inequalities.

Guiding the Problem String

The first three problems should go quickly and set the routine of identifying solutions of equations compared to inequalities. Model these three problems horizontally in a row on separate number lines. For the second set of three problems you will repeat the first three but represented in two dimensions on a coordinate grid. Their graphs should be underneath their corresponding one dimensional model to juxtapose their meanings and foster conversation. In each case, ask students to identify values that make the statements true, tag these as "solutions." Suggest a non-solution and a solution for students to evaluate.

The next problem, $2x - 6 = 0$, where the solution, $x = 3$, is modeled on a number line, while $y = 2x - 6$, where the solution is a set of ordered pairs that form the line, is modeled on the coordinate grid. The next two problems, $y > 2x - 6$ and $y \leq 2x - 6$ each represent a set of ordered pairs that form a part of a plane. The next problem is a system of equations where the solution is an ordered pair where the lines intersect. The last problem is a system of inequalities where the solution a set of ordered pairs that form a part of a plane.

Use a display grapher to support student work. Sketch the graphs as a record so students can compare solutions.

About the Mathematics

The number line is a useful model to represent solutions of one variable equations and inequalities. Single solutions of equations are represented with a point, while inequalities have an infinite number of solutions and are represented by a ray, line, or line segment. The coordinate grid is a model for two variable linear equations and functions; these have an infinite number of solutions and are also represented by lines. Two variable linear inequalities also have infinite numbers of solutions and are represented as parts of a plane by shaded areas of the graph. The special cases of vertical and horizontal lines have only one variable in the equation and are represented by $x = c$ and $y = c$, respectively, but have solutions with both an x and a y component.

Solutions to equations and inequalities are values which make the statement true. Systems of linear equations can have infinite solutions, no solutions, and exactly one solution. Systems of inequalities have solutions that satisfy both (or all) of the individual inequalities.

(continued)

Sample Interactions

Use the following as you plan how to elicit and model student strategies. This is not meant as a script, but as a view into the relationships involved and the intent of the problem string.

Teacher: *The first few problems today are really quick. If we start with the equation $x = -3$, an algebraic representation, how can we represent that on a number line?* **Student:** *You just draw a number line and put a dot on -3.* **Teacher:** *Yep, not bad, right?*	$x = -3$
Teacher: *Next up is $x \geq -3$. What are some values of x that would make this a true statement?* **Student:** 0. **Student:** 3 **Student:** *x could be -3.* **Teacher:** *Okay, I'm going to represent what you said on a new number line.* Teacher puts points on the values students suggested. **Teacher:** *How does that look? Did we represent all the values of x that make this a true statement?* **Student:** *No, there are a lot more numbers that can work for x.* **Teacher:** *I'll shade the number line from the -3 to the right and put an arrow on it. Why would we shade part of the number line instead of a point like in the first problem?* **Student:** *A line means all the points, even the points in between the integers. So -2.5 works too.*	$x \geq -3$
Teacher: *Nice, the next problem is $x < -3$. What are some values that make that statement true? What will that look like?* **Students:** -5. **Student:** -9. **Student:** *All the numbers to the left of -3.* **Teacher:** *So are you saying x is equal to -3?* **Student:** *No, it's only everything just less than -3.* **Teacher:** *I'll draw on open circle meaning everything just to the left of -3.*	$x < -3$

Algebra Problem Strings
©2017 Kendall Hunt Publishing

Teacher: *Each of those questions we answered in one dimension, sets of numbers on the number line. Now let's look at each of those statements in two dimensions. How can we model them on a coordinate grid? Let's start with x = −3. If x is −3, what is y? What are some points that make it a true statement? Would a table help us think about it?*

Student: *So, x has to be −3. It doesn't tell us what y is, so I guess y could be anything.*

Student: *Yeah, like (−3, 0) or (−3, 2).*

Teacher: *Let me put those on the graph. Is that all the points that will work? Will the point (2, −3) work? Does that make a true statement?*

Student: *No, because x has to be −3.*

Student: *But all the y values will work, as long as x is −3.*

Teacher: *To represent all of the points where x = −3, I'll draw a vertical line there.*

In 2 dimensions:

$x = -3$

Teacher: *The next one is x ≥ −3. What are some values that would make this a true statement?*

Student: *Well, x has to be greater than or equal to −3, so the point (0, 0) will work.*

Several other students offer valid points while the teacher records them on a coordinate grid. If no one offers a point with an x value of −3, the teacher can ask students to consider one.

Teacher: *Will (−3, 4) satisfy the statement?*

Student: *Oh yeah, x can be −3 because of the greater than or equal to symbol.*

Teacher: *Once again, did we show all the possible solutions?*

Student: *No, there are tons of points that will work. We need a solid line on x = −3 and shade in the right side of the graph.*

In 2 dimensions:

$x \geq -3$

The teacher repeats this conversation with the next problem, $x < -3$, making note of the dashed line separating the solution set from the rest of the grid.

In 2 dimensions:

$x < -3$

(continued)

Teacher: *The next one is 2x − 6 = 0. What value or values make this a true statement?* **Student:** *I solved it and got x = 3.* **Teacher:** *So we got just one answer here when before we were getting lots of solutions. Interesting.*	$2x - 6 = 0$
Teacher: *The next problem is y = 2x − 6. What changes from the previous? What values work for x and y here?* **Student:** *There is an infinite number of answers. It's a line.* **Teacher:** *So are you saying that any point on this line will be a solution to this function?* **Student:** *Yeah, even the points in between the integers—on to infinity.* **Teacher:** *How many solutions are there? And how is this different from the previous problem?* **Student:** *There wasn't a y in the other one, so we just got one answer. But when there is an x and a y there are lots of combinations that work.* **Student:** *And when it was in two dimensions or an inequality we got lots of points that worked.* **Teacher:** *I'm going to graph it on the display grapher. Yep, that looks like the line you were talking about.*	$y = 2x - 6$
Teacher: *And now I'll make a sketch of it here, because we might want to use the display grapher to graph future problems and this way we'll have a record of it.*	$y = 2x - 6$

Teacher: *Next, tell us about the solution to $y > 2x - 6$. What does it look like?*

Student: *I think it's $y = 2x - 6$, but it's a dotted line and shaded above the line.*

Teacher: *Someone else, explain why it's a dotted line.*

Student: *Because it's not equal, it's just greater than.*

Teacher: *How did you know whether to shade above or below? Turn and talk about this.*

Students talk.

Teacher: *What did you decide?*

Student: *We tried some points and the ones that work are all above the line.*

Student: *Since it says that y is greater than the line, I graphed points above the line.*

Teacher: *Does that make sense? Let's look at it on the display grapher.*

$y > 2x - 6$

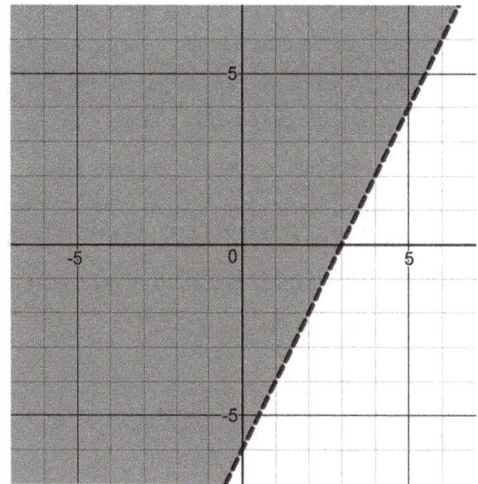

Teacher: *I'll make a sketch of it so we can refer back to it.*

$y > 2x - 6$

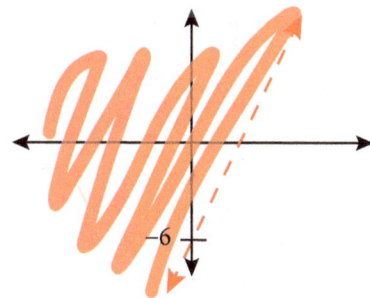

Teacher: *What if it changes just the tiniest bit, and there is a less than or equal to symbol now?*

Student: *Now the line is not dotted, it's solid. And you shade the other half of the plane, the part below the line.*

Teacher: *So now we shade the other part of the plane? Great work. I'll quickly sketch it.*

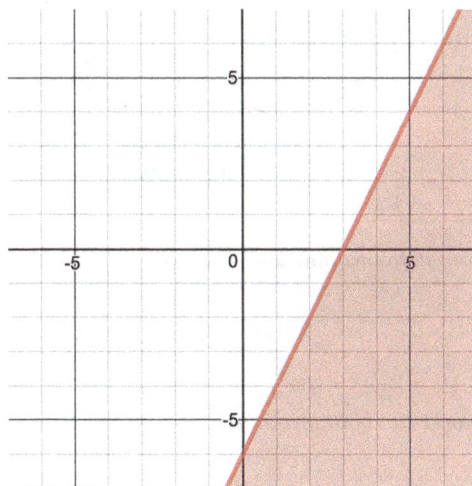

$y \leq 2x - 6$

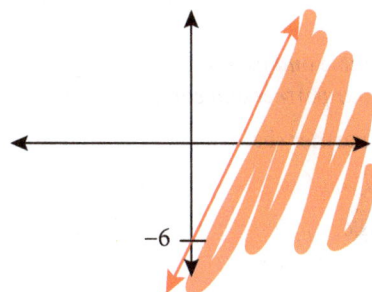

(continued)

Teacher: *This next problem is a system of equations. What is the solution to a system of equations? Turn and talk to your partner.*	$y = 2x - 6$ $y = \frac{1}{2}x$
Teacher: *Let's take a look at the display grapher. Is this what you were expecting? What is the solution?* **Student:** *Yes, it's the graph of two lines.* **Student:** *The solution is where they intersect.* **Teacher:** *So, the solutions to the two previous problems were all shaded, parts of planes. What is the solution here?* **Student:** *It's just a point. The point of intersection.* **Student:** *It looks like the point (4, 2).*	
Teacher: *Alright, the last problem is a system of two inequalities: $y > 2x - 6$ and $x \leq \frac{1}{2}x$. What points will work in both of those inequalities? What is the solution to a system of inequalities?* **Student:** *There should be an infinite number of solutions, just like in the other inequalities. We can graph the solutions.* **Teacher:** *Okay, talk about this one with your partner.* Partners talk while the teacher listens in.	$y > 2x - 6$ $y \leq \frac{1}{2}x$
Teacher: *I'm going to graph this on the display grapher. Tell us what you think the solution is and why.* **Student:** *The solution is where the shading overlaps.* **Teacher:** *Why?* **Student:** *The solution to the first inequality is everything above the dotted line. The solution to the second inequality is everything below the solid line, including the solid line. Then the solution is where they both are.* **Teacher:** *What are the solutions to the system? A point? A line? A bunch of points?* **Student:** *A bunch of points. It's really a section of a graph.*	

Teacher: *Create a table and list some points you think will work in both inequalities. When you have identified at least three, turn to your partner and compare your results.*

Students work and partner share.

Teacher: *What did you find?*

Student: *As long as our points were in the shaded region, they worked in both inequalities. That's kind of what the overlapping area means anyway.*

Teacher: *Nice insights. Great work today.*

Algebra Problem Strings
©2017 Kendall Hunt Publishing

Teacher: *How would you summarize some of the things that came up in this string today?*

Elicit the following:

- *Locations on a number line can represent values that make one variable equations true.*

- *Rays represent values that make one variable inequalities true.*

- *Lines represent values that make two variable or two dimensional equations true.*

- *Shaded areas (sections of planes) represent values that make two variable or two dimensional inequalities true.*

- *The intersection point of two lines is the solution to the system of those linear equations.*

- *Overlapping shaded areas (sections of planes) represent values that make systems of inequalities true.*

(continued)

Sample Final Display

Your display could look like this at the end of the problem string:

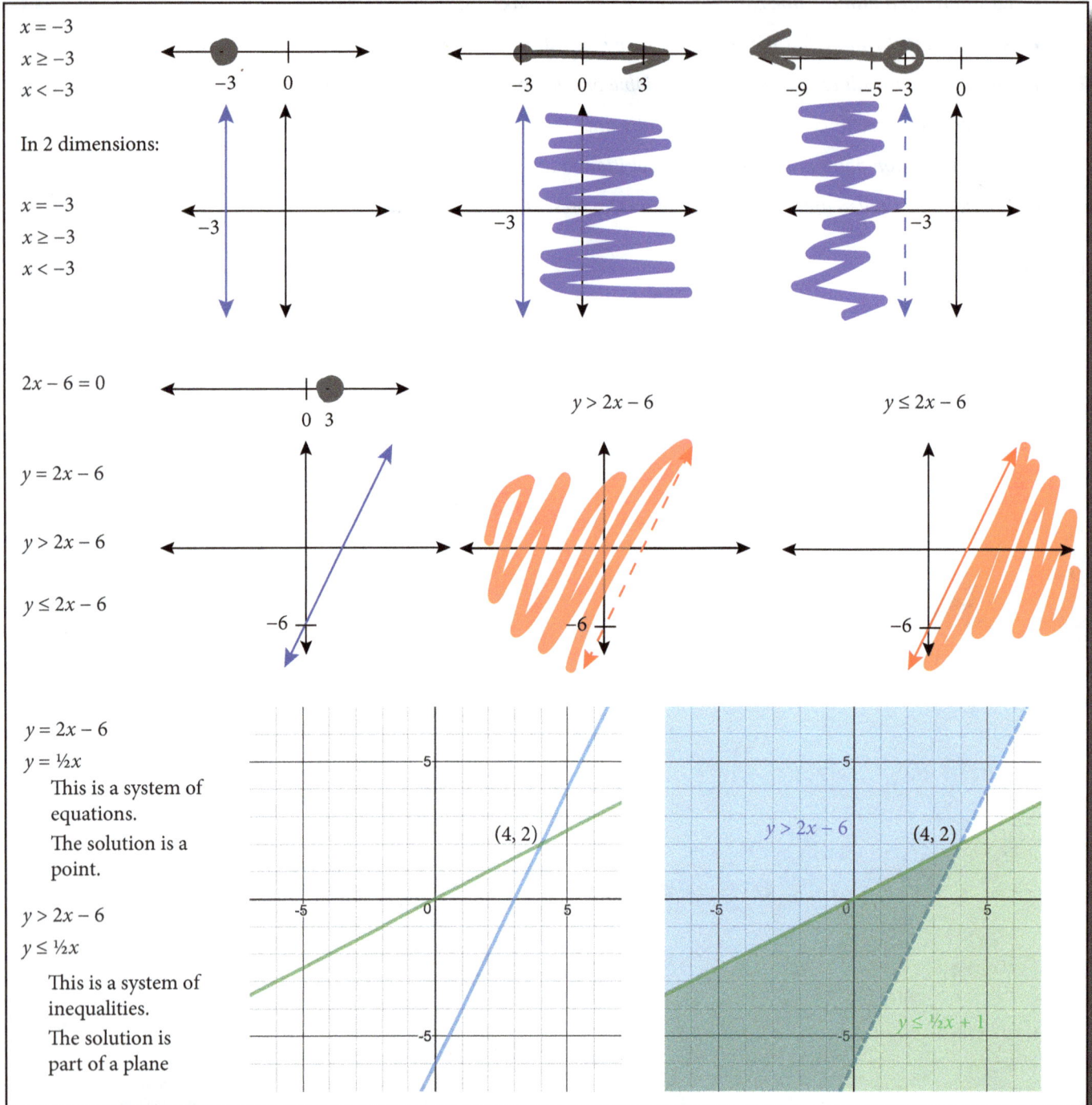

$x = -3$
$x \geq -3$
$x < -3$

In 2 dimensions:

$x = -3$
$x \geq -3$
$x < -3$

$2x - 6 = 0$

$y = 2x - 6$

$y > 2x - 6$

$y \leq 2x - 6$

$y > 2x - 6$

$y \leq 2x - 6$

$y = 2x - 6$
$y = \frac{1}{2}x$

This is a system of equations.
The solution is a point.

$y > 2x - 6$
$y \leq \frac{1}{2}x$

This is a system of inequalities.
The solution is part of a plane

(4, 2)

$y > 2x - 6$ (4, 2)

$y \leq \frac{1}{2}x + 1$

Algebra Problem Strings
©2017 Kendall Hunt Publishing

Facilitation Notes

This version of the problem string lists short notes for important teacher moves during the string. After you've done the string yourself and studied the relationships involved, you might make similar notes for the things you want a reminder of or deem important.

$x = -3$	The first three are quick.
	Graph on number lines horizontally in a row.
$x \geq -3$	What would these look like?
	Rational numbers? Points vs. rays?
$x < -3$	
In 2 dimensions:	Graph the rest of these on the display grapher and keep a sketch for reference.
$x = -3$	Now, in 2 dimensions. If x is -3, what is y?
	Sketch right below the corresponding problem in one dimension.
$x \geq -3$	What are some points that make it a true statement?
$x < -3$	Suggest a non-solution point.
$2x - 6 = 0$	Number line again. Solution is a point on the number line.
$y = 2x - 6$	What changes?
	Graph, ask for solutions. Solution is the whole line.
$y > 2x - 6$	What changes?
	Solution is a part of the plane, does not include the line.
$y \leq 2x - 6$	What changes? Solution is a part of a grid and includes the line.
$y = 2x - 6$	This is a system of equations. What is the solution?
$y = \frac{1}{2}x$	The solution is where the lines intersect, a point.
$y > 2x - 6$	This is a system of inequalities. What is the solution?
$y \leq \frac{1}{2}x$	Look at the overlap, where the inequalties share solutions of ordered pairs.
	Solution is a set of ordered pairs, a section of the grid.

Exponents 1: Definition

At a Glance	**Objectives**
	The goal of this problem string is for students to construct that exponents represent repeated multiplication.

4^3	**Placement**
5^3	This is the first in a series of seven strings designed to help students construct the properties of exponents. You can use this problem string to introduce or remind students of the meaning of an exponent, that it represents repeated multiplication of the same base.
$5 \cdot 5^2$	
$4^2 \cdot 4$	
$2^3 \cdot 2^2$	Use this problem string to introduce textbook Chapter 6 Exponents and Exponential Models.
$2^2 \cdot 2^4$	
x^m	

Guiding the Problem String

As you are facilitating this problem string, work to bring out the relationships and the idea of equivalence. This is not about students memorizing rules, but about students using what they know about multiplication to connect to the representation of exponents and powers. Encourage students to look for and verbalize patterns and equivalencies as you model the relationships.

The first two problems should be relatively quick and establish that $4^3 = 64$ and $5^3 = 125$. Now students will be able to refer to these relationships as they come up later in the string. The third problem is equivalent to the second problem and the fourth problem is equivalent to the first problem. Ask students to find equivalent statements. Use correct vocabulary as you repeat and model student suggestions. If needed, write the exponent of 1 where applicable. The fifth and sixth problems introduce a new base of 2. The sixth problem is equivalent to the first and fourth problems. The last problem is a chance for students to generalize the meaning of an exponent.

Throughout this string, you may be inclined to ask students to tell you the answer. Instead ask for equivalencies. Make it less about getting answers and more about relationship expressing.

About the Mathematics

One of the main ideas to bring out in this string is that students can either evaluate each power and then multiply the results, or simplify the expression to one power and then evaluate the power. For example, $2^2 \cdot 2^4 = 4 \cdot 16 = 64$ or $2^2 \cdot 2^4 = 2^6 = 64$.

Be purposeful to talk multiplicatively about the factors when discussing the number of factors represented by exponents. For example, in 2^3 the three 2s are multiplied. Do not say, "Three 2s." Three 2s is $2 + 2 + 2 = 3 \cdot 2$. On the other hand $2^3 = 2 \cdot 2 \cdot 2$ is three 2s multiplied by each other. Also, the exponent refers to the number of 2s $(2 \cdot 2 \cdot 2)$ as opposed to how many multiplications $(2 \cdot 2 \cdot 2 \cdot 2)$. In the power 2^3, the 3 is the exponent.

Sample Interactions

Use the following as you plan how to elicit and model student strategies. This is not meant as a script, but as a view into the relationships involved and the intent of the problem string.

Teacher: *To kick off this next chapter, let's do a short problem string. I don't think you'll need to record much, but if you want to jot down your thinking, you're welcome to. The first problem to think about is 4 cubed or 4 raised to the third. What is that equivalent to?* Think time. The teacher is purposeful about the varied use of "cubed" and "to the third."	4^3
Teacher: *What is 4 cubed, 4 raised to the third?* **Student:** *It's 64.* **Teacher:** *Did anyone get a different answer? No? How were you thinking about 4 cubed?* **Student:** *Four with the little 3 means 4 times 4 times 4. So I did 4 times 4 is 16 and 16 times 4 is 64.* **Teacher:** *We call that 4 cubed or 4 raised to the third.*	$4^3 = 64$ $4^3 = 64 = 4 \cdot 4 \cdot 4 = 16 \cdot 4 = 64$
Teacher: *Since we are looking for equivalent expressions, is there another way of writing 64 as a power, a number raised to an exponent?* **Student:** *I know that 8 squared is 64.*	$4^3 = 64 = 4 \cdot 4 \cdot 4 = 16 \cdot 4 = 64 = ?$ $4^3 = 64 = 4 \cdot 4 \cdot 4 = 16 \cdot 4 = 64 = 8^2$
Teacher: *The next problem is 5 cubed or 5 to the third. What is 5 cubed equivalent to?* Think time.	5^3
Teacher: *Okay, what is 5 cubed, 5 to the third, equivalent to?* **Student:** *It's 125.*	$5^3 = 125$
Teacher: *How?* **Student:** *It's the same thing, 5 times 5 times 5. So 5 times 5 is 25 and 25 times 5 is 125.*	$5^3 = 125 = 5 \cdot 5 \cdot 5 = 25 \cdot 5 = 125$
Teacher: *Let's keep looking at some powers. What is 5 times 5 squared equivalent to?* **Student:** *That's just 125.* **Teacher:** *You all seemed to think about that one pretty quickly. What were you thinking?* **Student:** *It's five times 25 and we just did that, 125.* **Student:** *It's also three 5s. We just did that 5 to the three.* **Teacher:** *We say five cubed or five to the third. Let's put some words to these parts. The 5 is called the base, the 3 is called an exponent, and the whole term, 5 cubed, is called a power.*	$5 \cdot 5^2$ $5 \cdot 5^2 = 5 \cdot 25 = 125 = 5^3$

(continued)

Teacher: *Okay, what is 4 squared times 4 equivalent to?*	$4^2 \cdot 4$
Student: *It's 16 times 4. That's 64.*	
Teacher: *Did anyone think about this as a bunch of 4s multiplied together?*	$= 16 \cdot 4 = 64$
Student: *I did. It's three 4s times each other, just like the first problem.*	
Teacher: *So you just noticed that it's 4 times 4 times 4 and we already found 4 cubed in the first problem? That is helpful.*	$= 16 \cdot 4 = 64 = 4 \cdot 4 \cdot 4 = 4^3$
Teacher: *The next expression for us to consider is 2 cubed times 2 squared. See if you can find 3 different expressions it is equivalent to?*	$2^3 \cdot 2^2$
Student: *It's 8 times 4.*	$= 8 \cdot 4 = 32 = (2 \cdot 2 \cdot 2) \cdot (2 \cdot 2) = 2^5$
Student: *That's 32.*	
Student: *There are three 2s times two 2s.*	
Student: *That's a total of five 2s times each other.*	
Teacher: *How can we write that with an exponent?*	
Student: *It's a big 2 with the little 5.*	
Student: *Two to the fifth.*	
Student: *The base is 2 and the exponent is 5.*	
Teacher: *Here's a new expression. Let's find some equivalents for two squared times 2 raised to the fourth.*	$2^2 \cdot 2^4$
Student: *It's 4 times 16. We've seen that before. It's 64.*	$= 4 \cdot 16 = 64 = (2 \cdot 2) \cdot (2 \cdot 2 \cdot 2 \cdot 2) = 2^6 = 4^3 = 8^2$
Student: *It's two 2s times four 2s.*	
Student: *It's six 2s.*	$2 + 2 + 2 + 2 + 2 + 2 = 6 \cdot 2$
Teacher: *Six 2s added together? What about the six 2s?*	
Student: *No, multiplied together, 2 to the 6th.*	$2 \cdot 2 \cdot 2 \cdot 2 \cdot 2 \cdot 2 = 2^6$
Student: *Two is the base and 6 is the exponent.*	
Student: *And we also know that it's all equal to 4 to the third.*	
Student: *And 8 squared.*	
Teacher: *There are a lot of equivalent ways of writing this. Nice.*	
Teacher: *So, it seems like we are getting a pretty good feel for what it means to have some base number, x, raised to some exponent m. What does that mean? Turn and talk to your partner.*	x^m
Students turn and talk while the teacher listens in.	

Algebra Problem Strings
©2017 Kendall Hunt Publishing

Teacher: *Start us off. What are you thinking?*	$x^m \qquad = x \cdot x \cdot \ldots x \cdot x$
Student: *It's like a whole bunch of x's.*	
Teacher: *Who can add on?*	
Student: *It's x's all times each other.*	
Teacher: *How many x's? Do we know?*	
Student: *It could be lots!*	
Student: *But I think it's m. Just like when we had 2 to the sixth there were six 2s. Now there are m x's.*	$x^m \qquad = \underbrace{x \cdot x \cdot \ldots x \cdot x}_{m\ x's}$
Teacher: *One way to say this is that if we see a base raised to an exponent, we know how many copies of that base are being multiplied together. Can someone say that the other way? What if we see a bunch of the same number being multiplied together?*	
Student: *Then we can write it more quickly using exponents.*	

Teacher: *How would you summarize some of the things that came up in this string today?*

Elicit the following:

- *The superscript, little numbers, are called exponents. The big, regular size, numbers are called bases. A base raised to an exponent is a called a power.*

- *Powers represent repeated multiplication of the same base.*

- *We can think about equivalent expressions by reassociating the factors.*

Sample Final Display

Your display could look like this at the end of the problem string:

$4^3 \qquad = 64 = 4 \cdot 4 \cdot 4 = 16 \cdot 4 = 64 = 8^2$

$5^3 \qquad = 125 = 5 \cdot 5 \cdot 5 = 25 \cdot 5 = 125$

$5 \cdot 5^2 \qquad = 5 \cdot 25 = 125 = 5^3$

exponent / base 5^3 power

$4^2 \cdot 4 \qquad = 16 \cdot 4 = 64 = 4 \cdot 4 \cdot 4 = 4^3$

$2^3 \cdot 2^2 \qquad = 8 \cdot 4 = 32 = (2 \cdot 2 \cdot 2) \cdot (2 \cdot 2) = 2^5$

$2^2 \cdot 2^4 \qquad = 4 \cdot 16 = 64 = (2 \cdot 2) \cdot (2 \cdot 2 \cdot 2 \cdot 2) = 2^6 = 4^3 = 8^2 \qquad 2 + 2 + 2 + 2 + 2 + 2 = 6 \cdot 2$

$\qquad\qquad\qquad\qquad\qquad\qquad\qquad\qquad\qquad\qquad\qquad\qquad\qquad 2 \cdot 2 \cdot 2 \cdot 2 \cdot 2 \cdot 2 = 2^6$

$x^m \qquad = \underbrace{x \cdot x \cdot \ldots x \cdot x}_{m\ x's}$

(continued)

Facilitation Notes

This version of the problem string lists short notes for important teacher moves during the string. After you've done the string yourself and studied the relationships involved, you might make similar notes for the things you want a reminder of or deem important.

4^3	What is four cubed, four raised to the third? Elicit $4 \cdot 4 \cdot 4 = 16 \cdot 4 = 8^2$. Model the associative property. If it comes up, 2^6.
5^3	What is five cubed, five raised to the third? Quick.
$5 \cdot 5^2$	Let's find some equivalencies. Elicit $5 \cdot (5 \cdot 5) = 5 \cdot 25$. Elicit that it's congruent to previous problem.
$4^2 \cdot 4$	Elicit $(4 \cdot 4) \cdot 4 = 16 \cdot 4$. Elicit that it's congruent to first problem.
$2^3 \cdot 2^2$	Elicit $8 \cdot 4 = (2 \cdot 2 \cdot 2) \cdot (2 \cdot 2)$. Elicit 2^5 (Did anyone think about it with one base and one exponent?)
$2^2 \cdot 2^4$	Elicit $4 \cdot 16 = (2 \cdot 2) \cdot (2 \cdot 2 \cdot 2 \cdot 2)$. Elicit 2^6 (Did anyone think about it with one base and one exponent?) Elicit that it's congruent to first and fourth problems.
x^m	So what does it mean when we see a number raised to an exponent? If not yet, define terms: base, exponent, power.

Algebra Problem Strings
©2017 Kendall Hunt Publishing

6.1 Division as Ratio

At a Glance	Objectives

At a Glance

$4 \times \underline{\quad} = 12$

$4 \times \underline{\quad} = 14$

$4 \times \underline{\quad} = 13$

$4 \times \underline{\quad} = 15$

$\times\underline{\quad}$

5, 30, _____

$\times\underline{\quad}$

5, 32.5, _____

Term #	Term
1	8 $\big)\times\underline{\quad}$
2	4

Term #	Term
1	8 $\big)\times\underline{\quad}$
2	6

Objectives

The goal of this string is to help students connect multiplication and division in such a way that they will be better able to realize that they can use division to find the common ratio in exponential data.

Placement

You can use this problem string before students are expected to write recursive rules for geometric sequences or exponential functions from data to help students realize they can use division to find the common ratio in exponential data.

This problem string can introduce textbook Lesson 6.1 Multiplicative Recursion to help students find the constant multiplier for geometric sequences or textbook Lesson 6.7 Fitting Exponential Models to Data to help students find the common ratio in the exponential function model.

Guiding the Problem String

The first four problems are meant to ground students in the connection between multiplication and division, but also to model student thinking about division as ratios. As you work with students, keep the discussion about equivalence. Don't let it be too pointed toward "what to do." Write 3.25 = 3 ¼ = ¹³⁄₄ and acknowledge them as equivalent. The division notation is social knowledge so treat it appropriately—"I can model your thinking this way." "We can write division like this, with fraction notation." The fifth and sixth problems are written as sequences, and the last two are in table format where the term number is paired with the term. These can serve as precursors to finding the common ratio of messy data in a table.

About the Mathematics

Sequences and division can be notated (by convention) in different ways. The notation is social knowledge and can be successfully told and shown to students. How to think about the relationships between multiplication, division, and ratios is logico-mathematical knowledge and needs to be worked out by students as they make connections by solving problems, looking for patterns, and using relationships.

Sample Interactions

Use the following as you plan how to elicit and model student strategies. This is not meant as a script, but as a view into the relationships involved and the intent of the problem string.

Teacher: *Today's problem string starts with this problem, an easy one: 4 times what is 12?*	$4 \times \underline{\quad} = 12$
Brief pause.	
Teacher: *And, I know that's not hard—what is that?*	$4 \times \underline{\ 3\ } = 12$
Student: *Three*	
Teacher: *Right, 4 times 3 is 12. So, if 4 people have $12, then they each had $3? Four times $3 is $12. I can also model that in this division notation, as 12 divided by 4 is 3.*	$\dfrac{12}{4} = 3$

(continued)

Teacher: *What if I told you that 4 times something is 14? Those 4 people had a total of $14. How much did each have?* Brief pause. *Did anyone get something more than $3? What did you find?* **Student:** *Three and a half.* **Teacher:** *How?* **Student:** *You need two more dollars. The four people split $2. So they each get 50 cents.* **Teacher:** *I might model your thinking like this. Again, we can write division like this, with fraction notation. You were thinking about 14 divided by 4 as 14 divided by 4 and then that leftover 2 divided by 4, so 3 and two-quarters, or 50 cents. And we can write that as ¹⁴⁄₄ or 3½ or 3.5.*	$4 \times \underline{\quad} = 14$ $4 \times \underline{3\frac{1}{2}} = 14$ $\frac{14}{4} = \frac{(12+2)}{4} = \frac{12}{4} + \frac{2}{4} = 3 + \frac{1}{2} = 3\frac{1}{2} = 3.5$
Teacher: *This time 4 times something is 13. Those 4 people had a total of $13. How much did each have?* **Student:** *You just need a dollar more, and so they each get one quarter.* **Teacher:** *I might model your thinking like this. You were thinking about 13 divided by 4 as 12 divided by 4 and then that leftover 1 divided by 4, so three and a quarter. And we can write that as ¹³⁄₄ or 3¼ or 3.25.*	$4 \times \underline{\quad} = 13$ $\frac{13}{4} = \frac{(12+1)}{4} = \frac{12}{4} + \frac{1}{4} = 3 + \frac{1}{4} = 3\frac{1}{4} = 3.25$
Teacher: *Next problem: 4 times what is 15?* **Student:** *Three and three-quarters. It's like you have three of those one-quarters from before.* **Student:** *Or I thought about sharing 3 dollars with 4 people, that is 3 over 4.* **Teacher:** *So, sharing $3 with 4 people is like 3 divided by 4, or three-fourths? Nice. Did anyone use 16 divided by 4? Could you?* **Student:** *Ahhh, yeah, you could have 16 divided by 4, that's 4, but it's one too many, so 4 minus a quarter is 3¾ or 3.75.*	$4 \times \underline{\quad} = 15$ $4 \times \underline{3\frac{3}{4}} = 15$ $\frac{15}{4} = \frac{(12+3)}{4} = \frac{12}{4} + \frac{3}{4} = 3 + \frac{3}{4} = 3\frac{3}{4}$ $\frac{15}{4} = \frac{(16-1)}{4} = \frac{16}{4} - \frac{1}{4} = 4 - \frac{1}{4} = 3\frac{3}{4}$
Teacher: *You've had problems before when we give you a list of numbers and you have to find the next number in the pattern, right? So, the next problem is one of those. Start with 5 and multiplicatively get to 30. What would that factor be? What times 5 is 30?* **Student:** *Six.* **Teacher:** *Again, that's an easy one, 6. So 5 times 6 is 30. Did anyone think of that as 30 divided by 5 equals 6? Could you? I'll write that too.*	$\overset{\times\_}{\frown}$ 5, 30, ____ $\overset{\times 6}{\frown}$ 5, 30, ____ $\frac{30}{5} = 6$

Algebra Problem Strings
©2017 Kendall Hunt Publishing

Teacher: *The next problem is in that same sequence format. The first term is 5 and you multiply that by something to get the next term, 32.5. So 5 times what is 32.5?* The teacher pauses, then elicits and models student responses.	$$\overset{\times\underline{\quad}}{\frown}$$ $$5, 32.5, \underline{\quad\quad}$$ $$\overset{\times 6\frac{1}{2}}{\frown}$$ $$5, 32.5, \underline{\quad\quad}$$ $$\frac{32.5}{5} = \frac{(30+2.5)}{5} = \frac{30}{5} + \frac{2.5}{5} = 6 + \tfrac{1}{2} = 6\tfrac{1}{2}$$
Teacher: *The next problem shows up in a table like this. If I told you that you get to the second term by multiplication, how could you get from the 8 to the 4?* Since the data could be additive or multiplicative, the teacher tells the students that it is multiplicative. **Students:** *Double. Times 2. One-half. Minus 4.*	Term # | Term 1 8 $)\times\underline{\quad}$ 2 4
Teacher: *How can we make sense of this? If you know this is multiplicative, how can you get from the 8 to the 4 with multiplication?* **Student:** *All I can think of is divided by 2.* **Student:** *Yeah, and that's like times one-half.* **Teacher:** *Eight times one-half is four? $8 \times \tfrac{1}{2} = 4$? Does it make it easier to think about if we use the commutative property, $\tfrac{1}{2} \times 8$, is that 4? Yes? What if I record the division? If it's 8 times something is 4, then 4 divided by 8 is that something. What is 4 divided by 8? I'll record it this way.* **Teacher:** *Ahhh, and ⁴⁄₈ simplifies to one-half which we can write as a fraction and a decimal.*	Term # | Term 1 8 $)\times\tfrac{1}{2}$ $8 \times \tfrac{1}{2} = 4$ $\frac{4}{8} = \frac{1}{2} = 0.5$ 2 4

Teacher: *So this is interesting. All of these other problems before resulted in a bigger number with multiplication. What do you notice about those multipliers?*

Student: *They are greater than 1.*

Teacher: *And now this is a fraction between 0 and 1? For this problem, 8 times a number between 0 and 1, 0.5, resulted in a smaller number than 8. How are you making sense of that?*

Student: *It's like when you find half of something or a third of something, it's less than the thing.*

Teacher: *Turn to your partner and discuss this. What does it mean to multiply by a number larger than 1? What does it mean to multiply by a number between zero and one?*

Students turn and talk while the teacher listens in.

(continued)

Teacher: *Okay, the last problem of the string shows up in another table like this:*		

Term #	Term	
1	8	$)\times \underline{\quad}$
2	6	

Teacher: *It's decreasing again! How can you get from 8 to 6 with multiplication? Did anyone use the previous problem to help?*

Think time.

Term #	Term	
1	8	$)\times \frac{3}{4}$
2	6	

Teacher: *What did you find? What number times 8 is 6?*

Student: *Three-fourths.*

Teacher: *Convince us that it's three-fourths!*

Student: *Half of 8 is 4 and a quarter of 8 is 2. Since it's 6, it's just ½ and ¼, so ¾.*

Teacher: *I might model your thinking like this: To get from 8 to 6, you could undo that by division, 6 divided by 8. Four divided by 8 you found by using the ⁴⁄₈ from before, so ½, and then thought about the left over 2—2 divided by 8 is ¼. So ½ and ¼ is ¾.*

Again, with division, we found the multiplier by dividing the term by the previous term. Great work. I wonder if these relationships might come in handy in this new chapter.

Term #	Term	
1	8	$)\times \frac{3}{4}$
2	6	

$8 \times \underline{\quad} = 6$

$\dfrac{6}{8} = \dfrac{4}{8} + \dfrac{2}{8} = \dfrac{1}{2} + \dfrac{1}{4} = \dfrac{3}{4} = 0.75$

Teacher: *How would you summarize some of the things that came up in this string today?*

Elicit the following:

- *Multiplication and division are related.*
- *To find the multiplier, you can divide a term by the previous term.*
- *If the given term is greater than the previous, the multiplier is greater than one.*
- *If the given term is less than the previous, the multiplier is between zero and one.*
- *Multiplying by one-half is equivalent to dividing by 2.*

Algebra Problem Strings
©2017 Kendall Hunt Publishing

Sample Final Display

Your display could look like this at the end of the problem string:

$4 \times \underline{\ 3\ } = 12$ $\qquad \dfrac{12}{4} = 3$

$4 \times \underline{\ 3\frac{1}{2}\ } = 14$ $\qquad \dfrac{14}{4} = \dfrac{(12+2)}{4} = \dfrac{12}{4} + \dfrac{2}{4} = 3 + \frac{1}{2} = 3\frac{1}{2} = 3.5$

$4 \times \underline{\ 3\frac{1}{4}\ } = 13$ $\qquad \dfrac{13}{4} = \dfrac{(12+1)}{4} = \dfrac{12}{4} + \dfrac{1}{4} = 3 + \frac{1}{4} = 3\frac{1}{4} = 3.25$

$4 \times \underline{\ 3\frac{3}{4}\ } = 15$ $\qquad \dfrac{15}{4} = \dfrac{(12+3)}{4} = \dfrac{12}{4} + \dfrac{3}{4} = 3 + \frac{3}{4} = 3\frac{3}{4}$ $\qquad \dfrac{15}{4} = \dfrac{(16-1)}{4} = \dfrac{16}{4} - \dfrac{1}{4} = 4 - \frac{1}{4} = 3\frac{3}{4}$

$\overset{\times 6}{\frown}$
$5,\ 30,\ \underline{\quad}$ $\qquad \dfrac{30}{5} = 6$

$\overset{\times 6\frac{1}{2}}{\frown}$
$5,\ 32.5,\ \underline{\quad}$ $\qquad \dfrac{32.5}{5} = \dfrac{(30+2.5)}{5} = \dfrac{30}{5} + \dfrac{2.5}{5} = 6 + \frac{1}{2} = 6\frac{1}{2}$

Term #	Term
1	8
2	4

$\Big) \times \frac{1}{2}$ $\qquad 8 \times \underline{\ \frac{1}{2}\ } = 4$ $\qquad \dfrac{4}{8} = \dfrac{1}{2} = 0.5$

Term #	Term
1	8
2	6

$\Big) \times \frac{3}{4}$ $\qquad 8 \times \underline{\ \frac{3}{4}\ } = 6$ $\qquad \dfrac{6}{8} = \dfrac{4}{8} + \dfrac{2}{8} = \dfrac{1}{2} + \dfrac{1}{4} = \dfrac{3}{4} = 0.75$

(continued)

Facilitation Notes

This version of the problem string lists short notes for important teacher moves during the string. After you've done the string yourself and studied the relationships involved, you might make similar notes for the things you want a reminder of or deem important.

$4 \times \underline{} = 12$	Quickly. Mention money context. Model 12/4.
$4 \times \underline{} = 14$	How do you know? Model 14/4.
$4 \times \underline{} = 13$	How do you know? Model 13/4.
$4 \times \underline{} = 15$	Model 13/4. Three of those 1/4s? Back one 1/4?
$\times\underline{}$ $\overarc{}$ 5, 30, $\underline{}$	What if it's written this way? Quickly.
$\times\underline{}$ $\overarc{}$ 5, 32.5, $\underline{}$	32.5 = 30 + 2.5 Write 1/2 x 5 = 2.5, x 1/2 and x 0.5

Term #	Term		
1	8	$\Big)\times\underline{}$	What if it's written this way? Wait, how can we get a smaller number? 8 to 4 by multiplication? Partner talk about multiplying by factors between zero and 1. Write "x 1/2" and "x 0.5"
2	4		

Term #	Term		
1	8	$\Big)\times\underline{}$	Linger. Decreasing again. Hmm... 8 to 6 by multiplication? Elicit using previous and simplifying. Multiplication can result in a smaller number, huh? Interesting.
2	6		

Algebra Problem Strings
 ©2017 Kendall Hunt Publishing

6.2 Exponents 2: Multiplying Like Bases

At a Glance	
	Objectives The goal of this problem string is for students to reason about equivalent expressions that deal with multiplying powers with like bases.
3^3 $3^2 \cdot 3$ $3^2 \cdot 3^2$ $2^3 \cdot 5^2$ $3^2 \cdot 5^2$ $2^2 \cdot 2^3$ $2^a \cdot 2^b$ $x^m \cdot x^n$ $ab^2 \cdot 3ab$ $2a^{15}b \cdot 3a^3 b^{10}$	**Placement** This is the second in a series of seven strings designed to help students construct the properties of exponents. You can use this problem string to build on the understanding started in problem string 6.0 to generalize the relationship of adding the exponents when multiplying powers with like bases. Use this problem string before or after textbook Lesson 6.2 Exponential Equations to help students reason about multiplying powers with like bases.
	Guiding the Problem String As you are facilitating this problem string, work to bring out the relationships and the idea of equivalence. This is not about students memorizing rules, but about students using what they know about multiplication to connect to the representation of powers. Encourage students to look for and verbalize patterns and equivalencies as you model the relationships. The first problem is a quick basis for the next two problems. Focus the conversation about how they are related and how you can use one to reason about another. The fourth and fifth problems are non-examples, supplied to help students face any over generalizing they might be doing, such as adding all exponents regardless of the base. The sixth problem, $2^2 \cdot 2^3$, is an opportunity to continue to build the generalization about multiplying like bases which students formalize for base 2 in the next problem, $2^a \cdot 2^b$, and base x in $x^m \cdot x^n$. The last two problems are opportunities to apply the generalization. Throughout this string, you may be inclined to ask students to tell you what to do. Instead, ask for equivalencies. Make it less about answer getting and more about relationship expressing.
	About the Mathematics One of the main ideas to bring out in this string is that students can either evaluate each power and then multiply the results, or simplify the expression to one power and then evaluate the power. For example, $2^2 \cdot 2^3 = 4 \cdot 8 = 32$ or $2^2 \cdot 2^3 = 2^5$. Be purposeful to talk multiplicatively about the factors when discussing the number of factors represented by exponents. For example, in 2^3 the three 2s are multiplied. Do not say, "Three 2s." Three 2s is $2 + 2 + 2 = 3 \cdot 2$. On the other hand $2^3 = 2 \cdot 2 \cdot 2$, three 2s multiplied by each other. Also, the exponent refers to the number of 2s ($2 \cdot 2 \cdot 2$) as opposed to how many multiplications ($2 \cdot 2 \cdot 2 \cdot 2$)

(continued)

Sample Interactions

Use the following as you plan how to elicit and model student strategies. This is not meant as a script, but as a view into the relationships involved and the intent of the problem string.

Teacher: *The first problem today is three cubed, or three raised to the third, or three raised to the exponent of 3. What is it equivalent to?* Brief think time.	3^3
Student: *It's 27.* **Teacher:** *How do you know?* **Student:** *Three times three is nine, times three is 27.* **Teacher:** *So, that's what this means? Three times three times three? Yes?*	$3^3 = 27$ $3^3 = 27 = 3 \cdot 3 \cdot 3 = 9 \cdot 3$
Teacher: *The next problem for us today is three squared times three. What is that equivalent to?* **Student:** *The same thing.* **Student:** *Yeah, it's 27 because it's 9 times 3.* **Teacher:** *So, I'm hearing two different ideas. One is that this is the same as the first problem, three 3s times each other. The other is that it's nine times three. I'll just record those equivalent ideas.*	$3^2 \cdot 3$ $3^2 \cdot 3 = 27 = (3 \cdot 3) \cdot 3 = 9 \cdot 3 = 3^3$
Teacher: *And the next problem is to find equivalencies for three squared times three squared.* Brief think time.	$3^2 \cdot 3^2$
Teacher: *What is equivalent to three to the second times three to the second?* **Students:** 81 **Teacher:** *Did anyone find three squared first? Yes?* **Student:** *Yeah, it's just 9 times 9, 81.*	$3^2 \cdot 3^2 = 81$ $= 81 = 9 \cdot 9$
Teacher: *Did anyone think about how many threes are being multiplied?* **Student:** *I did. There are two 3s in each three squared, so three to the four.*	$= 81 = 9 \cdot 9 = (3 \cdot 3) \cdot (3 \cdot 3) = 3^4$
Teacher: *And did anyone use the previous problem?* **Student:** *We already knew that three times three times three is 27, so I just multiplied 27 times 3.*	$= 81 = 9 \cdot 9 = (3 \cdot 3) \cdot (3 \cdot 3) = 3^4 = (3 \cdot 3 \cdot 3) \cdot 3 = 27 \cdot 3$

Algebra Problem Strings
©2017 Kendall Hunt Publishing

Teacher: *So, up here we had three 3s times each other and we wrote that as 3^3. Then we had three 3s times each other and we wrote that as $3^2 \cdot 3^1 = 3^3$. Then we had two 3s times each other, 3^2, times another two 3s times each other, 3^2, and we wrote that as 3^4. Turn to your partner and talk about any patterns you are noticing.* Students turn and talk to a partner while the teacher listens in.	$3^1 \cdot 3^1 \cdot 3^1 = 3^3$ $3^2 \cdot 3^1 = 3^3$ $3^2 \cdot 3^2 = 3^4$
Teacher: *What are you noticing?* **Student:** *It looks like the exponents are adding.* **Teacher:** *We can represent that as $3^{2+1} = 3^3$ and $3^{2+2} = 3^4$.*	$3^2 \cdot 3^1 \qquad 3^{2+1} = 3^3$ $3^2 \cdot 3^2 \qquad 3^{2+2} = 3^4$
Teacher: *Okay, the next expression for us to consider is two cubed times five squared. What is this equivalent to? Based on the work we just did, can we just add these exponents together? Turn and talk to your partner* Partners turn and talk while the teacher listens in.	$2^3 \cdot 5^2$
Teacher: *What did you decide?* **Student:** *We think it's 200. You can't just add the exponents.* **Teacher:** *Tell me about the 200.* **Student:** *It's just 8 times 25. That's like 4 times 50 or 2 times 100, so 200.* **Teacher:** *So you knew that 2^3 is 8 and 5^2 is 25 and you found 8 times 25. Can everyone think about 8 quarters? Is that the same as four 50s? Is that the same as two 100s? Nice work.* *But I thought there was something we just decided about adding exponents?* **Student:** *That's only when the bottom numbers are the same.* **Teacher:** *That's only when the base is the same?* **Student:** *Yes, because then you're just sort of counting the number of the base times itself. And that would be 10 to the 5th. That's a way bigger number than 200.* **Teacher:** *So, when you're keeping track of the number of copies of the base being multiplied, you can add the exponents like we did before. But when the bases are not the same?* **Students:** *Then you just have to figure each power.*	$2^3 \cdot 5^2$ $= 200 = 8 \cdot 25 = 4 \cdot 50 = 2 \cdot 100$ $\neq (2 \cdot 5)^{3+2} = (10)^5$
Teacher: *Well, so here is an expression when the exponents are the same. What about now? Is $3^2 \cdot 5^2 = (3 + 5)^2$? Can we add this way?* Think time.	$3^2 \cdot 5^2 \; ? \; (3 + 5)^2$

(continued)

Student: No, that doesn't make them equal. The left side is 9 times 25, 225. The right side is 8 squared, 64. They are not equal. **Student:** Before we were looking at the same base with different exponents being multiplied together. Then you can count the number of bases by adding those exponents. Here the bases are not the same. It doesn't matter that the exponents are.	$3^2 \cdot 5^2 \neq (3+5)^2$ $9 \cdot 25 \neq 64$
Teacher: How about this expression, $2^2 \cdot 2^3$? What are some equivalencies? Can you write this as one number? As an expression with one exponent?	$2^2 \cdot 2^3$
Teacher: What number is this equivalent to? **Students:** 32. **Teacher:** Why? **Student:** Because 4 times 8 is 32.	$2^2 \cdot 2^3 = 32 = 4 \cdot 8$
Teacher: And what if we wanted to write it as a 2 with one exponent? **Student:** It's two 2s times three 2s, that's five 2s. So 2 to the 5th. **Teacher:** Since it's five 2s all times each other, we can write that as 2 raised to the 2 plus 3, 2 raised to the 5th.	$2^2 \cdot 2^3 = 32 = 4 \cdot 8 = (2 \cdot 2) \cdot (2 \cdot 2 \cdot 2) = 2^5$ $2^2 \cdot 2^3 = 2^{2+3} = 2^5$
Teacher: What if all we know is that we have 2 raised to some number a times 2 raised to some number b? What does that mean? **Student:** It means there are a bunch of 2s times each other. There's a of them on the left and b of them on the right. **Teacher:** So how many total 2s times each other? **Student:** $a + b$.	$2^a \cdot 2^b$ $2^a \cdot 2^b = \underbrace{(2 \cdot 2 \cdot \ldots \cdot 2)}_{a \ 2s} \cdot \underbrace{(2 \cdot 2 \cdot \ldots \cdot 2)}_{b \ 2s} = \underbrace{2 \cdot 2 \cdot \ldots \cdot 2}_{(a+b) \ 2s} = 2^{a+b}$
Teacher: What if we had any random number x raised to some exponent m times the same number x raised to some exponent n. What else do we know? The teacher records the generalization as students discuss.	$x^m \cdot x^n = \underbrace{(x \cdot \ldots \cdot x)}_{m \ x's} \cdot \underbrace{(x \cdot \ldots \cdot x)}_{n \ x's} = \underbrace{x \cdot \ldots \cdot x}_{(m+n) \ x's}$
Teacher: Let's switch things up a bit. How does everything we've been talking about influence your thinking about this expression? What operation is going on in this expression? **Student:** Multiplication. **Teacher:** Let's multiply. What are you thinking? Think to yourself and when you are ready, compare your work with a partner to multiply ab^2 times $3ab$. Students work while the teacher circulates.	$ab^2 \cdot 3ab$

Algebra Problem Strings
 ©2017 Kendall Hunt Publishing

Teacher: *Will you two share your thinking please?*	
Student: *We talked to you about rearranging the order.*	
Teacher: *And what did you decide.*	
Student: *We decided that we could rearrange the order because it's multiplication. So we put the a's together and the b's together and got …*	
Teacher: *Hold on a bit. Let me get that up here. You put the a's together and the b's together and where is the 3?*	$ab^2 \cdot 3ab = 3 \cdot a \cdot a \cdot b^2 \cdot b$
Student: *It's out front just like coefficients for other variables. Then we see that there are two a's time each other, so a^2, and three b's times each other, so b^3.*	
Teacher: *So I can record that thinking like this.*	$= 3 \cdot a \cdot a \cdot b^2 \cdot b = 3a^{1+1}b^{2+1} = 3a^2b^3$

Teacher: *When we multiply factors like this so that we have each base represented once, that can be called simplifying the expression. All of these expressions are equivalent. We call the one on the right the simplified one.*	
Student: *That's kind of funny. It almost feels like the one with all of the a's and b's written out is the most simple. The one on the right looks like the most efficient.*	
Teacher: *That's a really good point. Simplified is a term that is used by convention—someone decided that's what we are going to call it. And there are other times we use simplified with different kinds of expressions that have nothing to do with exponents. So, while it's important to know what we mean by simplified, it might be more important to know when things are equivalent. And we're doing a great job with that today.*	

The teacher repeats with the last problem.	$2a^{15}b \cdot 3a^3b^{10}$ $= 2 \cdot 3 \cdot a^{15} \cdot a^3 \cdot b \cdot b^{10} = 6a^{15+3}b^{1+10} = 6a^{18}b^{11}$

Teacher: *How would you summarize some of the things that came up in this string today?*

Elicit the following:

- *Exponents mean repeated multiplication of the same base.*

- *If we are multiplying powers with the same base, we can count the number of copies of the base or we can add the exponents to find the total number of copies.*

- *We can rearrange factors in a multiplicative expression using the commutative property.*

(continued)

Sample Final Display

Your display could look like this at the end of the problem string:

3^3	$= 27 = 3 \cdot 3 \cdot 3 = 9 \cdot 3$
$3^2 \cdot 3$	$= 27 = (3 \cdot 3) \cdot 3 = 9 \cdot 3 = 3^3$ $3^{2+1} = 3^3$
$3^2 \cdot 3^2$	$= 81 = 9 \cdot 9 = (3 \cdot 3) \cdot (3 \cdot 3) = 3^4 = (3 \cdot 3 \cdot 3) \cdot 3 = 27 \cdot 3$ $3^{2+2} = 3^4$
$2^3 \cdot 5^2$	$= 200 = 8 \cdot 25 = 4 \cdot 50 = 2 \cdot 100$ $\neq (2 \cdot 5)^{3+2} = (10)^5$
$3^2 \cdot 5^2$	$\neq (3 + 5)^2$ $9 \cdot 25 \neq 64$
$2^2 \cdot 2^3$	$32 = 4 \cdot 8 = (2 \cdot 2) \cdot (2 \cdot 2 \cdot 2) = 2^5$ $2^2 \cdot 2^3 = 2^{2+3} = 2^5$
$2^a \cdot 2^b$	$= \underbrace{(2 \cdot 2 \cdot \ldots \cdot 2)}_{a \ 2s} \cdot \underbrace{(2 \cdot 2 \cdot \ldots \cdot 2)}_{b \ 2s} = \underbrace{2 \cdot 2 \cdot \ldots \cdot 2}_{(a+b) \ 2s} \quad = 2^{a+b}$
$x^m \cdot x^n$	$= \underbrace{(x \cdot \ldots \cdot x)}_{m \ x's} \cdot \underbrace{(x \cdot \ldots \cdot x)}_{n \ x's} = \underbrace{x \cdot \ldots \cdot x}_{(m+n) \ x's}$
$ab^2 \cdot 3ab$	$= 3 \cdot a \cdot a \cdot b^2 \cdot b = 3a^{1+1}b^{2+1} = 3a^2b^3$
$2a^{15}b \cdot 3a^3b^{10}$	$= 2 \cdot 3 \cdot a^{15} \cdot a^3 \cdot b \cdot b^{10} = 6a^{15+3}b^{1+10} = 6a^{18}b^{11}$

Algebra Problem Strings
 ©2017 Kendall Hunt Publishing

Facilitation Notes

This version of the problem string lists short notes for important teacher moves during the string. After you've done the string yourself and studied the relationships involved, you might make similar notes for the things you want a reminder of or deem important.

3^3	*What is this equivalent to? How do you know?*
$3^2 \cdot 3$	*Elicit using the previous and 9 times 3.*
$3^2 \cdot 3^2$	*Did anyone find 3^2 first?* *Did anyone think about how many 3s are being multiplied?* *Did anyone use the previous problem?* *Notice any patterns? Model adding the exponents.*
$2^3 \cdot 5^2$	*Based on the work we just did, can we just add these exponents together?* *Partner talk.* *Elicit $200 \neq 10^5$.*
$3^2 \cdot 5^2$	*Now the 2s are the same! So we can add 3 + 5?* *Base must be the same!*
$2^2 \cdot 2^3$	*Elicit using power first then multiply.* *Elicit finding the number of 2s being multiplied, record as $2 \cdot 2 \cdot 2 \cdot 2 \cdot 2$ and adding the exponents.*
$2^a \cdot 2^b$	*What if all we know is that we have 2 raised to some number a times 2 raised to some number b?* *How many 2s are being multiplied together?*
$x^m \cdot x^n$	*What if we had any random number x raised to some exponent m times the same number x raised to some exponent n. What else do we know? How many x's are being multiplied together?*
$ab^2 \cdot 3ab$	*How are you thinking about this expression? Let's multiply the factors.* *Can we reorder factors when multiplying?* *What does simplify mean?*
$2a^{15}b \cdot 3a^3b^{10}$	*How can we multiply these factors?*

6.3 Exponents 3: Multiplication

At a Glance	Objectives

Objectives
The goal of this problem string is for students to use patterns to reason about the relationships of powers raised to exponents.

$(3 \cdot 3)^2$

$(2^2)^3$

$(2^3)^2$

$(a^4)^3$

$(a^6)^2$

$(3^a)^3$

$(3^2)^b$

$(x^a)^b$

$(2a^2)^3$

$(-3ab^{10})^2$

Placement
This is the third in a series of seven strings in chapter 6 designed to help students construct the properties of exponents. You can use this problem string to build on student understanding to generalize the relationship of multiplying the exponents when raising a power to an exponent.

Use this problem string before textbook Lesson 6.3 Multiplication and Exponents to help students reason about raising powers to exponents.

Guiding the Problem String
The first problem is structured to suggest multiplying the $3 \cdot 3$ first so some students take up the strategy of evaluating inside the parentheses first in the rest of the problems. Throughout the rest of the string, elicit the strategy of evaluating the power and then raising to the exponent (power first strategy) and also the strategy of raising the power to the exponent first and then multiplying (exponent first strategy). The second and third problems use only numbers and are equivalent. This equivalence can help students realize that the commutative property of multiplication is at work in the power of a power relationship. The fourth and fifth problems are equivalent and use a variable as the base. These problems use different factor pairs of 12 $(4 \cdot 3, 2 \cdot 3)$ to continue to suggest that multiplication is at work. The sixth and seventh problems use a variable as an exponent giving students the opportunity to reason more abstractly about the number of copies of the base being multiplied. The eighth problem, $(x^a)^b$, is an opportunity for students to generalize the power of a power relationship. The last two problems are opportunities for students to use their emerging understanding of relationships to create equivalent expressions from more complicated expressions.

About the Mathematics
One of the main ideas in this string is to bring out that students can either evaluate each power and then raise the results to the exponent, or multiply the power by copies of itself according to the exponent. The equivalence of the results gives students an opportunity to make generalizations. For example:

$$(2^2)^3 = (4)^3 = 4 \cdot 4 \cdot 4 = 64$$
$$= 2^2 \cdot 2^2 \cdot 2^2 = 4 \cdot 4 \cdot 4 = 64$$

Be purposeful to talk multiplicatively about the factors when discussing the number of factors represented by exponents. For example, in 4^3 the three 4s are multiplied. Do not say, "Three 4s." Three 4s is $4 + 4 + 4 = 3 \cdot 4$. On the other hand $4^3 = 4 \cdot 4 \cdot 4$ is three 4s multiplied.

Algebra Problem Strings
©2017 Kendall Hunt Publishing

Sample Interactions

Use the following as you plan how to elicit and model student strategies. This is not meant as a script, but as a view into the relationships involved and the intent of the problem string.

Teacher: *To start our problem string today, what do you think of when you see this expression, $(3 \cdot 3)^2$?*	$(3 \cdot 3)^2$
Student: 81.	
Teacher: *Did anyone focus on the three times three first?*	$= 81 = (9)^2 = 9 \cdot 9$
Student: *Yeah, three times three is 9. So then it's 9 squared.*	
Teacher: *Did anyone think about the exponent first, the squared? Could you?*	
Student: *It's like, whatever is in the parentheses is times itself, squared. So three times three times another three times three.*	
Teacher: *I could write that like this.*	$= (3 \cdot 3)(3 \cdot 3) = 9 \cdot 9$
Teacher: *So, we can do what it tells us in the parentheses first or we can think about the power first? Interesting. Let's hang on to that idea.*	

Note: Students may have learned to simplify expressions through rote procedures that demand that the expression in the parentheses is simplified first, rather than through sense making. The idea that you can work with an exponent before the parentheses in these cases may be new. Invite students to make sense of both exponent first and power first strategies, since the exponent first strategy is helpful for making sense of multiplying exponents when you have a power of a power.

Teacher: *The next problem for us to consider today is $\left(2^2\right)^3$. What is this expression equivalent to?* Think time.	$\left(2^2\right)^3$
Teacher: *Is anyone thinking about the exponent first—thinking about cubing?* *Did anyone think about the power first, figuring the two squared?*	$= 64 = \left(2^2\right)\left(2^2\right)\left(2^2\right) = 4 \cdot 4 \cdot 4 = 16 \cdot 4$ $= \left(2^2\right)^3 = (4)^3$
Teacher: *So how many 2s were being multiplied in this expression? Six? How can we represent that?*	$= 2^6$

(continued)

Teacher: Okay, the next problem is $\left(2^3\right)^2$. What is this expression equivalent to? Is anyone thinking about the exponent first, thinking about squaring? Did anyone think about the power first, figuring the two cubed? Which strategy are you more inclined to think about, squaring first or figuring the two cubed first? Thinking about the exponent first or the power first? Turn and talk to your partner about the other strategy, the one you are not as inclined toward. Talk about it so your brain gets more used to it. Why can we either square first or figure the inside of the parentheses first?	$\left(2^3\right)^2$ $= 64 = \left(2^3\right)\left(2^3\right) = 8 \cdot 8$ $= \left(8^2\right) = 8 \cdot 8$
Teacher: A few of you were noticing that the last two problems were both equivalent to 64. Are the problems equivalent? How many 2s being multiplied does each problem represent? Either way, $\left(2^2\right)^3$ and $\left(2^3\right)^2$ represent six 2s multiplying? We can represent each of these as 2^6. Multiplication is commutative, two times three is equivalent to three times two. We can represent that.	$\left(2^2\right)^3 \overset{?}{=} \left(2^3\right)^2 \qquad 2 \cdot 3 = 3 \cdot 2$ $\left(2^3\right)^2 = 2^{3 \cdot 2} = 2^6 = 2^{2 \cdot 3} = \left(2^2\right)^3$
Teacher: Here's the next problem. How many a's are being multiplied here? How do you know? Did anyone think about the exponent first, cubing a^4 times a^4 times a^4? Did anyone think about the inside of the parentheses first, how many a's are being multiplied in a^4?	$\left(a^4\right)^3$ $= a^{12} = \left(a^4\right)\left(a^4\right)\left(a^4\right) = a^{4+4+4} = a^{3 \cdot 4}$ $(a \cdot a \cdot a \cdot a)(a \cdot a \cdot a \cdot a)(a \cdot a \cdot a \cdot a) = a^{12}$
Teacher: In this next problem, how many a's are being multiplied? How do you know? What are different ways we can think about this? Did anyone think about squaring first, a^6 times a^6? Focusing on the exponent first? Did anyone think about how many a's are being multiplied in a^6 first? Focusing on the power first? Which of these strategies seems more efficient? Why?	$\left(a^6\right)^2$ $= a^{12} = \left(a^6\right)\left(a^6\right) = a^{6+6} = a^{6 \cdot 2}$ $(a \cdot a \cdot a \cdot a \cdot a \cdot a)(a \cdot a \cdot a \cdot a \cdot a \cdot a) = a^{12}$

Algebra Problem Strings
 ©2017 Kendall Hunt Publishing

Teacher: *Let's look at these two problems. Are they equivalent?* *What does this have to do with factor pairs?* *Why?* *It sounds like you are saying that to find the number of a's being multiplied if a is raised to an exponent and that is raised to an exponent, you can multiply the exponent?* *Do you think that will always be true?*	$$\left(a^4\right)^3 \overset{?}{=} \left(a^6\right)^2$$ $$4\cdot 3 = 6\cdot 2 = 12$$
Teacher: *Okay, the next problem is $\left(3^a\right)^3$. How many 3s are being multiplied?* *How do you know?*	$$\left(3^a\right)^3$$ $$= 3^{3a} = \left(3^a\right)\left(3^a\right)\left(3^a\right) = 3^{a+a+a} = 3^{3\cdot a}$$
Teacher: *What are some equivalent expressions for $\left(3^2\right)^b$?* *Did anyone use our idea of multiplying the exponents?* *Did anyone think about 3^2 first, finding the number of 9s being multiplied?*	$$\left(3^2\right)^b$$ $$= 3^{2\cdot b} = 3^{2b} = \left(3^2\right)^b = 9^b$$ $$= \left(9\right)^b = 9^b$$
Teacher: *Okay, so if we have a power raised to an exponent, $\left(x^a\right)^b$, what else do we know?* *How many x's are being multiplied?* *Who can convince us?* *How do you know?*	$$\left(x^a\right)^b$$ $$= \underbrace{x^a \cdot ... \cdot x^a}$$ b copies of x^a $$= x^{a\cdot b} = x^{ab}$$
Teacher: *Let's have some fun and make these a bit more complicated. What are some equivalent expressions for $\left(2a^2\right)^3$?* *Did anyone think about the exponent 3 first?* *Did anyone think about multiplying the exponents?*	$$\left(2a^2\right)^3$$ $$= \left(2a^2\right)\left(2a^2\right)\left(2a^2\right) = 2\cdot 2\cdot 2\cdot a^2\cdot a^2\cdot a^2 = 8a^{2+2+2} = 8a^6$$ $$= 2^3\cdot\left(a^2\right)^3 = 8a^{2\cdot 3} = 8a^6$$
Teacher: *For our last problem today, what is this expression equivalent to?* *Did anyone think about the exponent 2 first?* *Did anyone think about multiplying the exponents?*	$$\left(-3ab^{10}\right)^2$$ $$= \left(-3ab^{10}\right)\left(-3ab^{10}\right) = (-3)(-3)aab^{10}b^{10}$$ $$= 9a^{1+1}b^{10+10} = 9a^2b^{20}$$ $$= (-3)^2(a)^2\left(b^{10}\right)^2 = 9a^2b^{10\cdot 2} = 9a^2b^{20}$$

Teacher: *How would you summarize some of the things that came up in this string today?*

Elicit the following:

- *We can find equivalent expressions of a power raised to an exponent by thinking about the exponent first.*

- *We can also find equivalent expressions of a power raised to an exponent by thinking about the power first.*

- *We can also find equivalent expressions of a power raised to an exponent by multiplying the exponents.*

(continued)

Sample Final Display

Your display could look like this at the end of the problem string:

$(3 \cdot 3)^2 \qquad = 81 = (9)^2 = 9 \cdot 9 \qquad = (3 \cdot 3)(3 \cdot 3) = 9 \cdot 9$

$(2^2)^3 \qquad = 64 = (2^2)(2^2)(2^2) = 4 \cdot 4 \cdot 4 = 16 \cdot 4 \quad = (2^2)^3 = (4)^3$

$\qquad\qquad (2^2)^3 = 2^6$

$(2^3)^2 \qquad = 64 = (2^3)(2^3) = 8 \cdot 8 \qquad = (8^2) = 8 \cdot 8$

$\qquad\qquad (2^3)^2 = 2^6$

$$(2^2)^3 \overset{?}{=} (2^3)^2 \qquad 2 \cdot 3 = 3 \cdot 2$$

$$(2^3)^2 = 2^{3 \cdot 2} = 2^6 = 2^{2 \cdot 3} = (2^2)^3$$

$(a^4)^3 \qquad = a^{12} = (a^4)(a^4)(a^4) = a^{4+4+4} = a^{3 \cdot 4} \qquad (a \cdot a \cdot a \cdot a)(a \cdot a \cdot a \cdot a)(a \cdot a \cdot a \cdot a) = a^{12}$

$(a^6)^2 \qquad = a^{12} = (a^6)(a^6) = a^{6+6} = a^{6 \cdot 2} \quad (a \cdot a \cdot a \cdot a \cdot a \cdot a)(a \cdot a \cdot a \cdot a \cdot a \cdot a) = a^{12}$

$$(a^4)^3 \overset{?}{=} (a^6)^2$$
$$4 \cdot 3 = 6 \cdot 2 = 12$$

$(3^a)^3 \qquad = 3^{3a} = (3^a)(3^a)(3^a) = 3^{a+a+a} = 3^{3 \cdot a}$

$(3^2)^b \qquad = (9)^b = 9^b \qquad = 3^{2 \cdot b} = 3^{2b} = (3^2)^b = 9^b$

$(x^a)^b \qquad = \underbrace{x^a \cdot \ldots \cdot x^a}_{b \text{ copies of } x^a} \qquad = x^{a \cdot b} = x^{ab}$

$(2a^2)^3 \qquad = (2a^2)(2a^2)(2a^2) = 2 \cdot 2 \cdot 2 \cdot a^2 \cdot a^2 \cdot a^2 = 8a^{2+2+2} = 8a^6$

$\qquad\qquad = 2^3 \cdot (a^2)^3 = 8a^{2 \cdot 3} = 8a^6$

$(-3ab^{10})^2 \quad = (-3ab^{10})(-3ab^{10}) = (-3)(-3)aab^{10}b^{10} = 9a^{1+1}b^{10+10} = 9a^2b^{20}$

$\qquad\qquad = (-3)^2(a)^2(b^{10})^2 = 9a^2b^{10 \cdot 2} = 9a^2b^{20}$

Algebra Problem Strings
©2017 Kendall Hunt Publishing

Facilitation Notes

This version of the problem string lists short notes for important teacher moves during the string. After you've done the string yourself and studied the relationships involved, you might make similar notes for the things you want a reminder of or deem important.

$(3 \cdot 3)^2$	What do you think when you see this expression? Elicit 3 · 3, then square. Elicit square, then 3 · 3.
$(2^2)^3$	What is this expression equivalent to? Elicit both exponent (cube) first and power (square) first strategies. How many 2s are being multiplied? How can we represent that?
$(2^3)^2$	What is this expression equivalent to? Elicit both exponent (square) first and power (cube) first strategies. Partner talk about strategy. How many 2s are being multiplied? How can we represent that? Are problems 2 and 3 equivalent? Multiplication is commutative!
$(a^4)^3$	How many a's are being multiplied? Elicit both exponent firt and power first strategies.. Which strategy do you think is more efficient?
$(a^6)^2$	How many a's are being multiplied? Elicit both exponent first and power first strategies. Which strategy do you think is more efficient? Are these last two problems equivalent? It sounds like you are saying that to find the number of a's being multiply, we can multiply the exponents. Will that always be true?
$(3^a)^3$	How many 3s? How can you convince us? Emphasize sense making.
$(3^2)^b$	Repeat.
$(x^a)^b$	If we have a power raised to an exponent, what else do we know? How can you convince us?
$(2a^2)^3$	Let's make it a bit more complicated. What are some equivalent expressions? Elicit exponent first strategy (cubing the 2a².) Elicit multiplying exponents. Stop modeling the power first strategy at this point in the string.
$(-3ab^{10})^2$	Repeat. Elicit exponent first strategy. Elicit multiplying exponents.

6.4 | Exponents 4: Equivalence

At a Glance

$$\left(3x^2 y\right)^3 = 9x^5 y^4$$

$$\left(2a^3 b^4\right)\left(4a^3 b^2\right) = 2^3 a^9 b^8$$

$$\left(4a^3 b^6\right)^4 = \left(8^2 a^6 b^{12}\right)^2$$

$$\left(-2x^{20} y^3\right)^3 = \left(-4x^{20} y^4\right)\left(2x^3 y^2\right)$$

Objectives

The goal of this problem string is for students to integrate and use the relationships of the definition of exponents, powers multiplied by powers of same bases and powers raised to exponents to justify equivalence of expressions.

Placement

This is the fourth in a series of seven strings designed to help students construct the properties of exponents. After students have begun developing the relationships of the definition of exponents, same base powers multiplied by powers, and powers raised to exponents, you can use this problem string to help students gain facility in recognizing and using those relationships.

Use this problem string to follow textbook Lesson 6.3 Multiplication of Exponents or to precede 6.5 Looking Back with Exponents.

Guiding the Problem String

This problem string is written as statements of equivalence, however the statements may be true or false. The first two problems are in a form where students may simplify the left expression to see if the expression on the right is equivalent. The third and fourth problems are such that neither side of the equation is simplified.

As you facilitate each problem, tell students that you are going to write a statement and the task is for them to evaluate if the statement is true or false. If the statement is false, make it true by changing the equal symbol (=) to the not-equal symbol (≠). The purpose of having students share their reasoning is for the teacher to model their reasoning to make it visible as students justify the relationships. When controversy arises, instead of relying on rules, ask students to justify by reasoning about relationships. This can often be done by expanding the term so that the powers are written as repeated multiplication. When students refer to an expanded power, help them become more efficient by suggesting the power representation, "Ahhh, so now we are sure that $\left(4a^3 b^6\right)^4$ is equivalent to a term with four factors of 4 times 12 factors of a times 24 factors of b and we can write that as $4^4 a^{12} b^{24}$. That is a lot more efficient than writing all of those factors out!"

About the Mathematics

Continue to be purposeful by referring multiplicatively to the factors when discussing the number of factors represented by exponents. For example, in a^3 the three a's are multiplied. Do not say, "Three a's." Three a's is $a + a + a = 3a$. On the other hand, $a^3 = a \cdot a \cdot a$ is three a's being multiplied.

Algebra Problem Strings
©2017 Kendall Hunt Publishing

Important Questions

Use the following as you plan how to elicit and model student strategies.

- *Is this statement true? Are the two expressions equivalent?*

- *How do you know?*

- *Did you simplify each side of the equation to see if you found the same simplified expressions?*

- *Did you expand each side of the equation to see if you found equivalent expanded expressions?*

How would you summarize some of the things that came up in today's string?
- *You can expand an expression so that you can see all of the factors.*

- *You can simplify an expression so that it is expressed in mathematical shorthand.*

- *You add can add exponents when same bases are multiplied.*

- *You can multiply exponents when a power is raised to an exponent.*

Sample Final Display

Your display could look like this at the end of the problem string:

$$\left(3x^2y\right)^3 \neq 9x^5y^4$$

$$\left(3x^2y\right)^3 = 27x^6y^3 \neq 9x^5y^4$$

$$\left(3x^2y\right)^3 = \left(3^1\right)^3\left(x^2\right)^3\left(y^1\right)^3$$
$$= 3^{1\cdot3}x^{2\cdot3}y^{1\cdot3}$$
$$= 27x^6y^3$$

$$\left(3x^2y\right)^3 = \left(3^1\right)^3\left(x^2\right)^3\left(y^1\right)^3$$
$$= (3\cdot3\cdot3)\left(x^2\cdot x^2\cdot x^2\right)(y\cdot y\cdot y)$$
$$= 27x^6y^3$$

$$\left(2a^3b^4\right)\left(4a^3b^2\right) \neq 2^3a^9b^8$$

$$\left(2a^3b^4\right)\left(4a^3b^2\right) = 2\cdot2\cdot2\cdot a^3a^3b^4b^2 = 2^3a^{3+3}b^{4+2} = 2^3a^6b^6$$

$$2^3a^6b^6 \neq 2^3a^9b^8$$

$$\left(4a^3b^6\right)^4 = \left(2^4a^6b^{12}\right)^2 \quad = 4^4a^{12}b^{24}$$

$$= 2^8a^{12}b^{24}$$

$$\left(4a^3b^6\right)^4 = 4^4a^{3\cdot4}b^{6\cdot4} = \left(2^2\right)^4a^{12}b^{24} = \left(2^4\right)^2\left(a^6\right)^2\left(b^{12}\right)^2$$

$$\left(2^4a^6b^{12}\right)^2 = 2^{4\cdot2}a^{6\cdot2}b^{12\cdot2} = \left(2^2\right)^4a^{12}b^{24} = 4^4\left(a^3\right)^4\left(b^6\right)^4$$

$$\left(-2x^{20}y^3\right)^3 \neq \left(-4x^{20}y^4\right)\left(2x^3y^2\right)$$

$$-8x^{60}y^9 \neq -8x^{23}y^6$$

$$\left(-2x^{20}y^3\right)^3 = (-2)^3\left(x^{20}\right)^3\left(y^3\right)^3 = -8x^{20\cdot3}y^{3\cdot3} = -8x^{60}y^9$$
$$= (-2)(-2)(-2)x^{20}x^{20}x^{20}y^3y^3y^3 = -8x^{60}y^9$$

$$\left(-4x^{20}y^4\right)\left(2x^3y^2\right) = (-4\cdot2)x^{20}x^3y^4y^2 = -8x^{23}y^6$$

(continued)

Facilitation Notes

This version of the problem string lists short notes for important teacher moves during the string. After you've done the string yourself and studied the relationships involved, you might make similar notes for the things you want a reminder of or deem important.

$\left(3x^2y\right)^3 = 9x^5y^4$	Here is a statement. Is it true? How can you use what you know about relationships? How can you convince us? Did anyone use the relationship of multiplying exponents of a power raised to an exponent? Did anyone cube 3x²y first, then expand?
$\left(2a^3b^4\right)\left(4a^3b^2\right) = 2^3a^9b^8$	Repeat.
$\left(4a^3b^6\right)^4 = \left(2^4a^6b^{12}\right)^2$	Repeat. Did anyone simplify both sides? Did anyone find equivalent expressions until you could write it like the other side of the equation?
$\left(-2x^{20}y^3\right)^3 = \left(-4x^{20}y^4\right)\left(2x^3y^2\right)$	Repeat. Did anyone simplify both sides? Did anyone choose just one factor to simplify in both expressions, found non-equivalence and those results convinced you the two expression were not equivalent? Which factor did you choose?

Algebra Problem Strings
©2017 Kendall Hunt Publishing

6.5 | Exponents 5: Division Property

At a Glance	Objectives
$$\frac{8}{4}$$ $$\frac{32}{4}$$ $$\frac{2\cdot2\cdot2\cdot2\cdot2}{2\cdot2}$$ $$\frac{3^4}{3}$$ $$\frac{2^m}{2^n}$$ $$\frac{x^m}{x^n}$$ $$\frac{2^5\cdot3}{2^2}$$ $$\frac{2^4 3^3}{12}$$	The goal of this problem string is for students to begin to generalize relationships with division of powers with like bases, where the exponents are positive.

Placement

This is the fifth in a series of seven strings for chapter 6 designed to help students construct the properties of exponents. You can use this problem string to develop student intuition about dividing powers with like bases before you introduce negative exponents.

Use this problem string either to introduce textbook Lesson 6.5 Looking Back with Exponents or to solidify the learning from the investigation The Division Property of Exponents.

Guiding the Problem String

The first two problems set the stage for division problems. They should go quickly. The third problem is equivalent to the second and can provide students the opportunity to think about the second problem, 32 divided by 4, as division of repeated multiplication of the same base. After solving the third problem, rewrite the first two problems in terms of exponents and nudge students to look for the relationship of the difference in the exponents of the numerator and denominator. The third problem is written as a base raised to exponents to further that conversation. For this and the rest of the problems, elicit two main strategies, thinking about multiplication or using the relationships of the exponents with like bases. Near the end of the string, encourage students to consider both strategies and wonder aloud which is more transparent and which is more efficient. For the last problem, encourage students to either simplify the multiplication first or subtract exponents.

If the more generalized relationship of including negative exponents arises, encourage students to continue to think about it and reassure students that you will be investigating that soon.

About the Mathematics

The word "cancel" is not helpful because it can be used in too many places to mean too many things and students often mix and match and strike out anything that looks similar, regardless of the operation at hand. Emphasize that a factor divided by itself is equivalent to 1 and that 1 times anything is that thing. Factors are not canceling or disappearing—you are using the equivalence of multiplying by 1 in a multiplication situation.

Sample Interactions

Use the following as you plan how to elicit and model student strategies. This is not meant as a script, but as a view into the relationships involved and the intent of the problem string.

Teacher: *This first problem might seem a bit easy but hang with me. What is eight divided by four?*	$$\frac{8}{4}$$

(continued)

Student: *Two.* **Teacher:** *Right, it's 2.*	$$\frac{8}{4} = 2$$
Teacher: *The next problem is 32 divided by 4. This line can mean to read this like a fraction, but it can also mean division, so 32 divided by 4. What is 32 divided by 4?* **Student:** *Eight.*	$$\frac{32}{4}$$ $$\frac{32}{4} = 8$$
Teacher: *The next problem is 2 times 2 times 2 times 2 times 2 all divided by 2 times 2. What is this equivalent to?* **Student:** *That's eight. It's just the same problem as 32 divided by 4.*	$$\frac{2 \cdot 2 \cdot 2 \cdot 2 \cdot 2}{2 \cdot 2} = \frac{32}{4} = 8$$
Teacher: *So, you might have noticed that it is equivalent to the second problem. Did anyone think about dividing out some of these 2s?* **Student:** *Yes, you can cancel two of the twos.* **Teacher:** *I don't find the word cancel very helpful. Are they disappearing? What is really happening? Turn to your partner and discuss what is happening with the twos.* Students talk to their partners while the teacher listens in.	$$\frac{2 \cdot 2 \cdot 2 \cdot 2 \cdot 2}{2 \cdot 2} = \frac{32}{4} = 8$$
Teacher: *What are you thinking about those twos?* **Student:** *It's division, right? Two divided by itself is 1.* **Teacher:** *I can model that thinking like this, where we think about the 2 divided by 2 and the next 2 divided by 2 as 1 times the rest of the 2s left. How many 2s are left multiplying?* **Student:** *There are three 2s left.* **Teacher:** *Let's write this problem more efficiently. How could we do that using exponents?* **Student:** *That's like 2 raised to the fifth over 2 squared.* **Teacher:** *That's like 2 raised to the fifth divided by 2 squared. And that is equivalent to two cubed? Yes?*	$$= 8 = \frac{2}{2} \cdot \frac{2}{2} \cdot \frac{2 \cdot 2 \cdot 2}{1} = 1 \cdot 1 \cdot 2 \cdot 2 \cdot 2 = 2^3 = \frac{2^5}{2^2}$$
Teacher: *Let's go back to the first problem. Can we write that in terms of 2 and exponents?* **Student:** *That is two to the third over 2 squared.* **Teacher:** *And I find the word "over" to be not very mathematical. What's going on? Division, again? So two to the third divided by 2 squared.*	$$\frac{8}{4} = 2^1 = \frac{2^3}{2^2}$$

Algebra Problem Strings
©2017 Kendall Hunt Publishing

Teacher: *And the second problem. Can we write that in terms of 2 and exponents? How do this exponent of 5 in the numerator and the exponent of 2 in the numerator relate to the exponent of 3 in this 2 raised to the third power?*	$$\frac{2^5}{2^2} = 2^{5-2} = 2^3$$
Student: *Well, 5 minus 2 is 3.*	
Teacher: *Sure enough. I wonder if that will always be true? Do you think the exponent in the numerator and denominator will have a relationship like this, subtraction? Hmmm...*	
Teacher: *Okay, here is the next problem. Lots of threes being multiplied. What are some thoughts about this problem?*	$$\frac{3^4}{3}$$
Student: *It's 27.*	$$= 27 = \frac{3 \cdot 3 \cdot 3 \cdot 3}{3} = \frac{3}{3} \cdot \frac{3 \cdot 3 \cdot 3}{1} = 1 \cdot 3 \cdot 3 \cdot 3 = 3^3$$
Teacher: *Tell me more.*	
Student: *Well there's four 3s on the top and one 3 on the bottom so that leaves three 3s.*	
Teacher: *But three 3s is 9—3 times 3 is 9.*	
Student: *I meant 3 times 3 times 3.*	
Teacher: *Who can restate that?*	
Student: *It's four 3s in the numerator and one 3 in the denominator. Like we said with the twos, three divided by itself is one so we just have three 3s being multiplied left in the numerator.*	
Teacher: *So, some of you were thinking about the number of threes in the numerator and denominator and finding the difference between them.*	
Teacher: *Was anyone thinking about what three to the fourth is?*	$$= \frac{81}{3} = 27 = 3^3$$
Student: *Yes, it's 81 and 81 divided by 3 is 27.*	
Teacher: *And 27, what is that in terms of three raised to an exponent?*	
Student: *That's three cubed.*	
Teacher: *So, how do these four 3s multiplying together and the one 3 in the denominator relate to the multiplying of the three 3s we just talked about?*	$$\frac{3^4}{3^1} = 3^{4-1} = 3^3$$
Student: *Subtraction again. Four minus one is three.*	
Teacher: *I'll write that. Four minus one is three and the difference between four and 1 is three. We can think of it either way.*	
Teacher: *If we've got some number like 2, raised to some exponent m divided by that same base 2 raised to a different exponent n, what are you thinking will happen to the twos?*	$$\frac{2^m}{2^n}$$

(continued)

Student: *The extra ones divide out, leaving ones.* **Teacher:** *So, how many twos will be left?* **Student:** *The difference between the m and n.* **Teacher:** *And we can write that with subtraction, like this.*	$$\frac{2^m}{2^n} = 2^{m-n}$$
Teacher: *In fact, if we had any base, like x, raised to an exponent divided by that same base, x, raised to a different exponent, how many x's would we have left over after the x's divided out?* **Student:** *We'd have x raised to the m minus n.*	$$\frac{x^m}{x^n}$$ $$\frac{x^m}{x^n} = x^{m-n}$$
Teacher: *I wonder how that might help us think about some equivalencies for this expression? What is this equivalent to?*	$$\frac{2^5 \cdot 3}{2^2}$$

Teacher: *Did anyone think about the number of 2s being multiplied?*

As students explain their thinking, the teacher models the relationships using multiplication.

$$\frac{2^5 \cdot 3}{2^2} = \frac{2 \cdot 2 \cdot 2 \cdot 2 \cdot 2 \cdot 3}{2 \cdot 2} = \frac{2}{2} \cdot \frac{2}{2} \cdot \frac{2 \cdot 2 \cdot 2 \cdot 3}{1} = 1 \cdot 1 \cdot 2 \cdot 2 \cdot 2 \cdot 3 = 2^3 3 = 8 \cdot 3 = 24$$

Teacher: *Did anyone think about the relationship between the exponents?* As students explain their thinking, the teacher models the relationships using subtraction. **Teacher:** *I wonder which of these ways of looking at these expressions seems more transparent, like you can see what's going on? Which of them feels more efficient?* **Student:** *When your write out all of the factors, I can see what is going on, but writing it with exponent subtraction is a lot faster.* **Teacher:** *We can call this process simplifying because there are no repeated factors in the numerator or the denominator.*	$= 2^{5-2} \cdot 3 = 2^3 \cdot 3 = 8 \cdot 3 = 24$
Teacher: *What about this expression? What is this equivalent to? After you have simplified this expression and found what it is equivalent to, see if you can use the other strategy to think about it. You can think about the multiplication first or think about subtracting exponents. Step out of your thinking and try to see if the other strategy also makes sense.*	$$\frac{2^4 3^3}{12}$$
Teacher: *Who usually thinks about subtracting the exponents first? Instead of telling us about your initial strategy, would one of you please tell us how you could use multiplication to figure this out?* As students explain their thinking, the teacher models the relationships using multiplication.	$= 36 = \frac{16 \cdot 27}{12} = \frac{(4 \cdot 4)(3 \cdot 9)}{4 \cdot 3} = \frac{4}{4} \cdot \frac{3}{3} \cdot \frac{4 \cdot 9}{1} = 1 \cdot 1 \cdot 36 = 36$

Algebra Problem Strings
©2017 Kendall Hunt Publishing

Teacher: *Who usually thinks about multiplying everything out? Instead of telling us about your initial strategy, would one of you please tell us how you could use subtraction with the exponents to figure this out?* As students explain their thinking, the teacher models the relationships using subtraction and exponents. **Teacher:** *Nice. We have two ways of reasoning about equivalencies when faced with expressions full of the same bases and exponents and division. Let's keep thinking about when each strategy might come in handy.*	$= \dfrac{2^4 \cdot 3^3}{2^2 \cdot 3} = 2^{4-2} \cdot 3^{3-1} = 2^2 \cdot 3^2 = 4 \cdot 9 = 36$

Teacher: *How would you summarize some of the things that came up in this string today?*

Elicit the following:

- *A number or variable divided by itself is equivalent to one.*

- *When dividing powers, you can rewrite each power in terms of multiplication to help determine where you have extra copies of the base being multiplied.*

- *When dividing powers with like bases, the difference between the exponents can be helpful to determine how many copies of the base are being multiplied.*

- *Writing powers expanded as multiplication of the factors helps us see what is going on, but takes time and space.*

(continued)

Sample Final Display

Your display could look like this at the end of the problem string:

$$\frac{8}{4} \qquad = 2^1 = \frac{2^3}{2^2}$$

$$\frac{32}{4} \qquad = 8$$

$$\frac{2 \cdot 2 \cdot 2 \cdot 2 \cdot 2}{2 \cdot 2} \qquad = 8 = \frac{2}{2} \cdot \frac{2}{2} \cdot \frac{2 \cdot 2 \cdot 2}{1} = 1 \cdot 1 \cdot 2 \cdot 2 \cdot 2 = 2^3 = \frac{2^5}{2^2} \qquad \frac{2^5}{2^2} = 2^{5-2} = 2^3$$

$$\frac{3^4}{3} \qquad = 27 = \frac{3 \cdot 3 \cdot 3 \cdot 3}{3} = \frac{3}{3} \cdot \frac{3 \cdot 3 \cdot 3}{1} = 1 \cdot 3 \cdot 3 \cdot 3 = 3^3 \qquad \frac{3^4}{3^1} = 3^{4-1} = 3^3$$

$$= \frac{81}{3} = 27 = 3^3$$

$$\frac{2^m}{2^n} \qquad = 2^{m-n}$$

$$\frac{x^m}{x^n} \qquad = x^{m-n}$$

$$\frac{2^5 \cdot 3}{2^2} \qquad = \frac{2 \cdot 2 \cdot 2 \cdot 2 \cdot 2 \cdot 3}{2 \cdot 2} = \frac{2}{2} \cdot \frac{2}{2} \cdot \frac{2 \cdot 2 \cdot 2 \cdot 3}{1} = 1 \cdot 1 \cdot 2 \cdot 2 \cdot 2 \cdot 3 = 2^3 3 = 8 \cdot 3 = 24$$

$$= 2^{5-2} \cdot 3 = 2^3 \cdot 3 = 8 \cdot 3 = 24$$

$$\frac{2^4 3^3}{12} \qquad = 36 = \frac{16 \cdot 27}{12} = \frac{(4 \cdot 4)(3 \cdot 9)}{4 \cdot 3} = \frac{4}{4} \cdot \frac{3}{3} \cdot \frac{4 \cdot 9}{1} = 1 \cdot 1 \cdot 36 = 36$$

$$= \frac{2^4 \cdot 3^3}{2^2 \cdot 3} = 2^{4-2} \cdot 3^{3-1} = 2^2 \cdot 3^2 = 4 \cdot 9 = 36$$

Algebra Problem Strings
©2017 Kendall Hunt Publishing

Facilitation Notes

This version of the problem string lists short notes for important teacher moves during the string. After you've done the string yourself and studied the relationships involved, you might make similar notes for the things you want a reminder of or deem important.

$\dfrac{8}{4}$	What is 8 divided by 4? Quick. How could we write 8 and 4 with exponents?
$\dfrac{32}{4}$	What is 32 divided by 4? Quick.
$\dfrac{2 \cdot 2 \cdot 2 \cdot 2 \cdot 2}{2 \cdot 2}$	What is this quotient? What is a number divided by itself? Write with exponents. Revisit problems 1 and 2 and write with exponents. How does a division problem written with exponents relate to the original and the answer?
$\dfrac{3^4}{3}$	What is this quotient? Elicit finding 3^4 and then dividing by 3. Elicit using relationships of exponents. What is happening? What pattern? Why? How does it make sense? Use removal (4 subtract 1) and difference (the difference between 4 and 1.)
$\dfrac{2^m}{2^n}$	So, then what is 2 raised to some number m divided by 2 raised to some number n?
$\dfrac{x^m}{x^n}$	Then what is any number x raised to some number m divided by that same x raised to some number n?
$\dfrac{2^5 \cdot 3}{2^2}$	What is this expression equivalent to? Elicit expanding and dividing. Elicit thinking about exponents and subtraction. Define simplify—no like bases in both the numerator and denominator.
$\dfrac{2^4 3^3}{12}$	Repeat. When might it be helpful to use exponents? Multiply everything out?

6.6 | Exponents 6: Negative Exponents

At a Glance	**Objectives**
	The goal of this problem string is to extend students' experience of dividing powers to dividing powers where the result has a negative exponent.

At a Glance

$$\frac{3}{3 \cdot 3 \cdot 3}$$

$$\frac{2^2}{2^3}$$

$$\frac{4^5}{4^3}$$

$$\frac{3^3}{3^5}$$

$$\frac{3^2 \cdot 2^2}{6^3}$$

$$\frac{x^m}{x^n}$$

$$m^{-a}$$

$$\frac{1}{m^{-a}}$$

Objectives

The goal of this problem string is to extend students' experience of dividing powers to dividing powers where the result has a negative exponent.

Placement

This is the sixth in a series of seven strings for chapter 6 designed to help students construct the properties of exponents. You can use this problem string to help students develop intuition about negative exponents after they have divided powers with like bases.

Use this problem string to introduce textbook Lesson 6.6 Zero and Negative Exponents or to solidify the learning after the investigation More Exponents.

Guiding the Problem String

This problem string consists of problems where the results are equivalent to powers with negative exponents. The first problem is a division problem where the numerator is smaller than the denominator to help students recall situations where quotients are between zero and one. It is written with repeated multiplication to provide the opportunity to connect to exponential notation and the relationship learned in problem string 6.5 about the difference of the exponents. The second problem is written as powers to invoke the exponent relationship, but model both the difference in the exponents and writing as the equivalent $\frac{4}{8} = \frac{1}{2}$, with the goal of connecting those two relationships. The third problem is equivalent to a power with a positive exponent to give all students the opportunity to reflect on all three of the problems and make more explicit the relationships where the quotient has positive or negative exponents and how that relates to the magnitude of the quotient. The fourth problem has a quotient with a negative exponent. Seek to help students connect the strategy of rewriting the powers as repeated multiplication with the division property of exponents. Refrain from making the division property a thing to do; rather keep students thinking about the relationships. The fifth problem has multiple bases to give students the opportunity to bring everything together. Because bases of 3 and 2 in the numerator are the factors of the 6 in the denominator, students have the opportunity to play with different ways of looking at the expression. Encourage creativity. Linger here and compare strategies, encouraging sense making. The last three problems are opportunities to generalize the relationships.

About the Mathematics

Be purposeful to use precise language. When referring to negative exponents, as in 3^{-2}, say "3 raised to the negative 2" or "3 to the negative 2 exponent" rather than "3 to the minus 2" (infers subtraction). Refrain from using positional words like *top* and *bottom*, rather use *numerator* and *denominator*, *dividend* and *divisor*. Ask "What is an equivalent expression?", rather than "What should we do?", emphasizing relationships rather than procedures.

While some texts refer to expressions with positive exponents as the *simplified* version of the expression, we use *simplified* here to mean that any common factors have been divided and thus the expression has no common factors in the numerator and denominator.

Important Questions

Use the following as you plan how to elicit and model student strategies.

- *What happens when you divide a number by a smaller number? How does the result (quotient) relate to the dividend when dividing by a divisor smaller than the dividend?*

- *What happens when you divide a number by a bigger number? How does the result (quotient) relate to the dividend when dividing by a divisor larger than the dividend?*

- *Which scenario (dividing by a smaller or larger number than the dividend) results in powers with positive exponents? Negative exponents?*

- *What does a negative exponent mean?*

- *How can we rewrite powers as repeated multiplication to help find equivalencies?*

- *How can we use the relationship between division of powers with like bases and the difference in their exponents to help find equivalencies?*

- *How can we rewrite powers with negative exponents as equivalent powers with positive exponents?*

How would you summarize some of the things that came up in this string today?

- *Division by a number smaller than the dividend results in a quotient larger than one.*

- *Division by a number larger than the dividend results in a quotient between zero and one.*

- *Negative exponents have everything to do with the magnitude (size) of the power and nothing to do with the positive or negative value of the power.*

- *We can divide expressions with exponents by subtracting the exponents if the expressions have the same base. If the exponent in the numerator is greater than the exponent in the denominator, the result is a positive exponent in the numerator. If the exponent in the numerator is less than the exponent in the denominator, the result is a negative exponent in the numerator.*

(continued)

Sample Final Display

Your display could look like this at the end of the problem string:

$$\frac{3}{3\cdot 3\cdot 3} \qquad =\frac{1}{9}=\frac{3}{27}=\frac{3}{3}\cdot\frac{1}{3\cdot 3}=\frac{3^1}{3^3}=\frac{1}{3^2} \qquad =3^{1-3}=3^{-2}$$

$$\frac{2^2}{2^3} \qquad =\frac{1}{2}=\frac{4}{8}=\frac{1}{2^1} \qquad =2^{2-3}=2^{-1}$$

$$\frac{4^5}{4^3} \qquad =16=4^2=4^{5-3}$$

$$\frac{3^3}{3^5} \qquad =\frac{3\cdot 3\cdot 3}{3\cdot 3\cdot 3\cdot 3\cdot 3}=\frac{1}{9}=\frac{1}{3^2}=\frac{3^1}{3^3} \qquad =3^{1-3}=3^{-2}$$

$$\frac{3^2\cdot 2^2}{6^3} \qquad =\frac{9\cdot 4}{216}=\frac{36}{216}=\frac{(3\cdot 3)(2\cdot 2)}{6^3}=\frac{(3\cdot 2)(3\cdot 2)}{6^3}=\frac{6^2}{6^3}=\frac{1}{6}$$

$$\frac{x^m}{x^n} \qquad =x^{m-n} \qquad \frac{\overbrace{x\cdot...\cdot x}^{m\ x\text{'s}}}{\underbrace{x\cdot...\cdot x}_{n\ x\text{'s}}}=\underbrace{x\cdot...\cdot x}_{(m-n)\ x\text{'s}} \qquad =(\underbrace{x\cdot...\cdot x}_{m\ x\text{'s}})\div(\underbrace{x\cdot...\cdot x}_{n\ x\text{'s}})=\underbrace{x\cdot...\cdot x}_{(m-n)\ x\text{'s}}$$

$$m^{-a} \qquad =\frac{1}{m^a}$$

$$\frac{1}{m^{-a}} \qquad =m^a$$

Algebra Problem Strings
©2017 Kendall Hunt Publishing

Facilitation Notes

This version of the problem string lists short notes for important teacher moves during the string. After you've done the string yourself and studied the relationships involved, you might make similar notes for the things you want a reminder of or deem important.

$\dfrac{3}{3 \cdot 3 \cdot 3}$	What is this equivalent to? Did anyone multiply the denominator first? Did anyone divide out a 3 first? Did anyone write the expression using exponents? What is 1/9 written with exponents? Which expression is simplified?
$\dfrac{2^2}{2^3}$	What is this equivalent to? Did anyone write this using exponents? Did anyone find the 2 squared and 3 cubed first?
$\dfrac{4^5}{4^3}$	In the last problem string, we found a pattern with like bases in the numerator and denominator. Using exponents, simplify. Go back to first problem and write using subtraction of exponents. Back to second problem and write using subtraction of exponents. What does the negative exponent mean? What is going on? Are positive or negative exponents simplified?
$\dfrac{3^3}{3^5}$	Did anyone multiply the powers out? Did anyone write the expression using exponents? Did anyone write the expression with a negative exponent?
$\dfrac{3^2 \cdot 2^2}{6^3}$	Did anyone multiply everything first? Did anyone rearrange the 3s and 2s? Could we subtract exponents from the beginning?
$\dfrac{x^m}{x^n}$	So if we have any number x raised to some exponent m divided by x to the n, what can we say?
m^{-a}	What is this equivalent to? How do you know?
$\dfrac{1}{m^{-a}}$	What is this equivalent to? How do you know?

Fitting Exponential Models to Data

At a Glance

$$y = 201(1-0.198)^x$$

$$y = 100(1-0.198)^x$$

$$y = 250(1-0.198)^x$$

$$y = 201(1-0.6)^x$$

$$y = 201(1-0.1)^x$$

"Years" elapsed	"Atoms" remaining
0	100
1	88
2	77

Objectives

The goal of this problem string is to solidify the learning about fitting exponential equations to data that took place in the Radioactive Decay Investigation in which students fit an exponential equation to exponential decay data.

Placement

You can use this problem string to help students solidify the meaning of the parameters of an exponential equation that models decay and how changes in the parameters affect the graph.

You could facilitate this string any time after the Radioactive Decay Investigation in textbook Lesson 6.7 Fitting Exponential Models to Data.

Guiding the Problem String

This problem string is based on a scenario where students investigated radioactive decay by simulating the data: students recorded the initial number of counters (201) and drew a central angle on a plate. For each trial "year," students dropped the counters on the plate, removing the number of counters in the angle (representing the decayed atoms). The remaining "atoms" represent the atoms that had not decayed. This data was collected and students found an exponential equation to model the data. The sample data used in the book was modeled with the first equation in the string, $y = 201(1-0.198)^x$.

Begin the problem string by reminding students about the context and asking students what the first equation means in terms of years and atoms. Then have students predict (remember) what the graph looks like. Bring out the rate of decay and the initial number of atoms. Display the graph. Add one equation at a time, asking students to predict the graph based on the changing parameters and describe the changes in the context. Take some time with problems 2 and 3 to work out why the changing "a" value not only vertically stretches or compresses the graph as we would expect $af(x)$ to do, but it also shifts the y-intercept. The fourth and fifth problems change the decay rate. Question students about the relationship between the decay rate and b in $y = a \cdot b^x$. The last problem is a small set of data for students to fit an equation to.

About the Mathematics

For the exponential equation $y = a \cdot b^x$, $1 - (\text{decay rate}) = b$.

The parameter a vertically stretches the graph of $y = b^x$ as we would expect, but it also stretches the y-intercept. Because the y-intercept of $y = b^x$ is (0, 1), the y-value of 1 is also multiplied by a. This is different than with linear equations where the value m only causes the line $y = x$ to rotate to create $y = mx$. This happens because the y-intercept of $y = x$ is (0, 0) and the y-value of 0 stretched (multiplied by a) is still 0. This problem string is written from the perspective that students have not yet learned transformations, so focus on identifying the connections between the context of the simulation and the algebraic representation of the function to which the data is being fit.

Algebra Problem Strings
©2017 Kendall Hunt Publishing

Important Questions

Use the following as you plan how to elicit and model student strategies.

- *How does this scenario, equation, and graph compare to the original? Why?*
- *Why does the "a" value shift the graph up or down? What is happening in the context of atoms?*
- *Why does the "a" value also vertically stretch or compress the graph? What is happening in the context of atoms?*
- *How does the decay rate relate to the "b" value in $y = a \cdot b^x$?*
- *What happens when b is between 0 and 1? In the equation, the graph, and the context?*
- *What does a high decay rate look like on the graph? A low decay rate? Why?*

How would you summarize some of the things that came up in this string today?

- *In the equation $y = a \cdot b^x$, the constant "a" is the y-intercept. It represents the initial number of "atoms" in this scenario.*

- *In the equation $y = a \cdot b^x$, the constant "b" is the multiplicative rate, or constant ratio, and represents the number of atoms remaining per year in this scenario. The rate "b" can be written as b = (1 − the decay rate of atoms per year).*

- *A high decay rate looks steep because more atoms are decaying each year which means there are less atoms remaining. A low decay rate looks more shallow because less atoms are decaying each year which means more atoms are remaining.*

Sample Final Display

Your display could look like this at the end of the problem string:

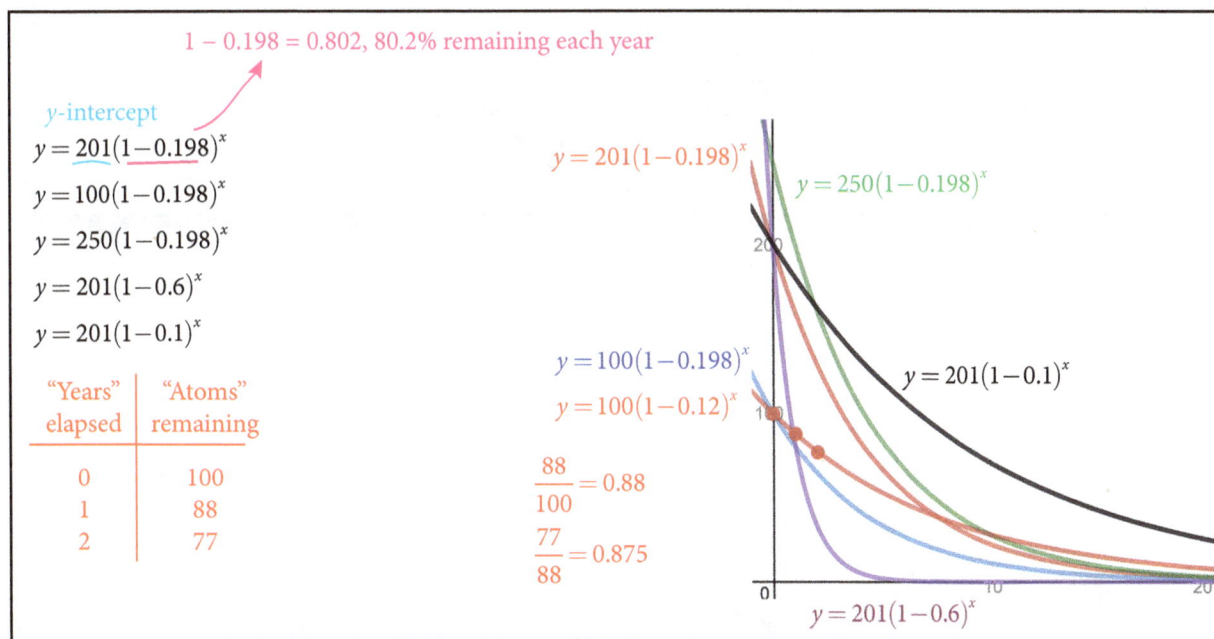

(continued)

Facilitation Notes

This version of the problem string lists short notes for important teacher moves during the string. After you've done the string yourself and studied the relationships involved, you might make similar notes for the things you want a reminder of or deem important.

$y = 201(1-0.198)^x$	Remember the "decaying atoms?" What is the 201 in terms of atoms? What is the 0.198? What did/would the graph look like? Display using the class grapher. Where on the graph do you see the 201? The 0.198? The 0.802?
	Present all the rest as different scenarios. Students predict the graphs and explain the changes.
$y = 100(1-0.198)^x$	What does this equation represent? How would the graph compare to the original? Partners predict. Display both graphs together. What changed? What didn't change? Partner talk.
$y = 250(1-0.198)^x$	Repeat. Predict first. Display all three graphs together. Why is the y-intercept changing?
$y = 201(1-0.6)^x$	Repeat. Now what changed? What didn't? Predict the graph. Look at 4 graphs. What do you notice? What does the 0.6 mean? What does 1 – 0.6 = 0.4 mean? How do these numbers relate to the central angle on the plate?
$y = 201(1-0.1)^x$	What's happening here? Repeat as the previous.

"Years" elapsed	"Atoms" remaining	
0	100	Here's some different data. Write a function to model this data. How did you find "a?" How did you find "b?" What is the decay rate? How does the function for this data compare to the other functions? How do the graphs compare? Which data had the highest starting number of atoms? Which data had the lowest starting number of atoms? Which data had the highest decay rate? Which data had the lowest decay rate?
1	88	
2	77	

Algebra Problem Strings
 ©2017 Kendall Hunt Publishing

6.8 | Simply Simplifying

At a Glance	Objectives

Objectives
The goal of this problem string is for students to develop efficient strategies for finding equivalent expressions for expressions involving positive and negative exponents.

At a Glance expressions:

$$\left(\frac{y^5}{y}\right)^{-3}$$

$$\left(x^{-3}\right)^{-4}$$

$$\left(\frac{x^{-2}}{x^{-3}}\right)^{-5}$$

$$\left(\frac{x^{-3}}{x}\right)^{-4}$$

$$\left(\frac{x^2 y^{-5}}{x^{-3} y^{-4}}\right)^{-1}$$

Placement
This is the last in a series of seven strings designed to help students construct the properties of exponents. You can use this problem string to help students solidify using the relationships of exponents as they develop efficient strategies.

Use this problem string any time after textbook Lesson 6.6 Zero and Negative Exponents. You can use it before or after textbook Lesson 6.8 Decreasing Exponential Models and Half-Life to help students solidify using the relationships of exponents as they develop efficient strategies.

Guiding the Problem String
The problems in this string are written to suggest two different strategies, "simplify the inside of the parentheses first" and "apply the outside exponent first." In the first problem it is efficient to simplify by working with the inside of the parentheses first. If any students apply the outside exponent first, elicit that strategy also. If not, the strategy of applying the outside exponent first will arise in the second problem. The third problem takes a bit more work to simplify inside the parentheses first, but is simplified fairly quickly by applying the outside exponent first. Have students debate which strategy is efficient for the fourth problem, which simplifies quickly by simplifying inside the parentheses first. The fifth problem is more complicated. Elicit both strategies and encourage students to consider a combination of the strategies.

About the Mathematics
Continue to be purposeful referring multiplicatively to the factors when discussing the number of factors represented by exponents. For example, in x^3 the three x's are multiplied. Do not say, "Three x's." Three x's is $x + x + x = 3x$. On the other hand $x^3 = x \cdot x \cdot x$ is three x's being multiplied.

Use precise language. When referring to negative exponents, as in x^{-2}, say "x raised to the negative 2" or "x to the negative 2 exponent" rather than "x to the minus 2" (infers subtraction). Refrain from using positional words like *top* and *bottom*, rather use *numerator* and *denominator*, *dividend* and *divisor*. Ask "What is an equivalent expression?", rather than "What should we do?", emphasizing relationships rather than procedures.

While some texts refer to expressions with positive exponents as the *simplified* version of the expression, we use *simplified* here to mean that any common factors have been divided and thus the expression has no common factors in the numerator and denominator.

(continued)

Sample Interactions

Use the following as you plan how to elicit and model student strategies. This is not meant as a script, but as a view into the relationships involved and the intent of the problem string.

Teacher: *Okay, we are going to start today's problem string with the expression I am writing on the board. What is this equivalent to? Find an equivalent expression such that there are no repeat factors.* Students work briefly while the teacher circulates, asking scaffolding questions as needed. The teacher is looking for students who are simplifying the expression inside the parentheses first and students who are applying the outside exponent (−3) first.	$\left(\dfrac{y^5}{y}\right)^{-3}$
Teacher: *Let's share some thinking. First, what simplified expression did you find?* **Student:** *I got y to the −12th.* **Student:** *I found 1 divided by y to the 12th.* **Teacher:** *What do we think? Which is correct?* **Student:** *They both are. They are the same.* **Teacher:** *They are equivalent? Convince us!*	$= y^{-12} = \dfrac{1}{y^{12}}$
Teacher: *I saw you working with the expression inside the parentheses first. Please tell us about that.* **Student:** *It's easy to see that y to the 5th divided by y is just y to the 4th. Then that's to the −3. So you multiply the 4 times −3 and that's y to the −12th.* **Student:** *Which is the same as 1 divided by y to the 12th.*	$= \left(y^4\right)^{-3} = y^{-12} = \dfrac{1}{y^{12}}$
Teacher: *But that's not how you approached it, right? Tell us about your thinking.* **Student:** *I was thinking about the relationship that you used at the end, that you can multiply the exponents. But I did it first. So it's y to the −15th divided by y to the −3rd. Then I flipped those because of the negative exponents.* **Teacher:** *You inverted the y to the −15th divided by y to the −3rd and got what?* **Student:** *That's y to the 3rd divided by y to the 15th. So there's y to the 12th left on the bottom.* **Teacher:** *So you're left with y to the 12th in the denominator. Okay, did anyone go a different direction once you got to y to the 3rd divided by y to the 15th?* **Student:** *I did. It's just y to the 3rd minus 15 and that's y to the −12th.* **Student:** *I did the same thing but earlier. Once I had y to the −15th divided by y to the −3rd, I just knew that was y to the −15 minus −3. That's y to the −15 plus 3 which is y to the −12th.*	$= \dfrac{y^{-15}}{y^{-3}} = \dfrac{y^3}{y^{15}} = \dfrac{1}{y^{12}}$ $= y^{3-15} = y^{-12}$ $= y^{-15-(-3)} = y^{-15+3} = y^{-12}$

Algebra Problem Strings
©2017 Kendall Hunt Publishing

Teacher: *Some people will expect you to have the resulting expression with no negative exponents in order to be the most simplified. As long as you have a no repeated factors, we'll call it simplified, but you should be able to justify either expression, y to the −12th or 1 divided by y to the 12th as equivalent.*	
Teacher: *The second problem is this one I am writing on the board. What is it equivalent to?*	$\left(x^{-3}\right)^{-4}$
Teacher: *What is this expression equivalent to?* **Student:** *This one is x to the positive 12th.*	$= x^{12}$
Teacher: *Did anyone simplify the expression inside the parentheses first? Tell us about that.* The teacher models student thinking as students explain two ways that they could have simplified the expression inside the parentheses first and then applied the outside exponent.	$= \left(\dfrac{1}{x^3}\right)^{-4} = \left(x^3\right)^4 = x^{12}$ $= \dfrac{1}{x^{3 \cdot 4}} = \dfrac{1}{x^{-12}} = x^{12}$
Teacher: *Did anyone just apply the outside exponent first? Tell us about that please.* The teacher models student thinking as students applying the outside exponent first.	$= x^{(-3)(-4)} = x^{12}$
Teacher: *Now that we can look at both of these ways of thinking, which strategy do you like for these problems, working inside the parentheses first or working with the outside exponent first?* **Student:** *For the first problem, inside worked great because it was easy to subtract positive exponents.* **Student:** *And outside for the second problem because of all of the negative exponents.*	
Teacher: *The third problem for us to simplify today is this one. What is it equivalent to?* Students work. The teacher circulates, looking for students who use a variation of either working with the inside expression first or applying the outside exponent first.	$\left(\dfrac{x^{-2}}{x^{-3}}\right)^{-5}$

(continued)

The teacher asks for the simplified equivalent expression and then models student thinking of simplifying the expression inside the parentheses first and then the strategy of applying the exponent first. **Teacher:** *Which of these strategies do you think was more efficient for this problem?* **Student:** *They seem just about the same to me.* **Student:** *I think that applying the outside exponent first looks just as efficient, but if you just multiply the exponents, you're almost done. The other strategy took more steps.* **Student:** *I didn't want to mess with subtracting negatives, not when I could multiply and get positives.* **Student:** *I agree. Once you multiplied the exponents, it is easy to see the x to the 10th divided by x to the 15th is just 1 divided by x to the 5th. Done.* **Teacher:** *So sometimes it might be more efficient to work with the expression in the parentheses first and sometimes to apply the outside exponent first?*	$= x^{-5} = \dfrac{1}{x^5}$ $= \left(\dfrac{x^3}{x^2}\right)^{-5} = (x)^{-5} = x^{-5}$ $= \left(x^{-2-(-3)}\right)^{-5} = \left(x^1\right)^{-5} = x^{-5}$ $= \dfrac{x^{(-2)(-5)}}{x^{(-3)(-5)}} = \dfrac{x^{10}}{x^{15}} = \dfrac{1}{x^5} = x^{-5}$
Teacher: *Here is the fourth problem. I wonder if one of the strategies might be more efficient here? Maybe try what comes natural first and then see if you can find something more efficient.*	$\left(\dfrac{x^{-3}}{x}\right)^{-4}$
The teacher repeats for this problem: circulating as students work, asking for the simplified equivalent expression, modeling student thinking of simplifying the expression inside the parentheses first, and then modeling the strategy of applying the exponent first. **Teacher:** *Now that we've seen them both, which strategy do you like better for this problem, the inside first or outside first?* **Student:** *Inside first. Once you did, then it's just one more thing to do. Doing the exponent first took more steps.* **Teacher:** *Turn to your partner and discuss what might make expressions so that it is more efficient to work inside the parentheses and what might make it more efficient to apply the exponent first.*	$= x^{16}$ $= \left(\dfrac{1}{x^3 x}\right)^{-4} = \left(\dfrac{1}{x^4}\right)^{-4} = \dfrac{1}{x^{-16}}$ $\dfrac{x^{(-3)(-4)}}{x^{1-4}} = \dfrac{x^{12}}{x^{-4}} = x^{12}x^4 = x^{16}$ $= x^{12-(-4)} = x^{16}$
Teacher: *This last problem is a bit more complicated. I wonder which strategy or maybe a combination of strategies might work well?* Students work, share their thinking, the teacher models it on the board, and the teacher draws out the summary comments that follow. See the Sample Final Display for some possibilities.	$\left(\dfrac{x^2 y^{-5}}{x^{-3} y^{-4}}\right)^{-1}$

Algebra Problem Strings
©2017 Kendall Hunt Publishing

Teacher: *How would you summarize some of the things that came up in this string today?*

Elicit the following:

- *We can simplify inside the parentheses first. This might be efficient when the expression in the parentheses simplifies quickly.*

- *We can apply the outside exponent first. This can be efficient when the exponents inside the parentheses are various different signs or when the outside exponent and inside exponents are negative.*

- *We can combine these strategies to help find equivalent expressions efficiently.*

- *We can invert powers inside the parentheses if they have negative exponents.*

Sample Final Display

Your display could look like this at the end of the problem string:

$$\left(\frac{y^5}{y}\right)^{-3} = y^{-12} = \frac{1}{y^{12}} \qquad = \left(y^4\right)^{-3} = y^{-12} = \frac{1}{y^{12}} \qquad = \frac{y^{-15}}{y^{-3}} = \frac{y^3}{y^{15}} = \frac{1}{y^{12}}$$

$$= y^{3-15} = y^{-12}$$

$$= y^{-15-(-3)} = y^{-15+3} = y^{-12}$$

$$\left(x^{-3}\right)^{-4} = x^{12} \qquad = \left(\frac{1}{x^3}\right)^{-4} = \left(x^3\right)^4 = x^{12} \qquad = x^{(-3)(-4)} = x^{12}$$

$$= \frac{1}{x^{3\cdot -4}} = \frac{1}{x^{-12}} = x^{12}$$

$$\left(\frac{x^{-2}}{x^{-3}}\right)^{-5} = x^{-5} = \frac{1}{x^5} \qquad = \left(\frac{x^3}{x^2}\right)^{-5} = (x)^{-5} = x^{-5} \qquad = \frac{x^{(-2)(-5)}}{x^{(-3)(-5)}} = \frac{x^{10}}{x^{15}} = \frac{1}{x^5} = x^{-5}$$

$$= \left(x^{-2-(-3)}\right)^{-5} = \left(x^1\right)^{-5} = x^{-5}$$

$$\left(\frac{x^{-3}}{x}\right)^{-4} = x^{16} \qquad = \left(\frac{1}{x^3 x}\right)^{-4} = \left(\frac{1}{x^4}\right)^{-4} = \frac{1}{x^{-16}} \qquad = \frac{x^{(-3)(-4)}}{x^{1\cdot -4}} = \frac{x^{12}}{x^{-4}} = x^{12}x^4 = x^{16}$$

$$= x^{12-(-4)} = x^{16}$$

$$\left(\frac{x^2 y^{-5}}{x^{-3} y^{-4}}\right)^{-1} = \frac{x^{-5}}{y^{-1}} = \frac{y}{x^5} \qquad = \left(\frac{x^2 x^3 y^4}{y^5}\right)^{-1} = \left(\frac{x^5}{y}\right)^{-1} = \frac{x^{-5}}{y^{-1}} = \frac{y}{x^5} \qquad = \left(\frac{x^{(2)(-1)} y^{(-5)(-1)}}{x^{(-3)(-1)} y^{(-4)(-1)}}\right) = \frac{x^{-2} y^5}{x^3 y^4} = \frac{y}{x^5}$$

$$= \left(x^2 x^3\right)^{-1}\left(\frac{y^{(-5)(-1)}}{y^{(-4)(-1)}}\right) = x^{-5} \cdot \frac{y^5}{y^4} = x^{-5} y$$

(continued)

Facilitation Notes

This version of the problem string lists short notes for important teacher moves during the string. After you've done the string yourself and studied the relationships involved, you might make similar notes for the things you want a reminder of or deem important.

$\left(\dfrac{y^5}{y}\right)^{-3}$	What is this equivalent to? Find an equivalent expression that has no repeated factors. Elicit simplify the expression in parentheses first. Elicit apply the outside exponent first (if students do this.) Mention that some may expect simplified to mean only positive exponents.
$\left(x^{-3}\right)^{-4}$	Repeat. The strategy of working with the expression inside sure seemed efficient for the first problem, but not so much for this one. Interesting.
$\left(\dfrac{x^{-2}}{x^{-3}}\right)^{-5}$	Repeat. Which strategy do you find more efficient for this problem? Why?
$\left(\dfrac{x^{-3}}{x}\right)^{-4}$	Repeat. Which strategy do you find more efficient for this problem? Why? Partner talk. What about an expression makes one strategy more efficient?
$\left(\dfrac{x^2 y^{-5}}{x^{-3} y^{-4}}\right)^{-1}$	Repeat. Which strategy do you find more efficient for this problem? Why? Did anyone try a combination of the strategies?

Algebra Problem Strings
©2017 Kendall Hunt Publishing

7.1 | Function Notation

At a Glance	Objectives
	The goal of this problem string is to help students learn the meaning of and construct the use of function notation and it's connection with ordered pairs.

At a Glance

$(2, 7)$

$f(-1) = 1$

$f(-4) = -5$

$(½, 4)$

$f(1) = $ ___

$f(\underline{}) = 0$

$f(x) = $ _____

Objectives
The goal of this problem string is to help students learn the meaning of and construct the use of function notation and it's connection with ordered pairs.

Placement
This problem string could be used to introduce students to function notation. Students should have prior experience writing equations of lines to be successful with the last three problems of the string.

You can use this problem string to begin textbook Chapter 7: Functions and Transformations to introduce function notation.

Guiding the Problem String
The first four problems should go quickly. Listen for students who are noticing that the four points might be colinear. If no one says anything, ask what they think or notice it aloud yourself. Then in the fifth question, ask students that if the function is a line, how could they use that to find the function value at 1. Take longer to let students wrestle with the fifth and sixth problems. Refer back to the previous points if needed to help ground them in the function notation. Even though it may appear that the point of the string is to write the equation of the line that contains the points, the emphasis is on the meaning and use of function notation. Keep asking students to clarify what the function notation for each question means. The last question is meant to help students realize that function notation can represent a point, but it can also represent sets of points that follow the rule. If things like the vertical line test come up, address them quickly, but keep the emphasis on the meaning and use of function notation.

About the Mathematics
Function notation is a social construct. By convention, the community of mathematicians have chosen to write function notation using potentially confusing notation, with parentheses that otherwise mean a grouping symbol and multiplication. The use of and facility with function notation is logico-mathematical and must be constructed through experience and making mental connections.

Sample Interactions

Use the following as you plan how to elicit and model student strategies. This is not meant as a script, but as a view into the relationships involved and the intent of the problem string.

Teacher: *To get us started today, put your pencils down and get your brain ready to visualize. What are some ways that mathematicians might represent (2, 7)?*	(2, 7)

(continued)

Student: *It's a point, so you could graph it.*

Student: *Put it in a table of x's and y's.*

The teacher graphs the point and puts it in a table. If no one mentions function notation, the teacher introduces it.

Teacher: *Mathematicians can represent the point (2, 7) by saying that for a function where the input, the x-value is 2, the function value or y-value is 7. We say this, "f of 2 is 7 or equals 7."*

(2, 7) $f(2) = 7$

x	y
2	7

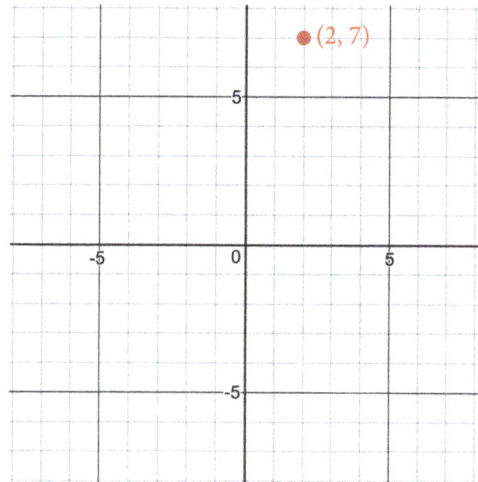

Teacher: *Let's go the other way. For the second problem, we have the function notation. So, for the same function, f, f of −1 is 1. How might mathematicians represent that? Can you picture it?*

$f(-1) = 1$

$f(-4) = -5$

(½, 4)

With student input, the teacher writes the point, puts it in the table, plots it on the graph, and then repeats quickly for the next two problem, $f(-4) = -5$ and then back to ordered pairs, (½, 4). The teacher listens for anyone who notices that the points might be colinear.

(2, 7) $f(2) = 7$

$f(-1) = 1$ (−1, 1)

$f(-4) = -5$ (−4, −5)

(½, 4) $f(½) = 4$

x	y
2	7
−1	1
−4	−5
½	4

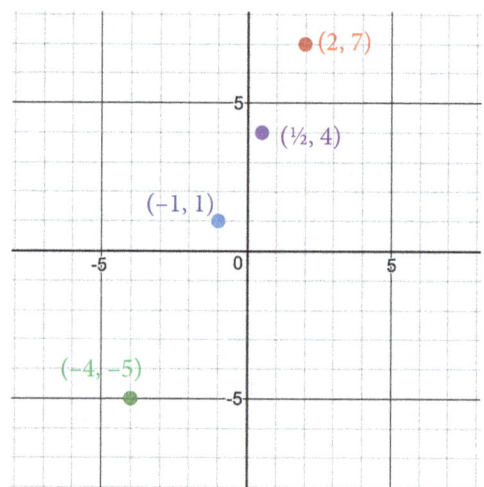

Teacher: *Those almost look linear. I wonder if those points are collinear, all lying on the same line? What do you think? Turn and talk to your partner. Are the points on the same line?*

Students turn and talk.

Teacher: *What do you think and why?*

Student: *The slope is the same between them. It's over 3 up 6 between (−1, 1) and (2, 7) and also from (−4, −5) to (−1, 1).*

Student: *Yes, the rate of change is 2. Between (−4, −5) and (2, 7), it's over 6 and up 12, still a slope of 2.*

Teacher: *I wonder if that might be helpful to know.*

$$\frac{3}{6} = \frac{6}{12}$$

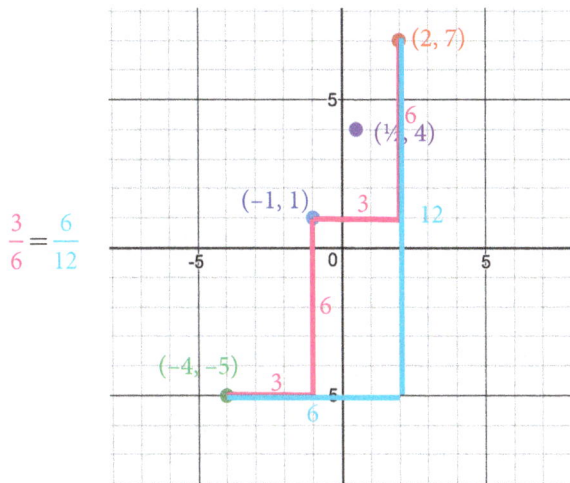

Teacher: *Okay, next problem, for the same function, what do you think f of 1 is? What does this even mean?*

Student: *If the input is 1, what is the output?*

Student: *If the x-value is 1, what is the y-value?*

Teacher: *Work on that. What is y if x is 1 for this function f?*

Students work and the teacher circulates, looking for students who estimate using the sketch of a line, students who use the rate of 2 and an established point, and students who find the equation of *f* and find *f*(1).

$$f(1) = \underline{\hspace{1cm}}$$

$$(1, \underline{\hspace{0.5cm}})$$

(continued)

Teacher: *I saw you pointing with you finger. Tell us about that.*

Student: *I just estimated where x is 1 and went up to the line that I sketched in. It looks like it's about 5.*

Teacher: *You were drawing in some other lines. How did that help?*

Student: *We knew that the rate of change is 2 so I went back one and down two from (2, 7) and got to (1, 5).*

Student: *I did the same kind of thing, but I went right 2 from the point (−1, 1) and up 4 to get to (1, 5).*

Teacher: *And I saw you do something totally different. Did I see an equation you were working with?*

Student: *Since it's a line, I wanted to find the equation. So, I knew y = b + 2x. I used the point (2, 7) and got 7 = b + 2(2) and solved for b and b is 3. So y = 3 + 2x. And (0, 3) is on the graph. When I plugged in 1 for x, I got y = 3 + 2 = 5.*

Teacher: *Since you brought it up, I'll add the point (0, 3) and the line y = 2x + 3 to the board.*

(2, 7)	$f(2) = 7$
$f(-1) = 1$	(−1, 1)
$f(-4) = -5$	(−4, −5)
(½, 4)	$f(½) = 4$
	(0, 3)
$f(1) = \underline{5}$	(1, 5)

x	$y = 2x + 3$
2	7
−1	1
−4	−5
½	4
1	5

$y = 3 + 2x$ $y = b + 2x, 7 = b + 2(2), b = 3$

$f(1) = 3 + 2(1) = 3 + 2 = 5$

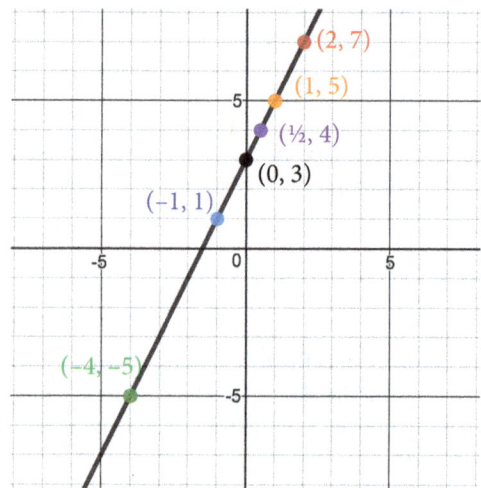

Teacher: *The next problem is kind of like the previous and kind of not. If I write f(___) = 0, what does that mean?*

$f(\underline{\quad}) = 0$

Student: *Now do we plug in 0?*

Student: *I don't think so. What you plug in goes inside the parentheses.*

Student: *Yeah, I think we need to find a y-value of 0.*

Teacher: *What about a y-value of 0?*

Student: *What is the x-value when the y-value is 0. It looks to me like it's in between −2 and −1.*

Teacher: *Right here? Nice. How could we get an exact answer?*

Student: *We know the slope is 2, but it seems like there must be an easier way.*

Student: *I think we can put the equation equals 0 and solve.*

Teacher: *Tell us more about that and I'll write it up here.*

The student talks about solving for *x* as the teacher represents it on the board. They plot the point and write in the −1.5 in the function notation.

x	$y = 2x + 3$
2	7
−1	1
−4	−5
½	4
1	5
1.5	0

(2, 7) $f(2) = 7$

$f(-1) = 1$ (−1, 1)

$f(-4) = -5$ (−4, −5)

(½, 4) $f(½) = 4$

(0, 3)

$f(1) = \underline{5}$ (1, 5) $y = 3 + 2x \quad y = b + 2x, 7 = b + 2(2), b = 3$

$f(1) = 3 + 2\,(1) = 3 + 2 = 5$

$f(\underline{-1.5}) = 0$ (1.5, 0) $f(?) = 2x + 3 = 0$

$2x = -3$

$x = -1.5$

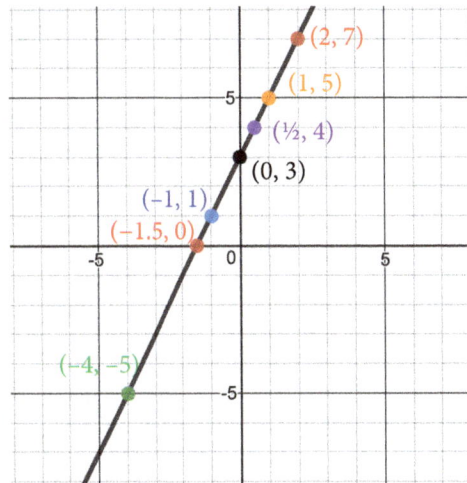

Teacher: *The last problem today is what is f of x? And what does that mean?*

Student: *I don't know. How do we plug an x in? What does that mean? It's not a number.*

Student: *I think it means that we do exactly that, plug an x in, that we write the equation of the line.*

Student: *That line stands for all of the values of x and y that are on that line, so yeah, f(x) = 2x + 3.*

Teacher: *Turn and talk about what you are thinking about f of x.*

Students turn and talk.

$f(x) = \underline{}$

(continued)

Teacher: *I heard lots of conversation about x's and y's and f. The function that we have been talking about this whole problem string is called f. And that function at any x value is 2 times that x plus 3 to get the y-value. So, I will add that to the table and the graph.*

(2, 7) $f(2) = 7$

$f(-1) = 1$ (−1, 1)

$f(-4) = -5$ (−4, −5)

(½, 4) $f(½) = 4$

(0, 3)

$f(1) = \underline{\ 5\ }$ (1, 5) $y = 3 + 2x \quad y = b + 2x, \ 7 = b + 2(2), \ b = 3$

$f(1) = 3 + 2\,(1) = 3 + 2 = 5$

$f(\underline{-1.5}) = 0$ (1.5, 0) $f(?) = 2x + 3 = 0$

$2x = -3$

$x = -1.5$

$f(x) = \underline{\ 2x + 3\ }$ (x, 2x + 3)

x	$y = 2x + 3$
2	7
−1	1
−4	−5
½	4
1	5
1.5	0
x	$2x + 3$

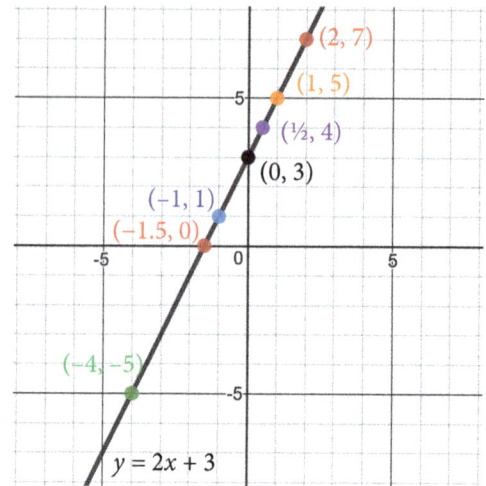

Teacher: *How would you summarize some of the things that came up in this string today?*

Elicit the following:

- *The name of this function is f.*
- *A function can have a lot of different points that have the same relationship.*
- *f(2) means the function value, the y-value, when x is 2.*
- *When you write a point in function notation, it's f of the x-value equals the y-value f(x-value) = y-value.*
- *When the function as specific input, it corresponds to a specific output, f(input) = output.*
- *f(x) is another way of referring to y-values.*
- *When you write out the function, f(x) = 2x + 3, that represents all of the (x, y) points that work in the function.*

Algebra Problem Strings
©2017 Kendall Hunt Publishing

Sample Final Display

Your display could look like this at the end of the problem string:

$(2, 7)$ $f(2) = 7$

$f(-1) = 1$ $(-1, 1)$

$f(-4) = -5$ $(-4, -5)$

$(½, 4)$ $f(½) = 4$

$(0, 3)$

$f(1) = \underline{\;5\;}$ $(1, 5)$ $y = 3 + 2x$ $y = b + 2x, 7 = b + 2(2), b = 3$

$f(1) = 3 + 2\,(1) = 3 + 2 = 5$

$f(\underline{-1.5}) = 0$ $(1.5, 0)$ $f(?) = 2x + 3 = 0$

$2x = -3$

$x = -1.5$

$f(x) = \underline{\;2x + 3\;}$ $(x, 2x + 3)$

x	$y = 2x + 3$
2	7
–1	1
–4	–5
½	4
1	5
1.5	0
x	$2x + 3$

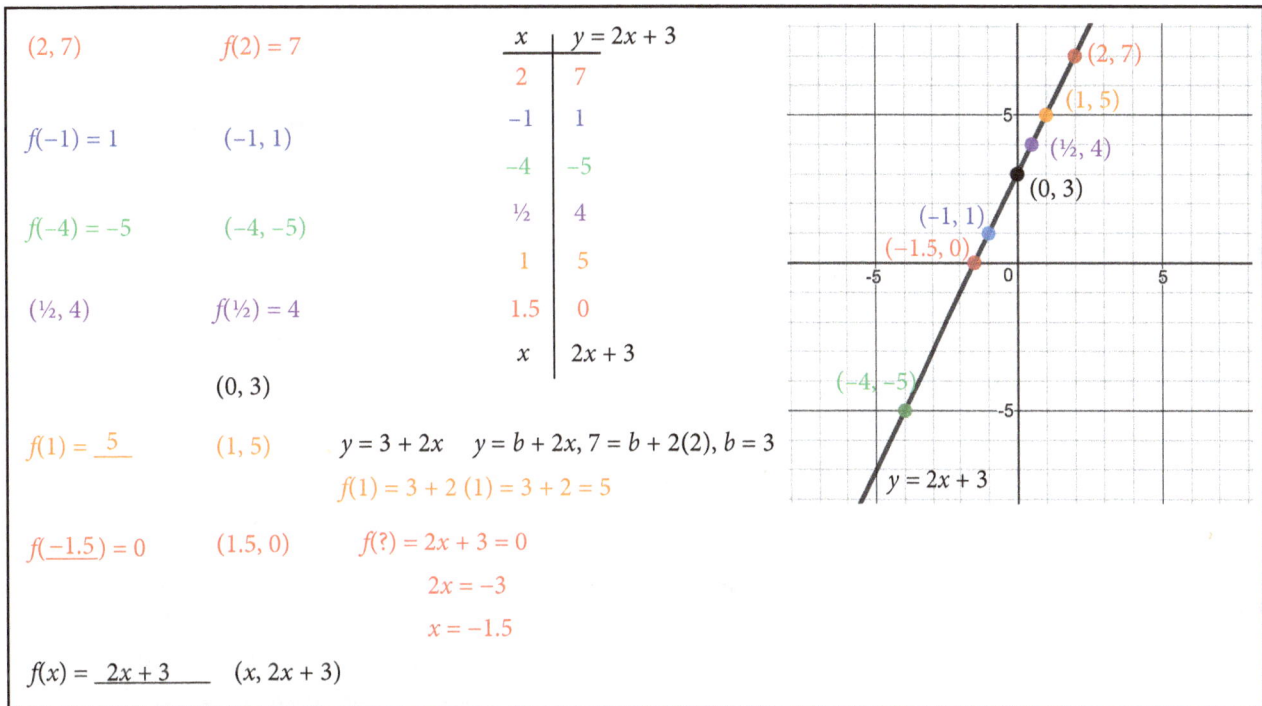

Facilitation Notes

This version of the problem string lists short notes for important teacher moves during the string. After you've done the string yourself and studied the relationships involved, you might make similar notes for the things you want a reminder of or deem important.

$(2, 7)$	What are some ways a mathematician might represent? Graph, table, function notation.
$f(-1) = 1$	Other direction. How might a mathematican represent? Ordered pair, graph, table.
$f(-4) = -5$	Repeat. Quick.
$(½, 4)$	What does this look like in function notation? Quick. Notice if anyone is thinking the points are colinear. Are the points colinear? Partner talk. Elicit and model slope triangles.
$f(1) = \underline{\quad}$	What does this mean? Elicit estimating with the graph. Elicit using the rate of change. Elicit finding and using the equation of the line.
$f(\underline{\quad}) = 0$	Similar, but different. What does this mean? How could we get an exact value? Elicit 2x+3 = 0, solve for x.
$f(x) = \underline{\qquad}$	What does this mean? What are all the y-values if we use x for the independent values?

7.2 Lines and Absolute Value

At a Glance	Objectives
	The goal of this problem string is for students to connect what they know about the graphs of linear equations to the graphs of related absolute value equations.

$$y = x$$
$$y = |x|$$
$$y = x + 3$$
$$y = |x| + 3$$
$$y = |x + 3|$$
$$y = 2x - 4$$
$$y = |2x - 4|$$
$$y = |2x| - 4$$
$$y = |x - 2|$$
$$y = |x| - 2$$

Placement
You can use this problem string to help students begin to consider the connection between the graphs of lines and related absolute value equations. This string's sample interaction assumes no prior experience with function transformations. If students have such experience, the conversation would include translations.

Use the problem string after you have introduced the absolute value function in textbook Lesson 7.2 Piecewise Functions and Absolute Value.

Guiding the Problem String
Use the class display grapher throughout this problem string to compare the graphs. Use different colors for each graph in the related sets to help students notice the relationships. Help students use the definition of the absolute value as the distance of a number from zero on a number line to help justify their predictions.

The first problem is the familiar linear parent function. Quickly use it to give students an opportunity to describe the relationship between it's graph and $y = |x|$. Spend a bit more time comparing the graphs of the next three problems related to $y = x + 3$, adding one problem at a time. Wonder with students about why the graphs pivot where they do. Repeat with the next set of three problems related to $y = 2x + 4$. Finish by asking students to create their own helper problem to help make sense of the graphs of the last two problems.

About the Mathematics
The absolute value of a number can be thought of as the distance that number is from zero on the number line. Because distance is never negative, the absolute value of a number is never negative. It is the size, or magnitude, of the number, without regard to wether the number is positive or negative.

Lesson 7.2 • Lines and Absolute Value (continued)

Sample Interactions

Use the following as you plan how to elicit and model student strategies. This is not meant as a script, but as a view into the relationships involved and the intent of the problem string.

Teacher: *This first problem is just to quickly establish what the line y = x looks like. Give me a brief description please.* **Student:** *It's the diagonal.* **Student:** *It has a slope of 1.* **Student:** *It goes through the origin (0, 0).* **Teacher:** *I'll put it in the display grapher. Look right?*	*y = x* *y = x*
Teacher: *We've just been talking about this function, called the absolute value function. What does it look like?* **Student:** *Like a v.* **Teacher:** *I'll put it in the display grapher also. How are these two graphs related?* **Student:** *Half is the same.* **Student:** *The absolute value is the same on the right but not on the left. On the left it goes up instead of down.* **Teacher:** *Another way to think about the absolute value function is that it represents the distance a number is from zero. How can you see that on the graph?* **Student:** *For all of the positive x's, it's that far, x, away from 0. But for the negative x's, the distance is positive.* **Teacher:** *Why does it change at zero?* **Student:** *At x = 0, it's 0 away from 0. That's the same as (0, 0) on y = x. But the negative x's are positive distance away from zero. Distance is positive.* **Teacher:** *So, the graph of absolute value of x kind of pivots there.*	*y = |x|* *y = |x|* *y = x*
Teacher: *This is an old friendly line that we've graphed before. What will the graph look like? Turn and tell your partner.* Students turn and talk while the teacher turns off the first two graphs and gets ready to display the new function.	*y = x + 3*

(continued)

Teacher: *Does this look like you were thinking? Yes?*

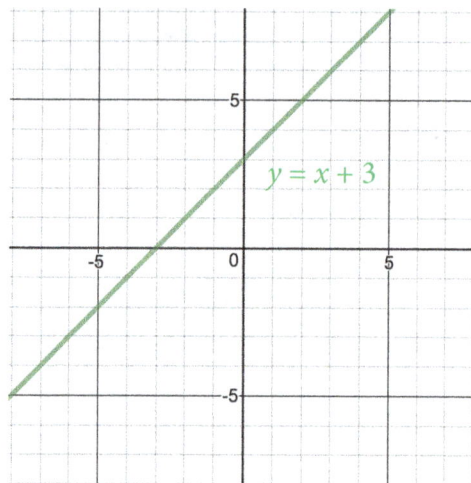

$y = x + 3$

Teacher: *Okay, now predict what you think the graph will look like for $y = |x| + 3$.*

Think time.

Teacher: *Here it is. What do you think? Why does it look like this? Turn and talk to your partner. How are these two graphs related? How are they different?*

Students turn and talk briefly while the teacher listens in.

Teacher: *What are you noticing?*

Student: *Again, the right half of the graph is the same as the line, but then it flips up.*

Student: *In the last one, it flipped up at 0. This time it flipped up at −3.*

Student: *If you plug in −3 to the purple function, you get 0 and the absolute value of 0 is 0.*

Student: *And if you plug in anything bigger than −3, you get positive numbers, so it's like the line. But if you plug in anything smaller, more negative, than −3, you get a negative number inside the absolute value, so that's why the graph goes up because it's a positive distance.*

$y = |x + 3|$

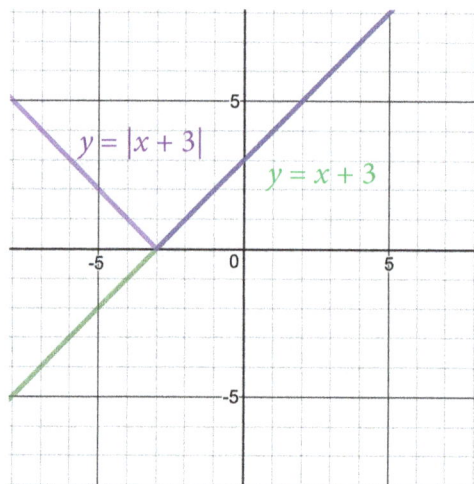

$y = |x + 3|$

$y = x + 3$

Algebra Problem Strings
©2017 Kendall Hunt Publishing

Teacher: *I wonder if any of that helps you make sense of what this graph might look like, $y = |x| + 3$? Turn and talk to your partner when you've had a chance to think about it and compare ideas.*

After a brief partner discussion, the teacher displays the graph.

Teacher: *How are you making sense of this graph?*

Student: *They all definitely have to do with the original line. It's like the absolute values are the line but then the line pivots somewhere.*

Teacher: *Where does it pivot? And why?*

Student: *I think it has to do with zero.*

Student: *Yeah, when y is zero.*

Student: *No, when x is zero. I'm not sure.*

Teacher: *What happens if you plug in $x = 0$? Or $y = 0$?*

Student: *I think it has to do with the stuff inside the absolute value being zero.*

Teacher: *Let's see how all of your ideas play out.*

$y = |x| + 3$

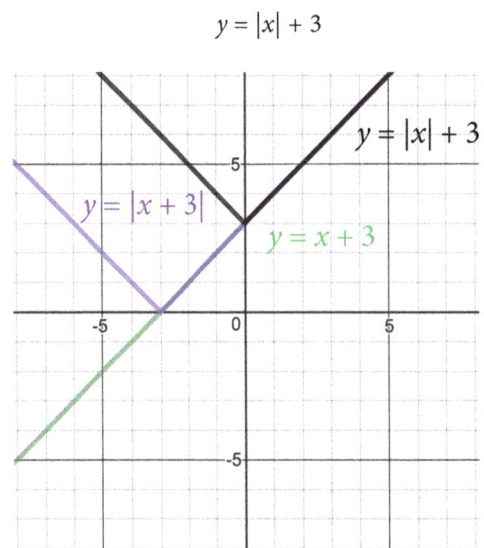

The teacher repeats with the next three problems, one at a time, asking students to briefly predict the graph and then justify the graph in terms of relationships.

Teacher: *Okay, what is going on?*

Student: *It's about where the absolute value is zero. When the stuff inside the absolute value is zero, then that's the turning point. After that, the graph is like the line. Before that, the graph is the line flipped up positive.*

Student: *Yeah, the pivot point is where the inside is zero. So $2x - 4$ is 0 at 2. And $2x$ is 0 at 0. Those are the places where the graph reflects up.*

Teacher. *Why?*

Student: *When you plug in x's that are bigger than the pivot point, the absolute value is positive. When you plug in a value that makes the absolute value zero, that's the pivot point. Then when you plug in values for x that are less, the absolute value is still positive, so the line reflects up.*

$y = 2x - 4$
$y = |2x - 4|$
$y = |2x| - 4$

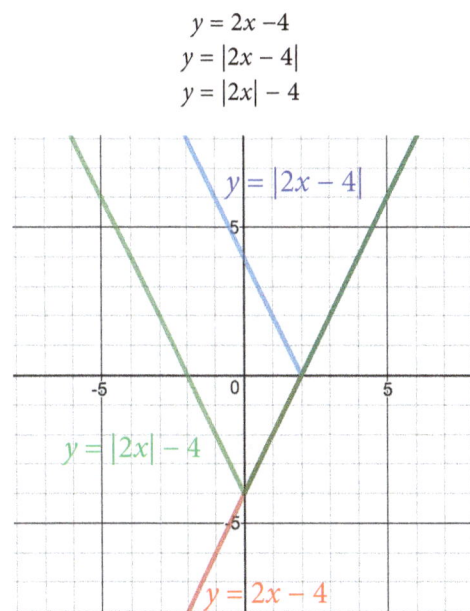

Teacher: *So, looking back at our string, I gave you a helper problem to think about the next two graphs. Then I gave you a helper problem for the fifth and sixth problems. For these last two problems, I did not make a helper problem. What helper problem can you think about to help with these graphs: $y = |x - 2|$ and $y = |x| - 2$?*

Brief think time.

$y = |x - 2|$

$y = |x| - 2$

(continued)

Teacher: *What is a good helper?* **Student:** *The line $y = x - 2$.* **Teacher:** *Thoughts on that? Agreed? Okay, then picture that graph. What does it look like? How will it help? Turn and talk to your partner and sketch what you think all three of the graphs will look like.* Students turn and talk while the teacher listens in.	$y = x - 2$
Teacher: *Okay, talk to me. Why do the graphs look like this?* **Student:** *It is all about where the stuff inside the absolute value is zero. That is the pivot point. When you go further to the left than that pivot point, the line reflects up.* **Student:** *That's because the absolute value of whatever is in there is always positive. So, the graph of the absolute value is the same as the line until the numbers inside the absolute value would be negative. Then the graph flips up.* **Teacher:** *We'll see more reflections like this in upcoming lessons. Nice work.*	

Teacher: *How would you summarize some of the things that came up in this string today?*

Elicit the following:

- *Absolute value is a way to describe the distance of the x-value from zero on a number line.*

- *The graph of the absolute value function $y = |x|$ is related to the line $y = x$.*

- *The graphs of functions like $y = |ax + b|$ are related to the graph of $y = ax + b$. They are the same when $ax + b \geq 0$, where $ax + b = 0$ is the pivot point, and they are the reflection over the x-axis when $ax + b < 0$.*

- *The graphs of functions like $y = |ax| + b$ are related to the graph of $y = ax + b$. They are the same when $ax \geq 0$, they pivot where $ax = 0$ is, and they reflect over the over the line $y = b$.*

Algebra Problem Strings
©2017 Kendall Hunt Publishing

Sample Final Display

Your display could look like this at the end of the problem string:

$y = x$

$y = |x|$

$y = x + 3$

$y = |x + 3|$

$y = |x| + 3$

$y = 2x - 4$

$y = |2x - 4|$

$y = |2x| - 4$

$y = |x| - 2$

$y = |x - 2|$

$y = x - 2$

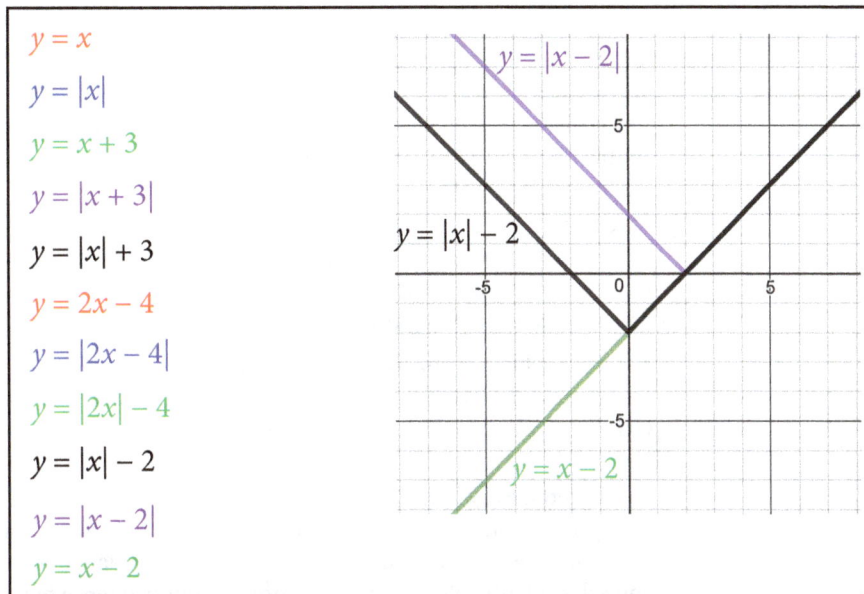

Facilitation Notes

This version of the problem string lists short notes for important teacher moves during the string. After you've done the string yourself and studied the relationships involved, you might make similar notes for the things you want a reminder of or deem important.

$y = x$	Quick. What does this look like? Show on display grapher.
$y = |x|$	Called the absolute value function. What does it look like? Show on display grapher. Discuss relationship. Clear display grapher.
$y = x + 3$	An old friend. What does this look like? Show on display grapher.
$y = |x + 3|$	Predict the graph. Pause. Show on display grapher. Why does the graph look like this? Justify using distance of a value from zero.
$y = |x| + 3$	Repeat. What matters? When x=0, y=0, something else?
$y = 2x - 4$	Quick. What does this look like? Show on display grapher.
$y = |2x - 4|$	Predict the graph. Pause. Show on display grapher. Why does the graph pivot and where? Where does 2x-4=0?
$y = |2x| - 4$	Predict the graph. Pause. Show on display grapher. Why does the graph pivot and where? Where does 2x=0?
$y = |x| - 2$ $y = |x - 2|$	The string has had helpers to help with the next problem(s). What function could help with these two? Graph suggested helpers. Elicit y=x-2. Use "reflection" to describe.

Connecting Linear, Absolute Value, and Quadratic Equations

At a Glance	Objectives
$x-3=2$ $\lvert x-3 \rvert = 2$ $(x-3)^2 = 4$ $x+4=3$ $\lvert x+4 \rvert = 3$ $(x+4)^2 = 9$	The goal of this problem string is to help students consider the connections between simple linear, absolute value, and quadratic equations.

Placement

You can use this problem string to introduce or strengthen connections between the parent absolute value function and the parent quadratic function in solving equations and graphing the functions.

Use this problem string to strengthen the connections students made between the parent absolute value function and the parent quadratic function in the investigation Graphing a Parabola in textbook Lesson 7.3 Squares, Squaring, and Parabolas.

Guiding the Problem String

This problem string is structured with two sets of problems, each set has a linear equation, a related absolute value equation, and a related quadratic equation. As you ask students to solve each equation, emphasize sense making by choosing students to share who are trying to reason about the solutions. If no one thinks about using graphs to find intersection points, ask students to consider the connections between the graphs. Keep all of the equations in a set on the display grapher simultaneously. Label the graphs and points and use color to help students make the connections.

About the Mathematics

The graphs of $y = f(x)$ and $y = g(x)$ can be used to find the solution to $f(x) = g(x)$. The solution(s) are the x-value(s) of the intersection point(s) of the graphs of $f(x)$ and $g(x)$.

Sample Interactions

Use the following as you plan how to elicit and model student strategies. This is not meant as a script, but as a view into the relationships involved and the intent of the problem string.

Teacher: *Today's problem string is going to be all about connections from things we've done to the new functions we have been learning. So, we'll start with an old friend. What is x if x subtract 3 is 2?*	$x - 3 = 2$
Student: *That's just 5.* **Teacher:** *And we've seen this modeled on an open number line. If x subtract 3 and 2 are in the same location, where is x?* **Student:** *It has to be 3 to the right, because you subtracted to get to x minus 3.* **Teacher:** *And we can model that using equations as well.*	$x = 5$ $x-3+3=2+3 \qquad x-3=2$ $\qquad x=5 \qquad\qquad +3 \ +3$ $\qquad\qquad\qquad\qquad\quad x=5$

Teacher: *We've also seen that modeled on an open coordinate grid. What does it look like to find where the line $y = x - 3$ intersects the line $y = 2$? Envision that in your mind. Picture each of those lines. Where do they intersect?*

Student: *They should intersect at 5.*

Teacher: *What do you mean at 5?*

Student: *It's two lines so they have to intersect at a point. It's where the lines have the same y-value, 2. So, I think it's the point $(5, 2)$.*

Teacher: *Let's take a look at those two lines. Where do they intersect?*

Student: *Yep, $(5, 2)$.*

$x - 3 = 2$

$x = 5$

$x - 3 + 3 = 2 + 3$

$x = 5$

$$\begin{aligned} x - 3 &= 2 \\ +3 \quad &+3 \\ \hline x &= 5 \end{aligned}$$

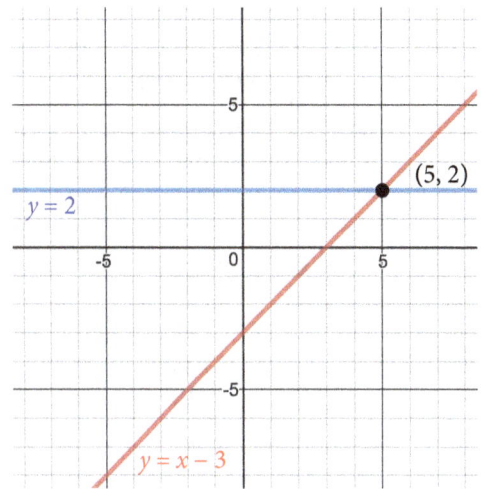

Teacher: *Okay, the next problem is what is x if the absolute value of the quantity x subtract 3 is 2? Solve that any way you want.*

Students work and the teacher circulates, looking for students who are solving algebraically and graphically.

$$|x - 3| = 2$$

Teacher: *I saw you writing equations. Please tell us about that.*

Student: *If x minus 3 is 2, then we already know that x is 5. If x minus 3 is −2, then x is 1.*

Teacher: *How do you know?*

Student: *Because if you add 3 to −2, that's 1.*

$x - 3 = 2 \qquad x - 3 = -2$

$x = 5 \qquad\quad x = 1$

(continued)

Teacher: *I saw you graphing. Tell us about that please.*	
Student: *We know the graph of y = |x − 3| will pivot at 3. So, there will be two intersections.*	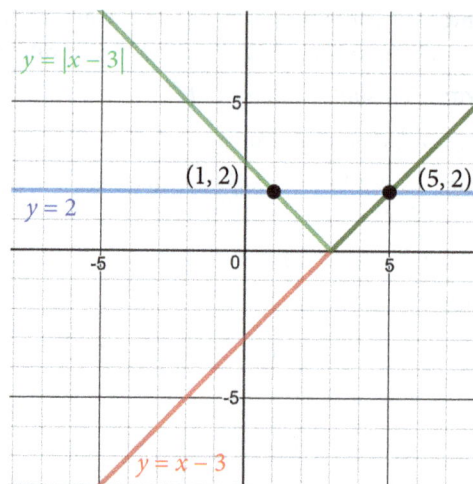
Teacher: *So, just visualizing the graphs helped you to think about the fact that there will be two answers?*	
Student: *Yes. We didn't really know where they were until we actually graphed them.*	
Teacher: *I'll put the graphs in the display grapher. So, here are the intersection points. So, what is the solution?*	
Student: *It's the x-values, 1 and 5.*	
Teacher: *How do you know it's the x-values, not the y-values of the points that are the solution?*	
Student: *Because we are setting the absolute value of x − 3 equal to y = 2, so we already know the y-values will be 2 when they intersect. We are looking for x when they have the same y-value.*	
Student: *The original problem only had an x in it, so we are looking for x-values.*	

Teacher: *The third problem today is to decide what is x, when x subtract 3 is squared and equals 4.*	$$(x-3)^2 = 4$$
I wonder what this one would look like?	
Students work and the teacher circulates, looking for students who are solving algebraically and graphically.	

Teacher: *I heard you two talking about something squared being 4. Could you tell us more about that please?*	$$(\quad)^2 = 4$$ $$(\quad) = 2$$
Student: *Yeah, we know that something squared is 4 so that thing is 2 because 2 times 2 is 4. So x − 3 is 2. Been there, done that. It's 5.*	$$x - 3 = 2$$ $$x = 5$$

Teacher: *But you two were graphing something. What were you graphing?*	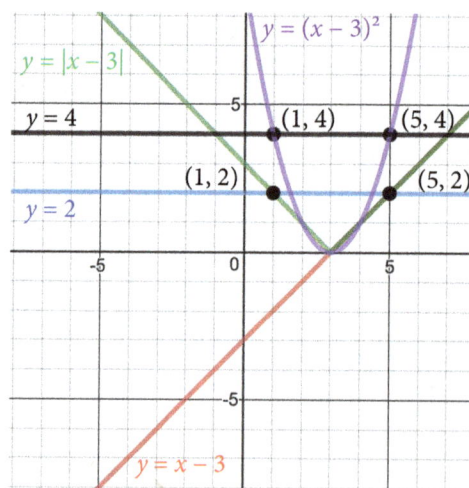
Student: *We thought about the parabola that we have just been learning about and graphed y = (x − 3)² and y = 4.*	
Student: *They intersect at x equals 1 and 5. We found two solutions, x = 1, 5.*	

Algebra Problem Strings
©2017 Kendall Hunt Publishing

Teacher: *Those of you who thought about a number squared, I'll just put parentheses to represent some number squared that equals 4 and you decided that it had to be 2. Do you have any other thoughts now? I can see you're smiling.* **Student:** *Right. We can also think about negative 2 squared, that's also 4. So, x minus 3 is negative 2 and we already solved that before, so x is 1. We agree that there are two solutions.*	$( )^2 = 4$ $( ) = -2$ $x - 3 = -2$ $x = 1$

Teacher: *I noticed that you just said twice that we had solved parts of this before. How are these problems related?*

Student: *The last two problems have the same answer.*

Student: *Answers.*

Teacher: *And where are those answers showing up in the graph.*

Student: *They are the same x-values.*

Teacher: *I wonder what this would look like if we graphed those x-values as a line, x = 1 and x = 5. Hmmm... what do you notice?*

Student: *That's interesting. They have the same intersection points.*

Student: *That's easier to see when you put the vertical lines on there.*

Teacher: *So, $|x-3| = 2$ and $(x-3)^2 = 4$ have the same solutions. Is that just coincidence? What do you think? Turn and talk to a partner.*

Teacher: *What are you thinking?*

Student: *The first two problems are related, because the absolute value is just like the line except it pivots at 3 and so you get two solutions.*

Student: *And we noticed that from the first problem to the third, you square the x minus 3 and you square the 2. So it makes sense that they would have the same solutions.*

$x - 3 = 2$
$x = 5$

$x - 3 + 3 = 2 + 3$
$x = 5$

$x - 3 = 2$
$+ 3 \ \ + 3$
$x = 5$

$|x - 3| = 2$
$x = 1, 5$

$x - 3 = 2 \quad x - 3 = -2$
$x = 5 \qquad x = 1$

$(x - 3)^2 = 4$
$x = 1, 5$

$( )^2 = 4$
$( ) = 2 \qquad ( ) = -2$
$x - 3 = 2 \quad x - 3 = -2$
$x = 5 \qquad x = 1$

(continued)

The teacher repeats with the last three problems, modeling student thinking and asking students to consider connections between the problems, the graphs, and the solutions.

$x + 4 = 3$

$x = -1$

$x + 4 - 4 = 3 - 4$ $x + 4 = 3$
 $\underline{-4 \quad -4}$
$x = -1$ $x = -1$

$|x + 4| = 3$

$x = -7, -1$

$x + 4 = 3$ $x + 4 = -3$
$x = -1$ $x = -7$

$(x + 4)^2 = 9$

$x = -7, -1$

$(\ )^2 = 9$
$(\ ) = 3$ $(\ ) = -3$
$x + 4 = 3$ $x + 4 = -3$
$x = -1$ $x = -7$

$y = (x+4)^2$

$(-7, 9)$ $(-1, 9)$ $y = 9$

$y = |x+4|$

$(-7, 4)$ $(-1, 4)$ $y = 3$

$y = x + 4$

$x = -7$ $x = -1$

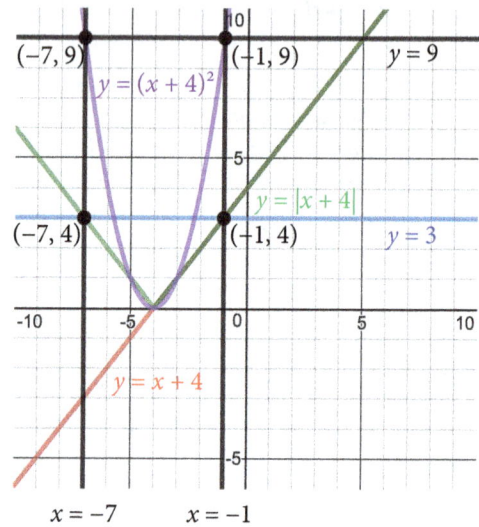

Teacher: *How would you summarize some of the things that came up in this string today?*

Elicit the following:

- We can find the solution to an equation like $|x - 3| = 2$ by finding the x-value of the intersection points of the graphs of $y = |x - 3|$ and $y = 2$.

- The graphs of functions like $y = |ax + b|$ are related to the graph of $y = ax + b$. They are the same when $ax + b \geq 0$, where $ax + b = 0$ is the pivot point, and they are the reflection over the x-axis when $ax + b < 0$.

- It looks like equations like $|x + a| = b$ are related to equations like $(x - a)^2 = b^2$ and that they have the same solutions.

Algebra Problem Strings
©2017 Kendall Hunt Publishing

Sample Final Display

Your display could look like this at the end of the problem string:

$x - 3 = 2$
$x = 5$

$x - 3 \quad 3 \quad x$

$x - 3 + 3 = 2 + 3$
$x = 5$

$x - 3 = 2$
$+ 3 \quad + 3$
$x = 5$

$|x - 3| = 2$

$x = 1, 5$

$x - 3 = 2 \qquad x - 3 = -2$
$x = 5 \qquad\quad x = 1$

$x - 3 \quad 3 \quad x$

$-2 \quad 3 \quad 1$

$(x - 3)^2 = 4$

$x = 5 \qquad x = 1$

$(\;)^2 = 4$
$(\;) = 2 \qquad (\;) = -2$
$x - 3 = 2 \qquad x - 3 = -2$
$x = 5 \qquad\quad x = 1$

$x + 4 = 3$
$x = -1$

$x \quad 4 \quad x + 4$

$-1 \quad 4 \quad 3$

$x + 4 - 4 = 3 - 4$
$x = -1$

$x + 4 = 3$
$- 4 \quad - 4$
$x = -1$

$|x + 4| = 3$

$x = -7, -1$

$x + 4 = 3 \qquad x + 4 = -3$
$x = -1 \qquad\quad x = -7$

$x \quad 4 \quad x + 4$

$-7 \quad 4 \quad -3$

$(x + 4)^2 = 9$

$x = -7, -1$

$(\;)^2 = 9$
$(\;) = 3 \qquad (\;) = -3$
$x + 4 = 3 \qquad x + 4 = -3$
$x = -1 \qquad\quad x = -7$

Graph with points $(-7, 9)$ and $(-1, 9)$ on $y = 9$; $y = (x + 4)^2$; $y = |x + 4|$; $y = 3$; points $(-7, 4)$ and $(-1, 4)$; $y = x + 4$; vertical lines $x = -7$ and $x = -1$.

(continued)

Facilitation Notes

This version of the problem string lists short notes for important teacher moves during the string. After you've done the string yourself and studied the relationships involved, you might make similar notes for the things you want a reminder of or deem important.

$x - 3 = 2$	We'll start with an old familar question. What is x? Picture, envision y=x−3 and y=2. Where do they intersect? Graph y=x−3 and y=2. Elicit that the solution to the equation is the x-value of the intersection point.
$\lvert x - 3 \rvert = 2$	Solve. Elicit algebraic solution. Elicit graphical solution. How does the graph help you know that there are 2 solutions to the equation?
$(x - 3)^2 = 4$	Solve. Elicit algebraic solution. Elicit graphical solution. How does the graph help you know that there are 2 solutions to the equation?
$x + 4 = 3$ $\lvert x + 4 \rvert = 3$ $(x + 4)^2 = 9$	Repeat with the next set of three equations, one at a time. How does the graph and solutions to the linear equations relate to the absolute value equations? How does the graph and solutions to the absolute value equations relate to the quadratic equations?

Algebra Problem Strings
©2017 Kendall Hunt Publishing

7.4 | Rate of Change

At a Glance	
(0, 0) (1, 1) (1, 1) (4, 2) (4, 2) (9, 3)	**Objectives** The goal of this problem string is for students to review finding the rate of change given 2 points.

Placement

You can use this problem string to review finding the slope of a line given 2 points while foreshadowing the graph of the square root function.

Use this problem string to quickly review finding the rate of change given two points in preparation for working with the average rate of change in textbook Lesson 7.4 Determining Rate of Change.

Guiding the Problem String

Project a blank grid on the board so that you can accurately plot the points in this problem string and then overlay the graph of the function at the end.

An alternative to starting with a stretched window is to start with a square window and just zoom out every time. If you do that, you'll need to redraw in the points and the connecting lines every time.

For each problem, ask students to find the rate of change for the line that contains the points. Nudge students toward thinking and reasoning to find the rates by asking students to share who plan ahead how to find the rate by thinking about the order to subtract or determining if the line is increasing or decreasing and using differences.

At the end of the string, ask students to stand back and look for relationships between the points. Overlay the graph of $y = \sqrt{x}$. Notice how the lines drawn between the points approximate the smooth radical function.

About the Mathematics

The points were purposefully chosen as points belonging to $y = \sqrt{x}$ so that students could gain experience finding the average rate of change between points that lie on a nonlinear curve.

We used a window of [−1, 10] by [−0.5, 3.5] so that when $y = \sqrt{x}$ is added at the end, students can see that the line segments approximate the curve, but do not match it. In a square window, it can be hard to distinguish the line segments from the function.

We created the graphs in Desmos by restricting the domain of each line. You can simply sketch in the lines, but if you want to use technology, here are the line segments: $y = x \{0 < x < 1\}$, $y = \frac{1}{3}(x-1) + 1\{1 < x < 4\}$, $y = \frac{1}{5}(x-4) + 2\{4 < x < 9\}$.

(continued)

Sample Interactions

Use the following as you plan how to elicit and model student strategies. This is not meant as a script, but as a view into the relationships involved and the intent of the problem string.

Teacher: *To get warmed up today, we're going to just refresh a bit on finding the rate of change between two points. What is the rate of change between these two points, (0, 0) and (1, 1)?*	(0, 0) (1, 1)
Teacher: *I saw some of you working on the formula to find the slope between two points. I also saw some of you not doing that. What were you thinking about?* **Student:** *It's just over 1 and up 1. That's a slope of 1.* **Student:** *It's like the line y = x, where the slope is 1.* **Teacher:** *Does anyone have any questions for them? Do you all understand how they were thinking? Remember, it's your job to understand or ask questions. No? Okay, I'll just plot those two points quickly and sketch in the line between them.*	
Teacher: *The second problem is to find the rate of change between these two points, (1, 1) and (4, 2)*	(1, 1) (4, 2)
Teacher: *I saw some of you subtracting, but thinking first. Tell us about that.* **Student:** *It's the difference in the y's divided by the difference in the x's. So 2 – 1 divided by 4 – 1, that's 1 divided by 3 which is one-third. Subtracting that way has no negatives.* **Student:** *Oh, nice! I subtracted 1 – 2 and 1 – 4, both with negative answers. Drat!* **Teacher:** *So, next time you want your brain to be inclined to think first and look to subtract with fewer negative results. Remember, that's why we share thinking, so that our brains have a chance to make connections that we can use later. So a ratio of up 1 to over 3? Yes? Okay.*	$m = \dfrac{2-1}{4-1} = \dfrac{1}{3}$ $m = \dfrac{1-2}{1-4} = \dfrac{-1}{-3}$

Algebra Problem Strings
©2017 Kendall Hunt Publishing

Teacher: *So, we should be able to start at (1, 1) and go over 3 and up 1 and run into the point (2, 4), right?*

And sure enough, we do. I'll sketch in the line between the two points.

Student: *That graph is really not square.*

Teacher: *How do you know?*

Student: *For one, the rectangles on the grid are really stretched, not close to squares. And for another that distance of 4 is almost the same as the distance of 1.*

Teacher: *Nice catch. So the graph is a bit skewed. I'm going to leave it that way for a reason that might become apparent with this next problem.*

Teacher: *The next problem is to find the slope between (4, 2) and (9, 3).*

The teacher erases the slope triangle.

(4, 2) (9, 3)

Teacher: *Who thought about the order to subtract? Tell us about that please.*

Student: *I subtracted 3 – 2 divided by 9 – 4, so 1 divided by 5 or one-fifth.*

Student: *I just knew that the slope was positive because of the points and I thought about the difference between 4 and 9 and 2 and 3 and I also got 1 to 5 or one-fifth.*

Teacher: *It sounds like you just envisioned the points and knew that the rate would be positive so then you just found the ratio of the differences.*

$$m = \frac{3-2}{9-4} = \frac{1}{5}$$

m is positive, $\dfrac{1}{5}$

Teacher: *I'll sketch that in. So the rate of 1 to 5, over 5 for every up 1, sure enough, gets us to that point (9, 3).*

Does anyone recognize any relationships between these points?

Student: *I think they are squared.*

Student: *Yeah, the x-values are squares of the y's.*

Student: *The y-values are just 0, 1, 2, 3.*

Teacher: *So, if you have an x-value of 4, how does that relate to the y-value of 2? If you have a x-value of 9, how does that relate to the y-value of 3? Turn to your partner and talk about these relationships.*

Students turn and talk while the teacher listens in.

Teacher: *What are you thinking?*

(continued)

Students: *It's like the squaring function we talked about yesterday.*

Student: *Except that the x's and y's are switched.*

Teacher: *What is the inverse, the undoing, of squaring?*

Student: *The square root.*

Teacher: *So, what if we graphed the square root of every number? Let's take a look at that graph, $y = \sqrt{x}$. What do you notice?*

Student: *It's like we were looking at the points on the function and finding the slope between them. The lines we drew kind of approximate the function. Cool.*

Teacher: *And that's a great lead in to today's lesson about finding average rates of change when the functions are not linear.*

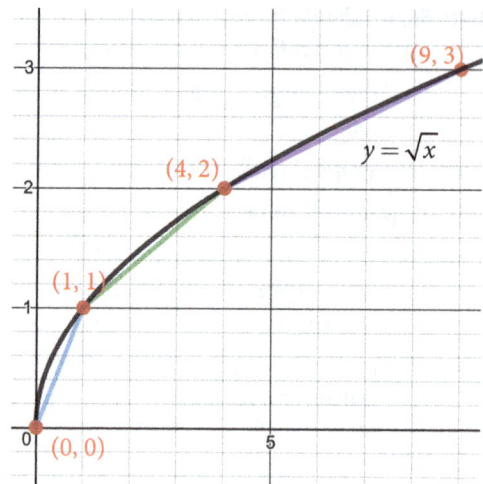

Teacher: *How would you summarize some of the things that came up in this string today?*

Elicit the following:

- *When finding the rate of change between two points, think before you subtract. Choose the order of subtraction wisely.*

- *The slope of the line between two points is the ratio of the difference between the y's to the difference between the x's.*

Sample Final Display

Your display could look like this at the end of the problem string:

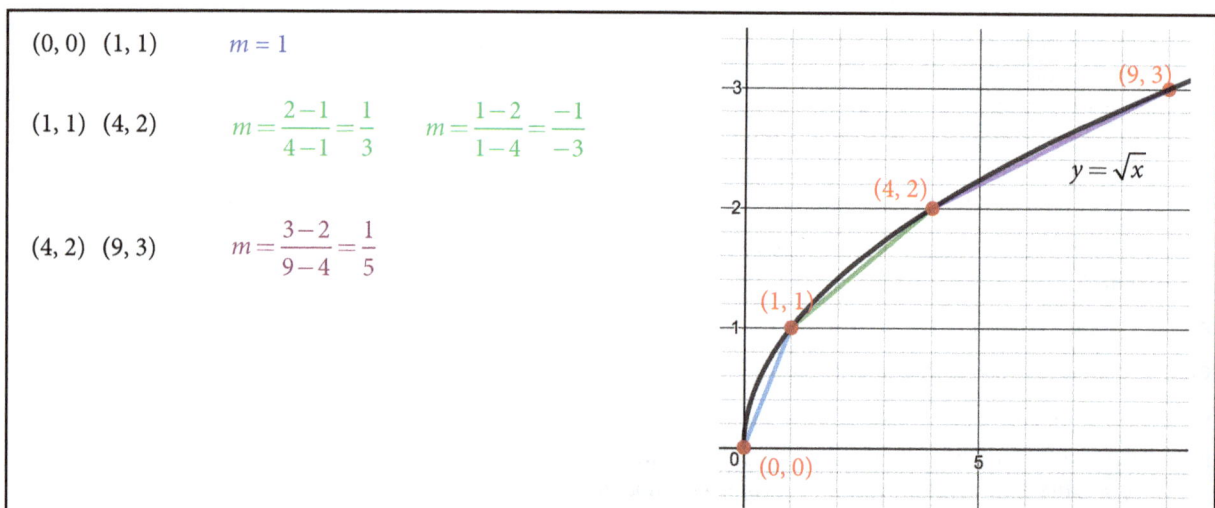

$(0, 0)\ (1, 1)$ $m = 1$

$(1, 1)\ (4, 2)$ $m = \dfrac{2-1}{4-1} = \dfrac{1}{3}$ $m = \dfrac{1-2}{1-4} = \dfrac{-1}{-3}$

$(4, 2)\ (9, 3)$ $m = \dfrac{3-2}{9-4} = \dfrac{1}{5}$

Algebra Problem Strings
©2017 Kendall Hunt Publishing

Facilitation Notes

This version of the problem string lists short notes for important teacher moves during the string. After you've done the string yourself and studied the relationships involved, you might make similar notes for the things you want a reminder of or deem important.

$(0, 0)$ $(1, 1)$	Project a grid on the display. To get warmed up for today, let's find the rate of change between these two points. Elicit just looking and thinking about y=x. Sketch the points and the line between.
$(1, 1)$ $(4, 2)$	Find the rate. Elicit subtracting in order. Elicit planning ahead the order to subtract. Sketch the points and the line between. Start from (1, 1) and draw slope triangle to confirm.
$(4, 2)$ $(9, 3)$	Find the rate. Elicit planning ahead the order to subtract. Elicit finding differences and considering the direction of the line. Sketch the points and the line between. Start from (4, 2) and draw slope triangle to confirm. What relationships do you see? Partner talk. Display the square root parent function with the display grapher.

7.5 Transforming Functions 1: Translating

At a Glance

$y = x^2$

$y = |x|$

$y = (x - 250)^2$

$y = |x| - 750$

$y = |x + 3000| - 1200$

$y = (x - 120)^2 + 425$

Objectives

The goal of this problem string is to develop students' facility with translations of functions. To do this students find appropriate viewing windows for translations of the parent functions $y = |x|$ and $y = x^2$. Using the power of technology, students can quickly test their assumptions by adjusting viewing windows. As they are pressed to defend their window choices, students build skill translating functions.

Placement

This is the first in a series of four problem strings that use finding appropriate viewing windows as a vehicle to build and strengthen students' understanding and facility with transformations. This string assumes students have been introduced to the parent absolute value and quadratic functions. You could use this problem string as students are learning about vertical and horizontal translations.

Use this problem string to help students solidify the learning in textbook Lesson 7.5 Translating Graphs.

Guiding the Problem String

The first two problems are intended to ground students in the behavior of the parent functions $y = |x|$ and $y = x^2$ and to establish some parameters for appropriate viewing windows. If students are familiar with absolute value and quadratic functions, focus these questions on what it means to find a good viewing window. The rest of the problems are designed to be well outside of the beginning window for most graphing software. Encourage students to predict the window using what they know. Allow students to swipe or zoom, but encourage them to predict first before they just start guessing. Then press them for justification when they do find the function, asking questions like, "*Why is the function there?*" and "*How can you use your knowledge of transformations to defend your choice?*"

When you gather students to collectively find an appropriate window, use the window setting feature, asking students to justify choices based on transformations.

About the Mathematics

There are infinite possibilities for appropriate viewing windows for each function. It is not important that students find the same windows. It is important that the windows are appropriate (showing the important features with not a lot of extra space). It is most important that students begin to use their knowledge of transformations to guide their thinking and justify their choices.

The graphs of these functions represent continuous, infinite relationships. As you transform functions, do not refer to functions as the letters *U* or *V* or use language that would suggest that the transformations are moving around a static shape. The parent parabola is not a static shape that eventually goes vertical, but an infinite set of points that keeps increasing as *x* increases and symmetrically increases as *x* decreases. In this string, each of those points are being translated, creating a transformed, continuous function.

Sample Interactions

Use the following as you plan how to elicit and model student strategies. This is not meant as a script, but as a view into the relationships involved and the intent of the problem string.

Teacher: *To start off today, let's visualize the graph of the parent function $y = x^2$. Turn and talk about what you picture in your mind.*

Students turn and talk while the teacher listens in.

Teacher: *I'm going to graph it with our display grapher. Is this a good viewing window for this function?*

Student: *Sure, you can see the whole function.*

Student: *Well, not the whole function since it goes on and on forever, but you get the gist of it.*

Teacher: *What do you mean the gist of it?*

Student: *You can see the important parts.*

Teacher: *What are the important parts of a quadratic function?*

Student: *The shape. The x-intercepts. The y-intercepts. The vertex.*

Teacher: *And we can see all of that so we think it's a good viewing window, or an appropriate viewing window. Good can be subjective. Can we agree that a good viewing window shows the important parts of a function?*

Student: *We don't really need all of that blank area at the bottom.*

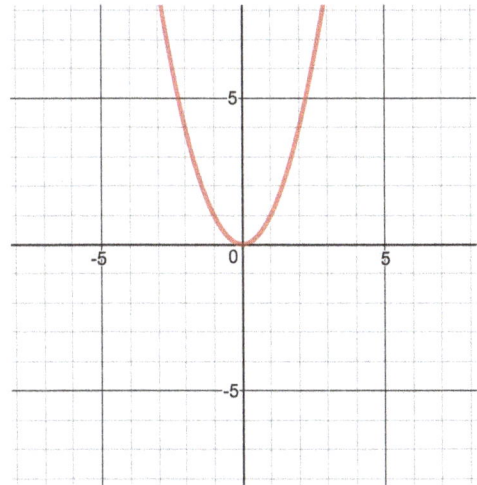

Teacher: *Is anyone bothered by all of the extra space in the bottom of the graph?*

Student: *You could get rid of that and fill the window with more of the function.*

Teacher: *Like how?*

Student: *Make the y's start at 0 or maybe a little lower, like −1.*

Student: *Now it feels like we could see less of the x's so that it would spread out a bit.*

Student: *What do you mean?*

Student: *There's too much space on the sides. We could zoom in more on the sides.*

Teacher: *How would you do that?*

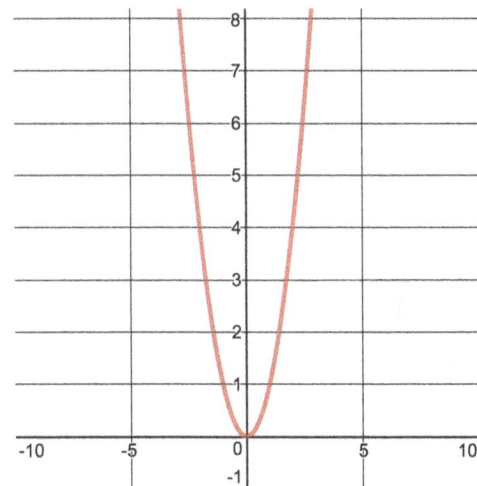

(continued)

Student: *Right now, we only see the parabola in between −4 and 4. You could change the x's to −4 and 4.*

Student: *Yes, that looks better.*

Teacher: *Why do you say that it looks better?*

Student: *We got rid of the extra and the parabola looks good, not squashed or stretched or missing parts.*

Teacher: *Could we have other windows that we could call good? Let's just say that there could be lots of good windows and when we find a window, it just needs to be good enough, showing the important parts, but not a lot of empty space.*

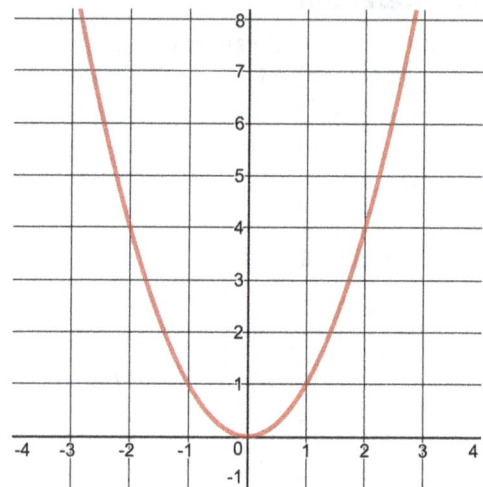

Teacher: *What is a good viewing window for y = |x|? First, envision what the graph looks like.*

Now I'll start us off by going back to the home window and putting it in the display grapher. Is this a good viewing window?

Student: *If we want to get rid of the extra space, cut off the bottom.*

Teacher: *How?*

Student: *The y's should only go to like −1.*

$$y = |x|$$

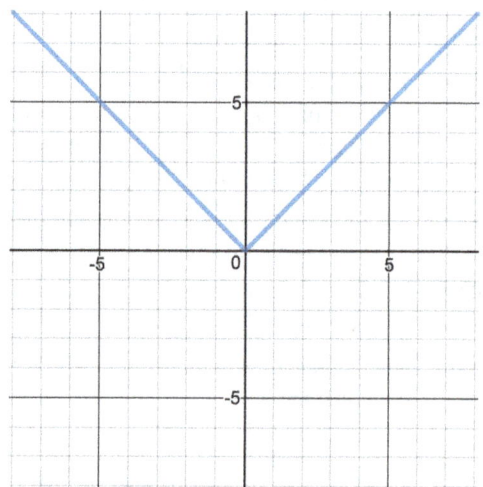

Teacher: *What do we think of this window?*

Student: *It's good.*

Teacher: *I'll switch back and forth between the two windows. What do you notice?*

Student: *I notice that there is less empty space?*

Teacher: *Did anyone notice that the graph got steeper?*

Student: *Yes, at first the lines were at a 45 degree angle. Now they are steeper.*

Student: *And also that the grid rectangles are stretched. They were more square before.*

Teacher: *So, in the home window, the rectangles are square. We call that a square window. It represents the graph without it being stretched or squished. Should we add something about square windows to our list of good windows?*

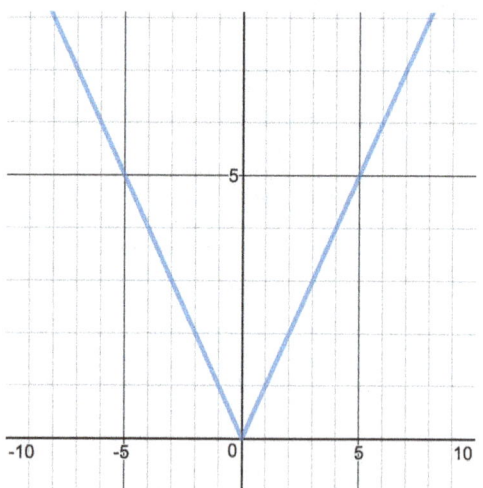

Algebra Problem Strings
©2017 Kendall Hunt Publishing

Student: *Is that what the zoom square button is for?*

Teacher: *Shall we check that out? Here is the graph after we've chosen zoom square. What do you notice?*

Student: *The squares are squares!*

Student: *And the lines look like 45 degrees again.*

Student: *But now we have extra space!*

Teacher: *So, we might have to choose between a square window or seeing too much extra blank space.*

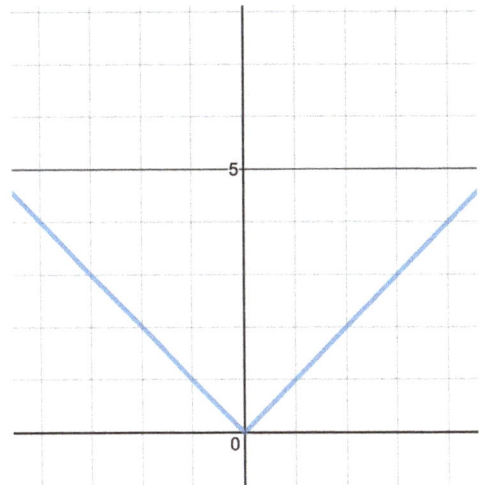

Teacher: *Let's have some fun finding good viewing windows for functions. The next problem is to find a good viewing window for $y = (x-250)^2$. You can use your devices to find it, but try to reason about it, rather than just randomly guessing.*

The teacher circulates and looks for students who are reasoning and students who are swiping or zooming in and out.

If the students do not have devices on which to graph, ask students to visualize where they could find the graph. Then work together as a class to find an appropriate window.

Teacher: *Who thinks they've found a good window?*

Student: *I think mine's fine. It's just over to the right so I can see 250.*

Teacher: *How did you find it?*

Student: *I just swiped over and there it was.*

$$y = (x-250)^2$$

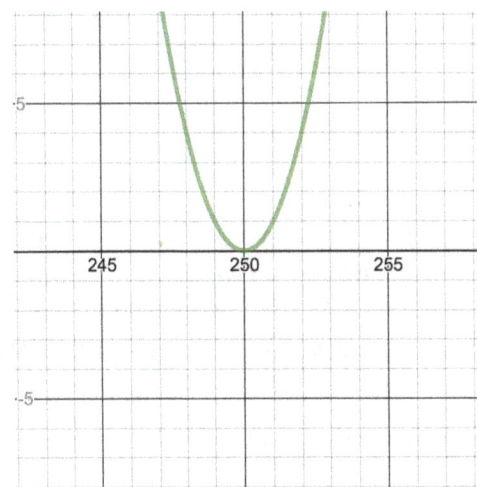

Teacher: *So, with a bit of luck, you ran into it? Can we make sense of why we can see the function in this window?*

Student: *You've got x – 250 in the parentheses.*

Teacher: *Right, we've replaced x with x – 250.*

(continued)

Student: *Yeah, so the function has to be 250 to the right of where the parent function was. But, I would lose the bottom part again. We don't need that extra space.*

Student: *And while you're at it, close in on the x's too. You've got too much horizontal space, just use from 245 to 255.*

Teacher: *Like this? What do you all think? I see lots of nodding. Why look to the right?*

Student: *When it's x −250, the graph shifts right.*

Teacher: *So, when you replace x with x − 250, the graph shifts, or translates, right. Let's make a note of that.*

$y = x^2$

$y = |x|$

$y = (x - 250)^2$ Look right!

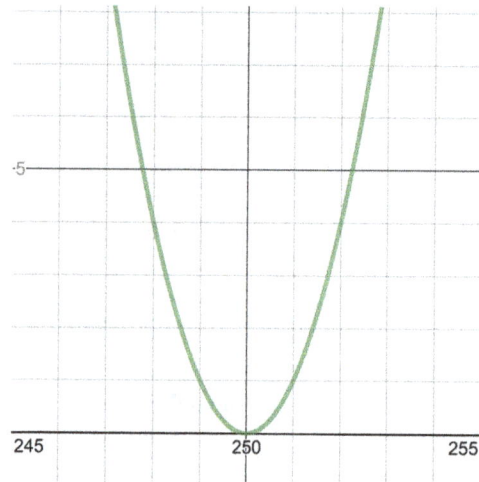

Teacher: *For this next problem, I want you to think and predict first, before you just start swiping or zooming. The next problem is to find an appropriate viewing window for $y = |x| - 800$ in as few moves as possible. Be ready to explain why you are looking where you are looking.*

The teacher circulates, asking students to defend their choices or justify a window if they just happen on it by swiping.

Teacher: *When you are ready, turn and share your window and defend your choices.*

Students turn and talk and the teacher circulates.

Algebra Problem Strings
©2017 Kendall Hunt Publishing

Teacher: *Let's come back together as a group and find a viewing window that we can agree is good enough. I noticed that you two were swiping and then you gave up on that. Would you tell us about that please?*

Student: *We thought it might be down but after swiping down we still didn't see it, so we changed the window by putting in numbers.*

Teacher: *What window did you decide on?*

The teacher starts with their window and then with student input finds a suitable window.

Teacher: *What could we make a note of here, with this function and why?*

Student: *Look down! Because you have the original |x| but down 800.*

$y = x^2$

$y = |x|$

$y = (x - 250)^2$ Look right!

$y = |x| - 800$ Look down!

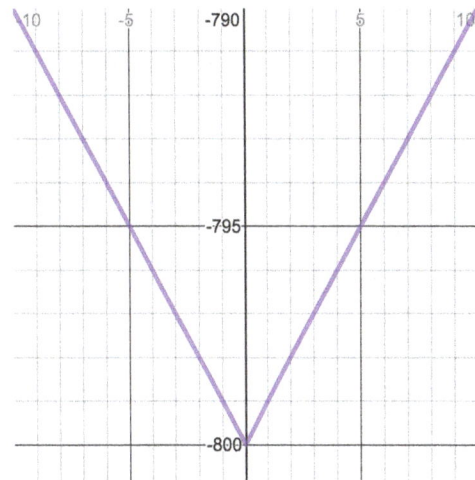

This is only one possible appropriate viewing window for this function. It is less important that students find the same windows. It is important that the windows are appropriate (showing the important features with not a lot of extra space.)

Teacher: *For the next problem, I want you to work with your partner. One of you use the grapher. The other partner makes suggestions. You have to agree before the graphing partner can input the ideas. So, there should be lots of discussing and agreeing before the partner making the changes actually makes any changes. Got it? Find an appropriate viewing window for $y = |x + 3000| - 1200$. Remember, talk and agree first, then make the changes.*

The teacher circulates as students work, encouraging good partner work and pressing for justification. Then the teacher calls the class together and they decide on an appropriate viewing window such as the following. The class suggests adding the note to "Look left and down!"

$y = x^2$

$y = |x|$

$y = (x - 250)^2$ Look right!

$y = |x| - 800$ Look down!

$y = |x + 3000| - 1200$ Look left and down!

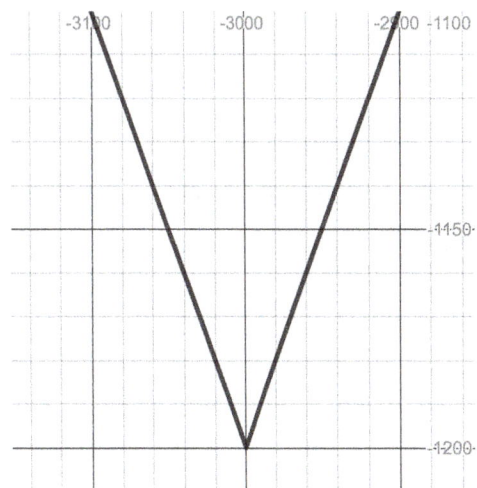

(continued)

> **Teacher:** *The last problem today is to find a suitable viewing window for* $y = (x-120)^2 + 425$. *What's going on with this function? How can that help you find a good viewing window? This time, change who is holding the grapher. Remember, you have to agree before that person makes any changes. Go!*
>
> See the Sample Final Display for a possible viewing window and note for this problem.

Teacher: *How would you summarize some of the things that came up in this string today?*

Elicit the following:

- *Translations really do shift functions around! Translations are additive changes.*

- *Using our knowledge of translations, we can find a window that shows the important features of the function that has been translated.*

- *If you add to or subtract from the function, the y-values change additively.*

- *If you replace x with x plus or minus something, the function shifts left or right. The important features now happen at different x-values.*

- *There can be lots of different appropriate viewing windows for the same function, but they should all include the important features of the function and not extra space.*

Sample Final Display

Your display could look like this at the end of the problem string:

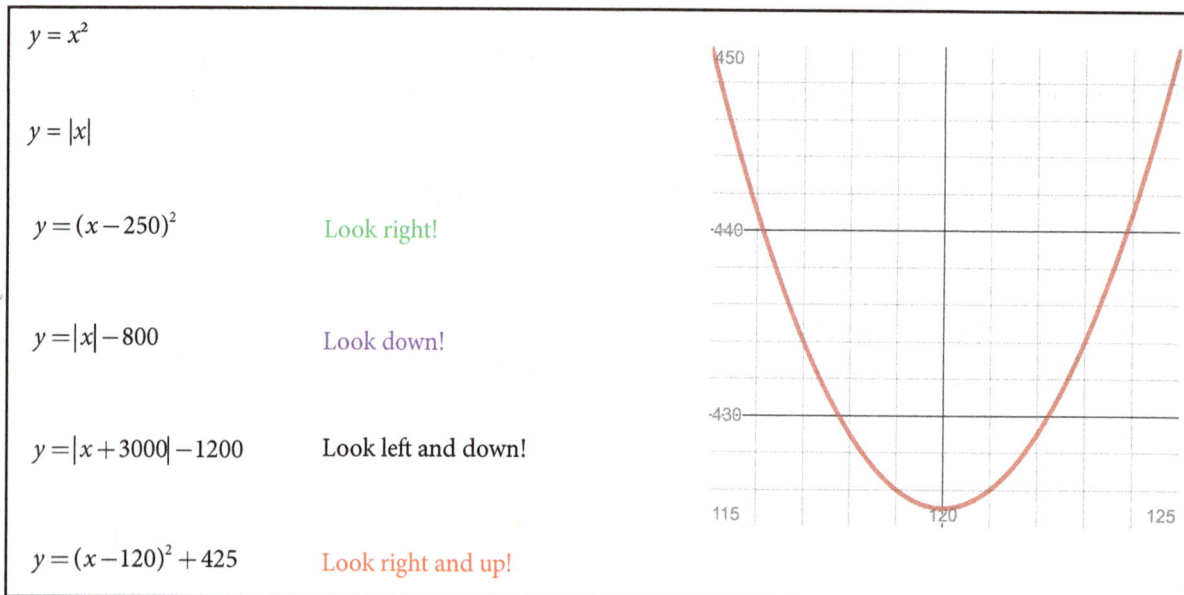

$y = x^2$	
$y = |x|$	
$y = (x-250)^2$	Look right!
$y = |x| - 800$	Look down!
$y = |x+3000| - 1200$	Look left and down!
$y = (x-120)^2 + 425$	Look right and up!

Algebra Problem Strings
©2017 Kendall Hunt Publishing

Facilitation Notes

This version of the problem string lists short notes for important teacher moves during the string. After you've done the string yourself and studied the relationships involved, you might make similar notes for the things you want a reminder of or deem important.

$y = x^2$	Visualize: what does the graph look like? Display with display grapher. What makes a good, appropriate viewing window? Shows important features, no extra space.
$y = \lvert x \rvert$	Visualize: what does the graph look like? Find a good viewing window. Discuss square windows. We might have to choose between no extra space and square windows.
$y = (x - 250)^2$	Find a good viewing window. Work together as a class. "Replace x with x-250." Make a note "Look right".
$y = \lvert x \rvert - 800$	Predict! Find a good viewing window in as few moves as possible. Turn and defend choices. Find an appropriate viewing window as a class. Press for justification. Note "Look down."
$y = \lvert x + 3000 \rvert - 1200$	Partner up. One holds grapher. Only make changes when both agree. Talk, defend, change, adjust. Find an appropriate viewing window as a class. Press for justification. Note "Look left and down."
$y = (x - 120)^2 + 425$	Repeat. Note "Look right and up." What do we know about translations?

Transforming Functions 2: Reflecting

At a Glance

$y = 2^x$

$y = -2^x$

$y = 2^{-x}$

$y = -2^{-x}$

$y = -2^x + 100$

$y = -2^{x+100}$

$y = -x^2$

$y = -|x|$

$y = -|x - 400| - 600$

$y = -(x + 2000)^2 + 5000$

Objectives

The goal of this problem string is to develop students' facility with reflections of functions. To do this students find appropriate viewing windows for reflections of several parent functions. Using the power of technology, students can quickly test their assumptions by adjusting viewing windows. As they are pressed to defend their window choices, students build skill reflecting functions.

Placement

This is the second in a series of four problem strings that use finding appropriate viewing windows as a vehicle to build and strengthen students' understanding and facility with parent functions and transformations. You could use this problem string as students are learning about reflections.

This problem string could come during the work of textbook Lesson 7.6 Reflecting Points and Graphs.

Guiding the Problem String

Use the first question to establish the behavior of the parent function. The next three problems play with reflecting the exponential function over the y-axis, the x-axis and then both axes. These should go fairly quickly, but take the time to watch a few points being reflected to reinforce that you are transforming a whole set of relationships, not a static shape. The remaining problems combine translations and reflections with various functions. You can change up the rhythm of the problem string by asking students to predict first, having students work with a partner where they must agree before changing anything, having students work individually and then comparing windows with a partner, or finding an appropriate window together as a class. The emphasis should always be on pressing for justification of the viewing window based on transforming the parent function.

See problem string 7.5 for a sample dialogue of a problem string with a similar structure.

About the Mathematics

The graphs of these functions represent continuous, infinite relationships. As you transform functions, do not refer to the functions using language that would suggest that the transformations are moving around a static shape. The parent exponential function is not a static shape that eventually goes horizontal to the left and vertical to the right, but an infinite set of points that keeps getting closer to zero as x decreases and increases as x increases. In this string, each of those points are being reflected across an axis, creating a transformed, continuous function.

Algebra Problem Strings
©2017 Kendall Hunt Publishing

Important Questions

Use the following as you plan how to elicit and model student strategies.

- *What is the general behavior of the parent function?*

- *What happens to the y-values of f(x) in the transformation −f(x)? Which variable is being multiplied by −1? How does that affect the graph?*

- *What happens to the function f(x) when you replace x with −x, so f(−x)?*

- *Are the transformations −f(x) and f(−x) additive or multiplicative?*

- *What changes, the x- or y-values, when you reflect a function over the x-axis? How can we represent that using function notation?*

- *What changes, the x- or y-values, when you reflect a function over the y-axis? How can we represent that using function notation?*

- *Which transformations do you consider first? Does it matter?*

How would you summarize some of the things that came up in this string today?

- *Reflections are transformations where you reflect the function over a line. This happens with either −f(x) or f(−x).*

- *When a function f(x) reflects over the x-axis, all of the y-values take on the opposite sign. What y-values were once positive are now negative. What y-values were once negative are now positive. This is represented by −f(x).*

- *When a function f(x) is reflected over the y-axis, all of the y-values that used to correspond to x, now correspond to −x. This is represented by f(−x).*

- *We can combined reflections and translations. We are wondering if order matters. We are thinking it makes sense to reflect first then shift.*

(continued)

Sample Final Display

Your display could look like this at the end of the problem string:

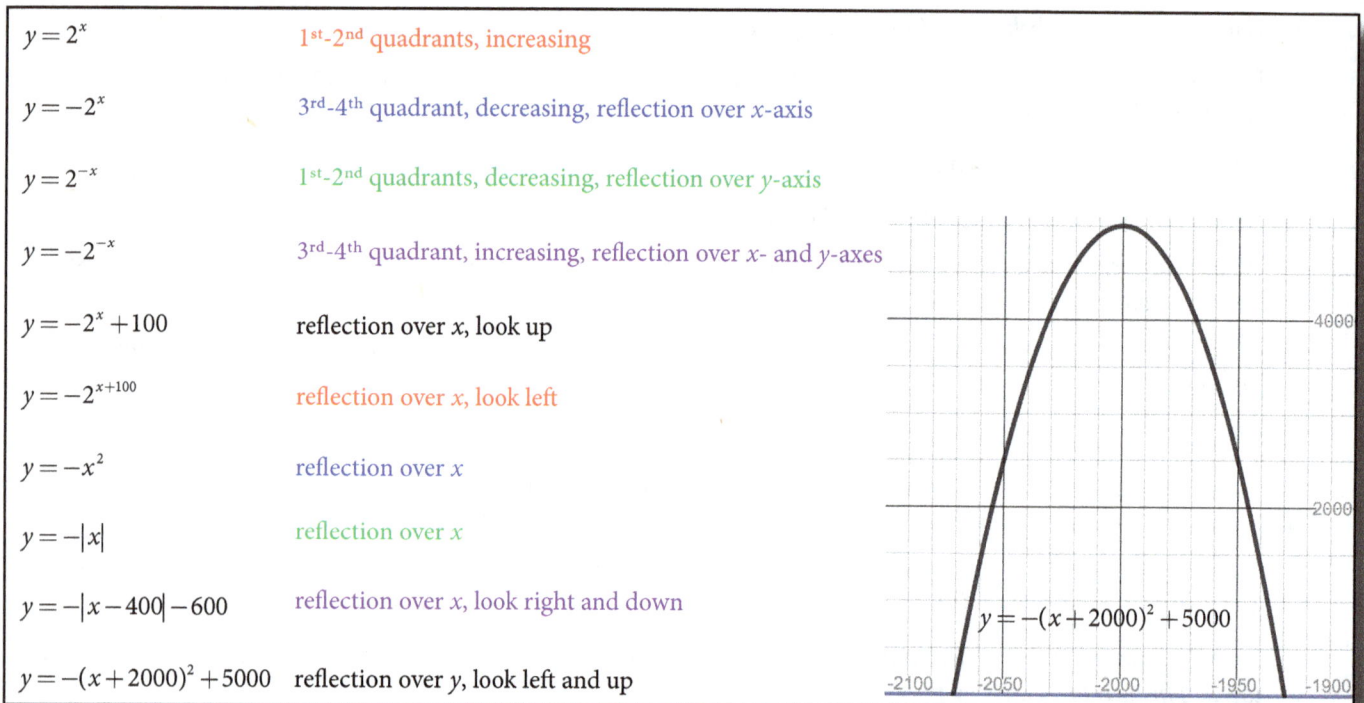

$y = 2^x$	1st-2nd quadrants, increasing
$y = -2^x$	3rd-4th quadrant, decreasing, reflection over x-axis
$y = 2^{-x}$	1st-2nd quadrants, decreasing, reflection over y-axis
$y = -2^{-x}$	3rd-4th quadrant, increasing, reflection over x- and y-axes
$y = -2^x + 100$	reflection over x, look up
$y = -2^{x+100}$	reflection over x, look left
$y = -x^2$	reflection over x
$y = -|x|$	reflection over x
$y = -|x - 400| - 600$	reflection over x, look right and down
$y = -(x+2000)^2 + 5000$	reflection over y, look left and up

$y = -(x+2000)^2 + 5000$

Facilitation Notes

This version of the problem string lists short notes for important teacher moves during the string. After you've done the string yourself and studied the relationships involved, you might make similar notes for the things you want a reminder of or deem important.

$y = 2^x$	Visualize: what does the graph look like? Display with display grapher. How would you describe the window and graph? Note "1st-2nd quadrants, increasing." Review appropriate viewing windows.
$y = -2^x$	Find a good viewing window. Note "3rd-4th quadrants, decreasing, reflection over x-axis."
$y = 2^{-x}$	Replace x with -x, the opposite of x. Repeat. What do we know about reflections? Note "1st-2nd quadrants, decreasing, reflection over y-axis."
$y = -2^{-x}$	Predict! How can you use what you know? Note "3rd-4th quadrants, increasing, reflection over x-, y-axes."
$y = -2^x + 100$	Partner up. One holds grapher. Only make changes when both agree. Talk, defend, change, adjust. Find an appropriate viewing window as a class. Press for justification. Note "Reflection over x, look up."
$y = -2^{x+100}$	Repeat. Note "Reflection over x, look left."
$y = -x^2$	Repeat. Quick. Note "Reflection over x."
$y = -|x|$	Repeat. Quick. Note "Reflection over x."
$y = -|x - 400| - 600$	Repeat. Note "Reflection over x, look right and down."
$y = -(x+2000)^2 + 5000$	Repeat. Note "Reflection over x, look left and up."

Algebra Problem Strings
©2017 Kendall Hunt Publishing

Transforming Functions 3: Stretching and Compressing

At a Glance

$y = x^2$

$y = 0.01x^2$

$y = 1000x^2$

$y = 10(x - 250)^2$

$y = \frac{1}{4}x^2 - 500$

$y = -20x^2 - 10$

$y = -\frac{1}{5}(x - 30)^2 + 150$

Objectives

The goal of this problem string is to develop students' facility with dilating functions. To do this students find appropriate viewing windows for vertical stretches and compressions of parent functions. Using the power of technology, students can quickly test their assumptions by trying and adjusting. As they are pressed to defend their window choices, students build skill dilating functions.

Placement

This is the third in a series of four problem strings that use finding appropriate viewing windows as a vehicle to build and strengthen students' understanding and facility with transforming parent functions. You could use this problem string as students are learning about vertical dilations.

This problem string could come during the work of textbook Lesson 7.7 Stretching and Shrinking Graphs.

Guiding the Problem String

The first problem is the parent function, $y = x^2$ as a reference for the rest of the string. The next two problems deal with first vertically compressing the quadratic function by a scale factor of 0.01 and then vertically stretching the quadratic function by a scale factor of 1,000. Take some time to follow how a few points change location multiplicatively to reinforce that you are transforming a whole set of relationships, not a static shape. The remaining problems are combinations of translating, reflecting, and dilating the quadratic parent function. You can change up the rhythm of the problem string by asking students to predict first, having students work with a partner where they must agree before changing anything, having students work individually and then comparing windows with a partner, or finding an appropriate window together as a class. The emphasis should always be on pressing for justification of the viewing window based on the transformations.

About the Mathematics

The graphs of these functions represent continuous, infinite relationships. As you work with transformations, do not refer to the function as the letter U or use language that would suggest that the transformations are squishing a static shape. Words like skinnier, taller, fatter, and smaller all connote a static shape. The quadratic parent function is not a static shaped U that ends, but an infinite set of points that keep increasing as x increases and decreases. In this string, the distances from the center of dilation, the origin, of those points are being dilated, a multiplicative change, creating a transformed, continuous function. Use words like, "the y-values are getting bigger faster" to describe vertical stretches or "the y-values are getting bigger more slowly" to describe vertical compressions. Points represent a location in space and have no size.

(continued)

Sample Interactions

Use the following as you plan how to elicit and model student strategies. This is not meant as a script, but as a view into the relationships involved and the intent of the problem string.

Teacher: *We are going to find some more viewing windows today. The first problem is $y = x^2$. Picture in your mind what this function looks like.*

Now show me physically what it would look like. Look around, do you agree?

Let's put that in our display grapher as a reference. What do you think? Does it match your prediction?

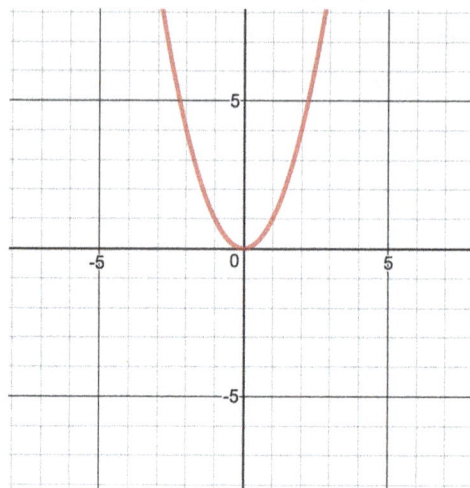

Teacher: *The next problem is $y = 0.01x^2$. Predict first. Envision. What do you think this will look like in this same window?*

Turn and tell your partner what you are thinking.

I'll put it in the display grapher.

What is a better viewing window, so that we have less empty space?

Student: *Let's cut off some of the bottom.*

Teacher: *Okay.*

Student: *And come in a little so we don't have extra space on the sides.*

$$y = 0.01x^2$$

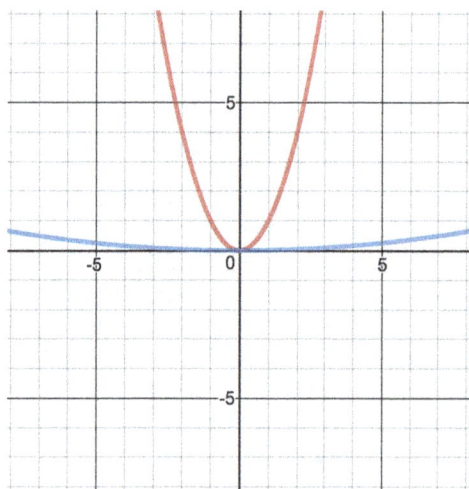
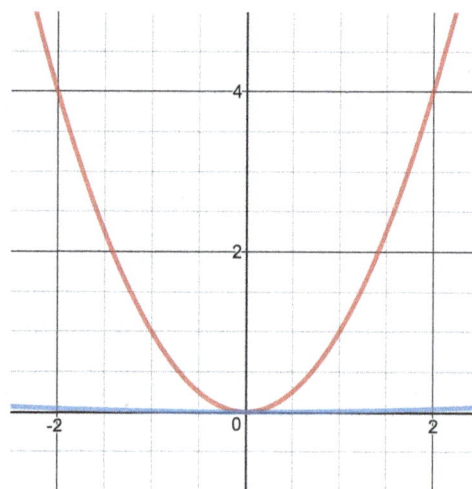

Algebra Problem Strings
©2017 Kendall Hunt Publishing

Teacher: *What's going on? Why does the blue $y = 0.01x^2$ look like that?*

Student: *It's flatter, squished.*

Teacher: *Based on this viewing window, the function appears flatter. Let's take a look at why. I wonder if it is helpful to look at a few points. If we go over 1, then it's up 1 squared, right. So, (1, 1) and back (−1, 1). But in the new function, what is the y-value when you go over 1?*

Student: *It's 0.01.*

Student: *It's one-hundredth. That's small.*

Teacher: *What about over 2?*

Student: *In the original, that's up 4, so (2, 4) and (−2, 4). In the new function, it's over 2, up 0.01 times 4.*

Student: *Or 4 times one-hundredth. Like 4 pennies. That's 0.04, not very high up.*

Teacher: *So, as we let x increase or decrease, the y-values are increasing slower than the original function.*

Student: *Yeah, it's like each y-value is only one-hundredth what it used to be. They are getting bigger slower.*

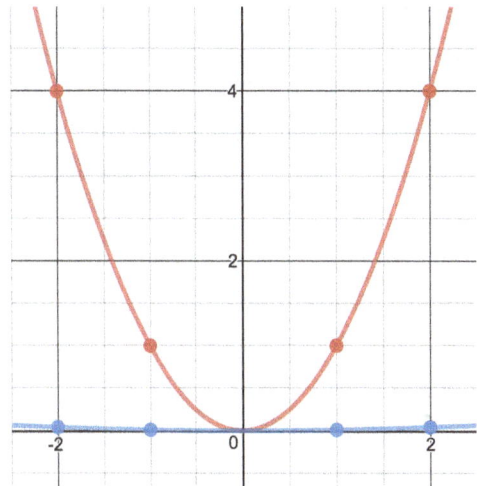

Teacher: *So, what would be a better viewing window for this blue $y = 0.01x^2$. I'll turn off the parent function and let's just focus on finding a good window for $y = 0.01x^2$.*

Students work and the teacher circulates, looking for students who focused on changing the y's and students who focused on changing the x's.

Teacher: *It seems to me like some of you took drastically different approaches. Let's start with your group. Tell me what window you had and I'll put it in the display grapher.*

Student: *We kept the −2.5 to 2.5 on the x's but changed to −0.01 to 0.07 on the y's.*

Teacher: *Got it. Did anyone else take this approach? I see some nods. What were you thinking?*

Student: *I could see that since the y-values were getting big so slowly, I decided to see less y-values, kind of zoom in on the y's. When I made them small enough, I could see the function pretty well.*

Teacher: *Tell me about the x's in your window.*

Student: *We just left them alone. They are the same as they were.*

Teacher: *So you concentrated on changing the y-values. Does that make sense to everyone?*

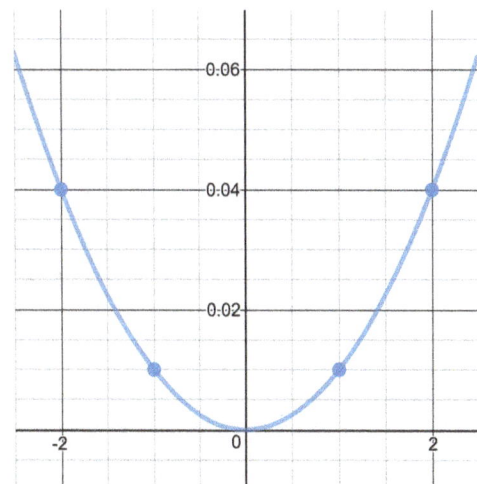

(continued)

Teacher: *But that is not what some of you did. Tell us about that.*

Student: *We did just the opposite. When it was so flat, we thought about going way far out so that it could grow.*

Teacher: *What do you mean, way far out? Tell me your window.*

Student: *We kept the y's the same, but went −25 to 25 on the x's.*

Student: *Whoa.*

Teacher: *Why do you say that?*

Student: *The windows are so different but the graph ends up looking very similar. That's interesting.*

Student: *One group changed the x's and the other the y's.*

Student: *I am not sure I understand the difference in what they did.*

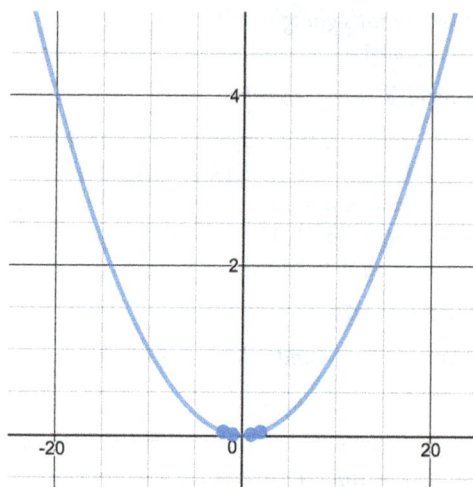

Teacher: *Can someone help us understand how you went from that short, squatty graph to the nice looking quadratic function that filled the screen.*

Student: *It's like they zoomed in really close on the y's and then stretched the y's up to fill the window.*

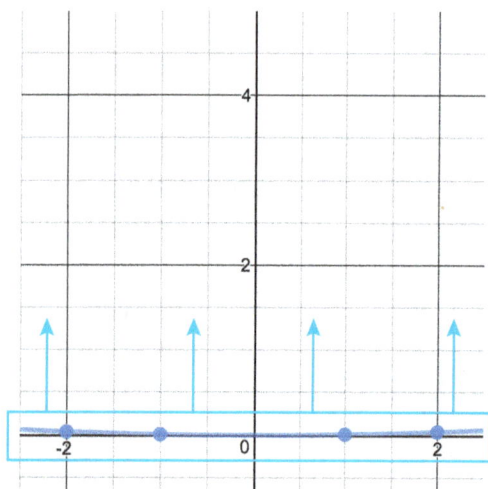

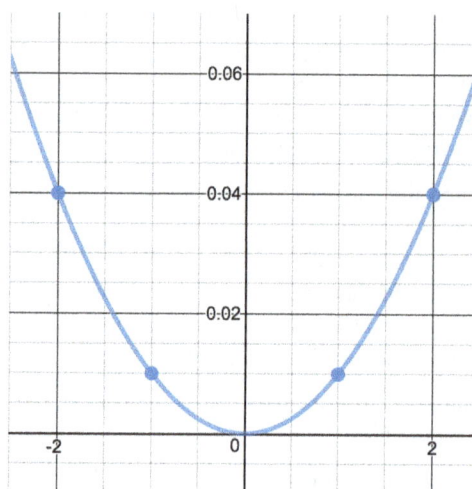

Algebra Problem Strings
©2017 Kendall Hunt Publishing

Student: *And this group went way outside the x's, far out, and then squished the whole thing in to fit into the window.*

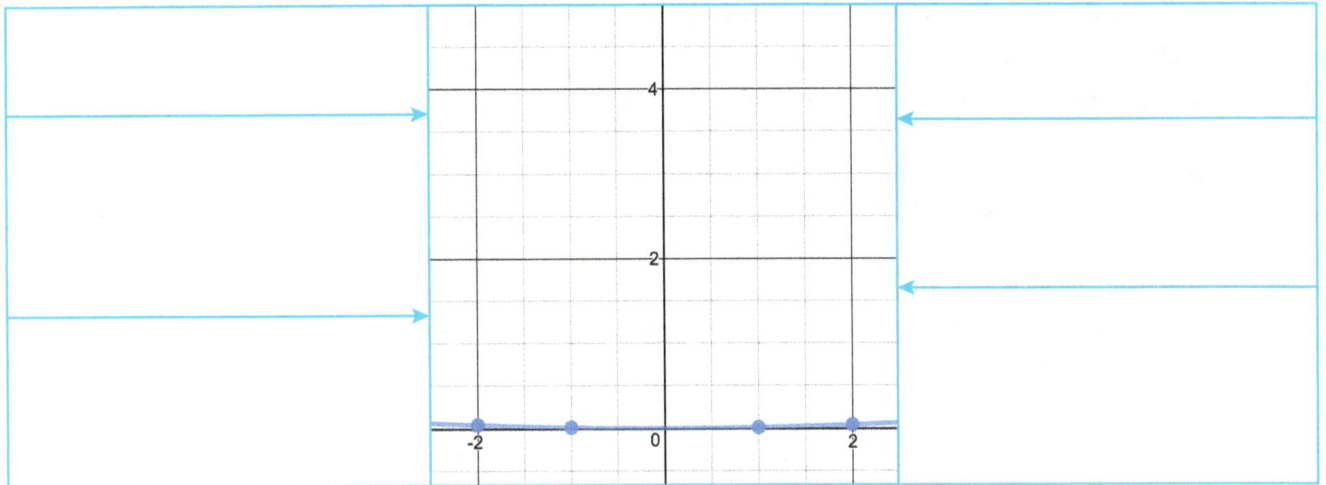

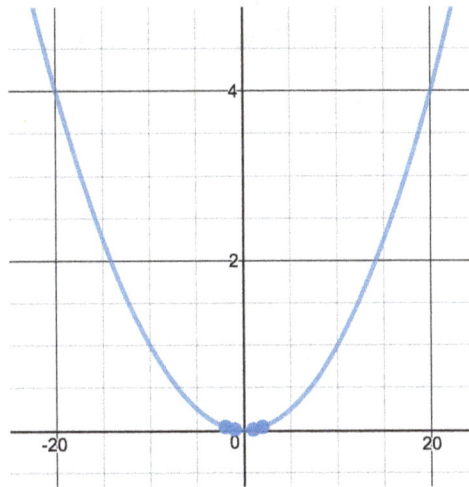

Teacher: *Great work. I'm going to put the parent function back up and our next problem, y = 1,000x² in the viewing window that we decided was good for the parent. Tell me about this graph.*	$y = 1{,}000x^2$

(continued)

Student: *That looks really skinny.*

Teacher: *Let's talk about why it looks that way. What's going on?*

Student: *Well, it's 1,000 times the y-values.*

Teacher: *Who can add on to that? No one? I wonder if looking at a couple of points might help. What's going on at x = 1?*

Student: *In the parent function, that's the point (1, 1). In the new one, that's the point (1, 1,000). Ahhh, that's way up there. It's off our graph.*

Student: *And that's the same for (−1, 1). It's now up at (−1, 1,000).*

Teacher: *It sounds like the y-values are getting bigger faster. Those y-values have really been stretched, haven't they?*

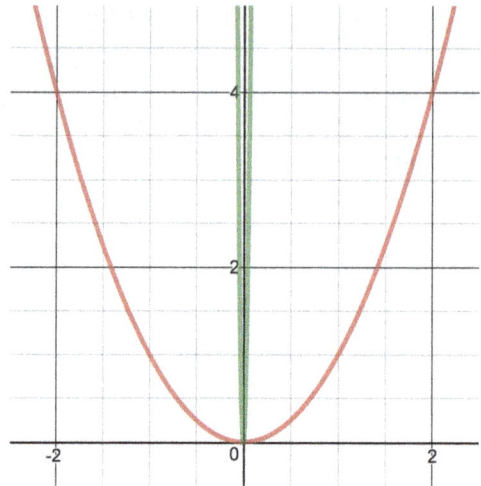

Teacher: *Go ahead and work to find a good viewing window for this function.*

Students work. The teacher chooses students to share who have either zoomed in on the *x*'s or the *y*'s and facilitates a conversation around their choices.

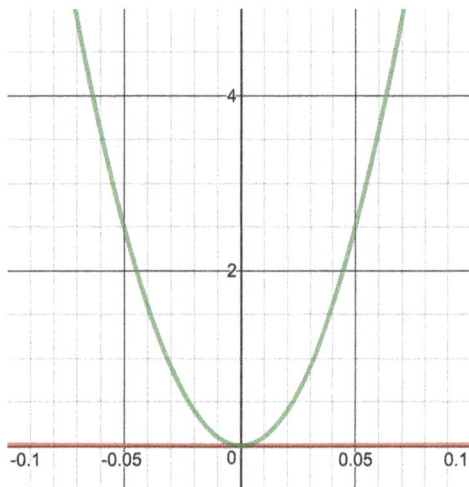

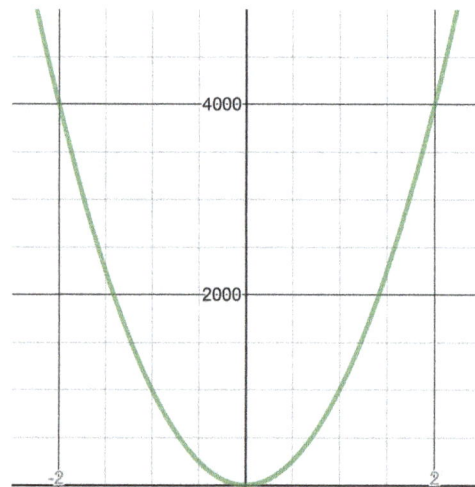

Teacher: *Thinking about the last two problems, why did one vertically compress and the other vertically stretch?*

Student: *I think it has to do with fractions. If it's bigger or smaller than 1.*

Student: *The very first graph was being multiplied by 1, so it makes sense that multiplying it by something less than one would give y-values smaller than the parent function. That's a compression.*

Student: *Less than 1 but still positive?*

Teacher: *I think you are talking about values between zero and one, those fractions would produce a vertical compression.*

Student: *Yeah, and bigger than one makes a vertical stretch.*

Algebra Problem Strings
©2017 Kendall Hunt Publishing

The teacher gives students the next four problems in the string one at a time, $y = 10(x-250)^2$, $y = \frac{1}{4}x^2 - 500$, $y = -20x^2 - 10$ and $y = -\frac{1}{5}(x-30)^2 + 150$ (for a sample graph, see the Final Display), and facilitates a conversation around combining transformations.

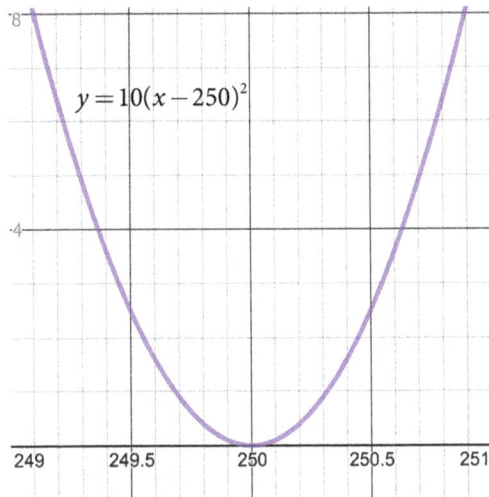

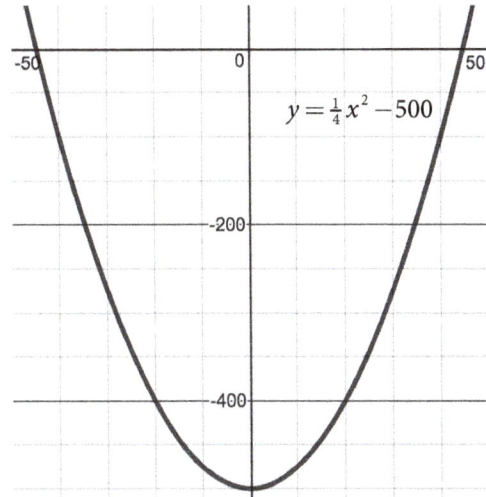

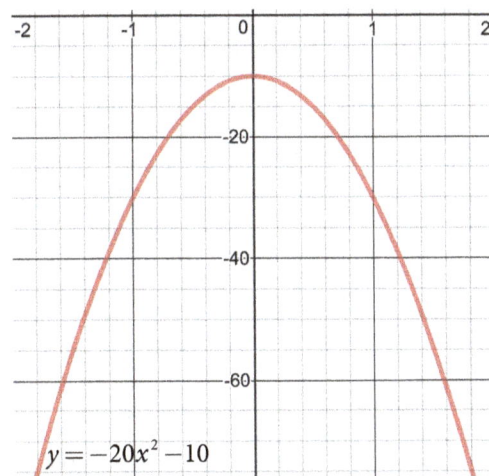

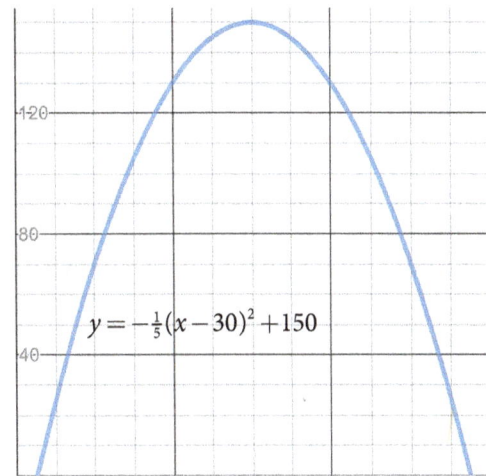

$y = 10(x-250)^2$

$y = \frac{1}{4}x^2 - 500$

$y = -20x^2 - 10$

$y = -\frac{1}{5}(x-30)^2 + 150$

Teacher: *How would you summarize some of the things that came up in this string today?*

Elicit the following:

- *When you multiply a function by a value, it's multiplying all of the y-values, scaling the y-values.*

- *For af(x), when the scale factor a is greater than 1, the result is a vertical stretch because the y-values are getting bigger faster than in the parent function.*

- *For af(x), when the scale factor a is between 0 and 1, the result is a vertical compression because the y-values are getting bigger more slowly than in the parent function.*

(continued)

Sample Final Display

Your display could look like this at the end of the problem string:

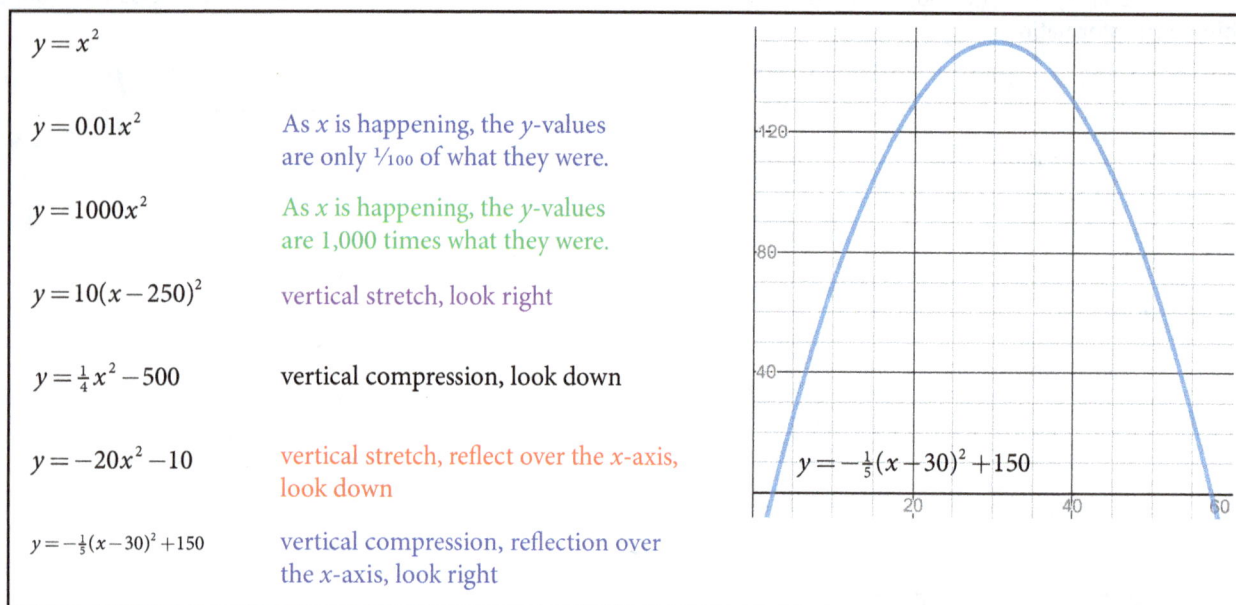

$y = x^2$		
$y = 0.01x^2$	As x is happening, the y-values are only $1/100$ of what they were.	
$y = 1000x^2$	As x is happening, the y-values are 1,000 times what they were.	
$y = 10(x-250)^2$	vertical stretch, look right	
$y = \frac{1}{4}x^2 - 500$	vertical compression, look down	
$y = -20x^2 - 10$	vertical stretch, reflect over the x-axis, look down	
$y = -\frac{1}{5}(x-30)^2 + 150$	vertical compression, reflection over the x-axis, look right	

$$y = -\tfrac{1}{5}(x-30)^2 + 150$$

Facilitation Notes

This version of the problem string lists short notes for important teacher moves during the string. After you've done the string yourself and studied the relationships involved, you might make similar notes for the things you want a reminder of or deem important.

$y = x^2$	Visualize: what does the graph look like? Display with display grapher.
$y = 0.01x^2$	What does 0.01 mean? What's happening to the y-values? Compare to parent when x=1, 2. Compare windows, both zooming in on x's or xooming in on y's. Note "As x is happening, the y-values are only 1/100th of what they were."
$y = 1000x^2$	What's happening to the y-values? Compare to parent when x=1, 2 Compare windows, both zooming in on x's or zooming in on y's. Note "As x is happening, the y-values are 1,000 times what they were."
$y = 10(x-250)^2$	Partner up. One holds grapher. Only make changes when both agree. Talk, defend, change, adjust. Find an appropriate viewing window as a class. Press for justification. Note "Vertical stretch, look right."
$y = \frac{1}{4}x^2 - 500$	Repeat. Note "vertical compression, look down."
$y = -20x^2 - 10$	Repeat. Note "vertical stretch, reflection over x, down."
$y = -\frac{1}{5}(x-30)^2 + 150$	Repeat. Note "vertical compression, reflection over x, look right."

Algebra Problem Strings
©2017 Kendall Hunt Publishing

Transforming Functions 4: Combinations

At a Glance

$$y = -0.1|x - 35| - 200$$

$$y = 50(x + 400)^2 + 87$$

$$y = 5\sqrt{-x} + 15$$

Objectives

The goal of this problem string is to develop students' facility with combining the transformations of translations, reflections, and dilations of functions. To do this students find appropriate viewing windows for the transformations of the functions $y = |x|$, $y = x^2$, $y = \sqrt{x}$, and $y = a^x$. Using the power of technology, students can quickly test their assumptions by trying and adjusting. As they are pressed to defend their window choices, students build skill transforming functions.

Placement

This is the fourth in a series of four problem strings that use finding appropriate viewing windows as a vehicle to build and strengthen students' understanding and facility with transforming parent functions. You could use this problem string after students have learned about translation, reflections, and vertical dilations.

This problem string could come during the work of textbook Lesson 7.8 Using Transformations to Model Data.

Guiding the Problem String

These problems are meant to bring together the functions and combinations of transformations. The first two problems are with the more familiar absolute value and quadratic functions. The third problem is with the less familiar square root function which can serve as an assessment to determine how well students are internalizing the effects of transformations given any parent function. When you give students the third problem, wonder aloud if it might be helpful to find the graph of the parent function first, before students begin to try to find the given function. The fourth problem gives students another shot at reflecting over the y-axis.

You can change up the rhythm of the problem string by asking students to predict first, having students work with a partner where they must agree before changing anything, having students work individually and then comparing windows with a partner, or finding an appropriate window together as a class. The emphasis should always be on pressing for justification of the viewing window based on the transformations of the parent function.

See problem string 7.5 for a sample dialogue of a problem string with a similar structure.

About the Mathematics

These problems combine translations, reflections, and dilations with quadratic, square root, and absolute-value functions.

(continued)

Important Questions

Use the following as you plan how to elicit and model student strategies.

- *What effect is $f(x) + b$ on the graph of $f(x)$? Why? How does that help you find an appropriate viewing window?*

- *What effect is $f(x + b)$ on the graph of $f(x)$? Why? How does that help you find an appropriate viewing window?*

- *What effect is $-f(x)$ on the graph of $f(x)$? Why? How does that help you find an appropriate viewing window?*

- *What effect is $f(-x)$ on the graph of $f(x)$? Why? How does that help you find an appropriate viewing window?*

- *What effect is $af(x)$ on the graph of $f(x)$ when $a > 1$? Why? How does that help you find an appropriate viewing window?*

- *What effect is $af(x)$ on the graph of $f(x)$ when $0 < a < 1$? Why? How does that help you find an appropriate viewing window?*

- *Which of these transformations are additive and are shape preserving? Which are multiplicative and shape changing?*

How would you summarize some of the things that came up in this string today?
- *Translations are transformations $f(x) + b$ where you shift the function $f(x)$ right for $b < 0$ and left for $b > 0$.*

- *Reflections are transformations where you reflect the function $f(x)$: $-f(x)$ over the x-axis and $f(-x)$ over the y-axis.*

- *Dilations are transformations $af(x)$ where you vertically stretch $f(x)$ by a scale factor of a when $a > 1$, and vertically compress $f(x)$ by a scale factor of a when $0 < a < 1$.*

Sample Final Display

Your display could look like this at the end of the problem string:

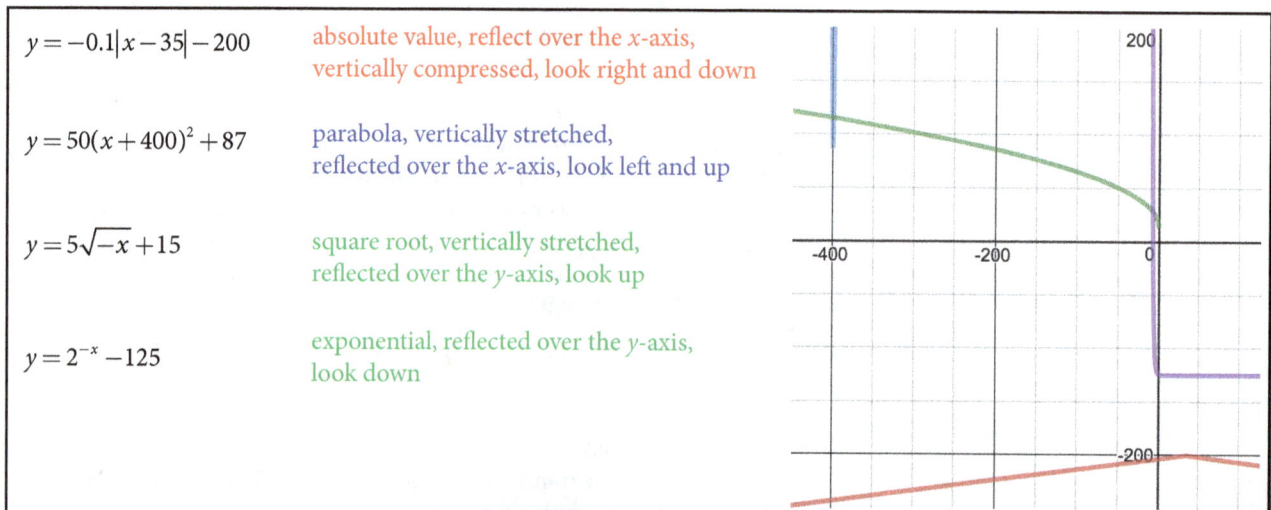

$y = -0.1\lvert x - 35 \rvert - 200$	absolute value, reflect over the x-axis, vertically compressed, look right and down
$y = 50(x + 400)^2 + 87$	parabola, vertically stretched, reflected over the x-axis, look left and up
$y = 5\sqrt{-x} + 15$	square root, vertically stretched, reflected over the y-axis, look up
$y = 2^{-x} - 125$	exponential, reflected over the y-axis, look down

Algebra Problem Strings
©2017 Kendall Hunt Publishing

Facilitation Notes

This version of the problem string lists short notes for important teacher moves during the string. After you've done the string yourself and studied the relationships involved, you might make similar notes for the things you want a reminder of or deem important.

$y = -0.1\|x - 35\| - 200$	Partner up. One holds grapher. Only make changes when both agree. Talk, defend, change, adjust. Find an appropriate viewing window as a class. Press for justification. Note absolute value, vertically compressed, look right and down.
$y = 50(x + 400)^2 + 87$	Repeat. Note parabola, vertically stretched, look left and up.
$y = 5\sqrt{-x} + 15$	Repeat. Note square root, vertically stretched, reflected over the y-axis, look up.
$y = 2^{-x} - 125$	Repeat. Note exponential, reflected over the y-axis, look down. For fun, find a window to see all four at once!

7.9 | The Rational Parent Function

At a Glance

Workers	Hours
20	6
10	
40	
	24
15	
1	
	h
w	

# of Volunteers	Time (hours)
250	3
500	
	2
	6
	25
10	
	h
v	

Objectives

The goal of this problem string is to connect what students have learned about inverse variation with the graph of the rational parent function. Students can bring in their emerging understanding of transformations to find a good viewing window.

Placement

This problem string is designed to refer to the learning that students did in problem strings 2.5 Reasoning About Inverse Variation 1 and 2.6 Reasoning About Inverse Variation 2. These problem strings have students using work problems to reason proportionally inversely. Use the problem string to help students connect that learning to the graph of the rational parent function.

You can use this problem string to introduce or support textbook Lesson 7.9 Introduction to Rational Functions.

Guiding the Problem String

These scenarios were developed in problem strings 2.5 and 2.6. If you did not facilitate these problem strings, consider doing only half of this string or this string over two class times.

For the first problem, remind students about the scenario (or develop it as necessary) and quickly find the missing number of workers or hours, reasoning inverse proportionally and modeling in the ratio table. Then ask students to generalize how to find the number of workers needed if you must complete the job in h hours. Graph the ordered pairs and the function in the display grapher. Expand the viewing window to see the function in the third quadrant. Repeat for the next scenario.

As students determine the missing values, encourage them to use what they already know. Bring out the constant of variation by noticing the products. Help students connect the constant of variation with the related rational function.

About the Mathematics

Dilations of the rational parent function are related to the inverse variation that students have already studied. These functions are often represented by the division equation below, but should also be understood as related to the multiplication equation, where k is the constant of variation and is referred to as the dilation scale factor.

$$xy = k \qquad y = \frac{k}{x}$$

Lesson 7.9 • The Rational Parent Function (continued)

Sample Interactions

Use the following as you plan how to elicit and model student strategies. This is not meant as a script, but as a view into the relationships involved and the intent of the problem string.

Teacher: *Today, we're going to pull back on a problem string we did quite a while ago. Remember when we talked about the apple orchard that needed the whole orchard cleared. We knew it took 20 workers 6 hours to do the whole job. And then we figured out some times and numbers of workers. Let's figure a few of those again.*

The teacher writes the information on the board and models a strategy for some of the entries.

Note: For a sample dialog and modeling of student strategies, see problem strings 2.5 and 2.6.

Workers	Hours
20	6
10	12
40	3
5	24
15	8
1	120
120	1

Teacher: *So, what if we wanted to figure out the number of workers if we knew that it would take h hours? What is the relationship between the numbers of workers and the number of hours?*

Student: *If you multiply them together, you get 120.*

Teacher: *Does everyone agree? Yes? But how does that help us find the number of workers if we know the number of hours?*

Student: *Well, if you divide 120 by 6, you get 20. And if you divide 120 by 10, you get 12. So, if you divide 120 by h, you get w.*

Teacher: *What do you think about that?*

Student: *Yes, that makes sense.*

Teacher: *And what if we wanted to find the number of hours it will take if we have w workers?*

A similar conversation takes place.

Workers	Hours
20	6
10	12
40	3
5	24
15	8
1	120
120	1
$\frac{120}{h}$	h
w	$\frac{120}{w}$

Teacher: *I wonder what a graph of that would look like? Let's plot the ordered pairs first.*

Let's have the number of workers be x and the number of hours be y. What values make sense for x?

Student: *One worker up to 120 workers.*

Teacher: *And what values make sense for y, for the number of hours?*

Student: *It could take as little as 1 hour and as many as 120 hours.*

Teacher: *Before we look at the graph, predict first what you think the plot will look like.*

Ready? Here it is. How does it compare to your mental image? Why does the graph have this pattern?

Student: *As the number of workers gets bigger, the number of hours gets smaller and vice versa.*

(continued)

Algebra Problem Strings
©2017 Kendall Hunt Publishing

359

Teacher: *Predict what you think the graph of* $y = \dfrac{120}{x}$ *will look. Here it is. What do you think?*

Turn to your partner and discuss this function and it's graph.

Students turn and talk briefly while the teacher listens in.

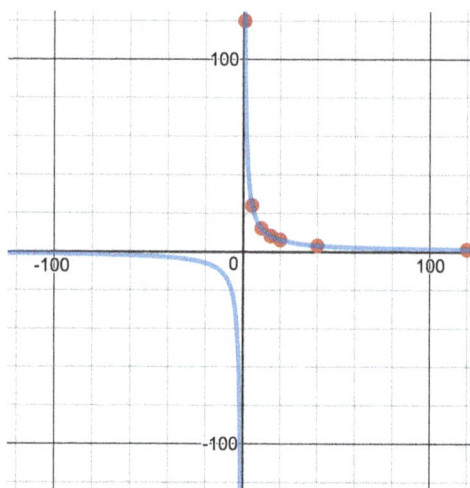

Teacher: *What are you noticing?*

Student: *Well, it fits the points.*

Student: *As one variable gets bigger, the other gets smaller. It fits the orchard, but then goes beyond. It's not limited to a whole number of people.*

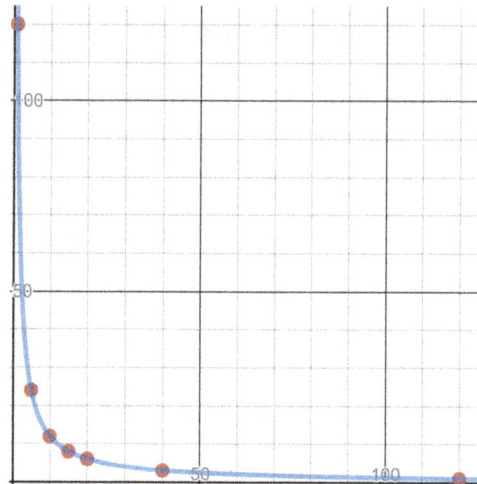

Teacher: *So, speaking of going beyond, let's open up this window a little. What do you notice now?*

Student: *Is that something different? Did you put another function in?*

Teacher: *No, it's just these points and this one function. How are you making sense of this?*

Student: *So, both x and y are negative.*

Student: *The shape is sort of the same but flipped.*

Teacher: *Turn and talk to your partner. Make sense of this function, 120 divided by x, and its graph.*

Students turn and talk briefly while the teacher listens in.

Teacher: *Let's see if looking at another familiar scenario can be helpful. Do you remember that we also worked with a scenario where we had volunteers stuffing envelopes? We knew that it took 250 volunteers 3 hours to get all of the envelopes stuffed. Let's run through a few of the numbers we figured out.*

The teacher asks the values, one a time, and models a strategy for some of the problems. This goes quickly. The teacher wonders aloud about the pattern and how it relates to the previous apple orchard scenario.

# of Volunteers	Time (hours)
250	3
500	1.5
375	2
125	6
30	25
10	75
v	$\dfrac{750}{v}$
$\dfrac{750}{h}$	h

Algebra Problem Strings
©2017 Kendall Hunt Publishing

Teacher: *Let's plot these points and look at this function.*

The teacher works with students to find a good viewing window for the ordered pairs (number of volunteers, number of hours) and graphs the function $y = \dfrac{750}{x}$.

The teacher crafts a conversation about the connection between the ordered pairs and the graph of the function.

As an extension, the teacher could ask about the graph of the function as x gets really large or as x gets closer to 1. These values are outside of the problem situation, but belong to the graph of the function.

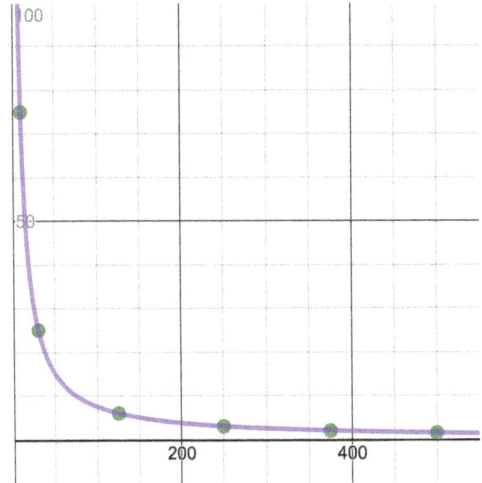

Teacher: *Anyone wondering what the function looks like in the rest of the plane? Let's open up the window.*

What are you thinking? How are you making sense of this graph?

Students turn and talk briefly while the teacher listens in.

Teacher: *Tell us what you think about the third quadrant.*

Student: *It makes sense for the equation because a positive 750 divided by a negative x is a negative y.*

Student: *The third quadrant is where both x and y are negative.*

Student: *But there aren't negative numbers of workers or negative numbers of hours.*

Student: *We can't have a half of a volunteer either. The curve is the function, not what makes sense in the situation.*

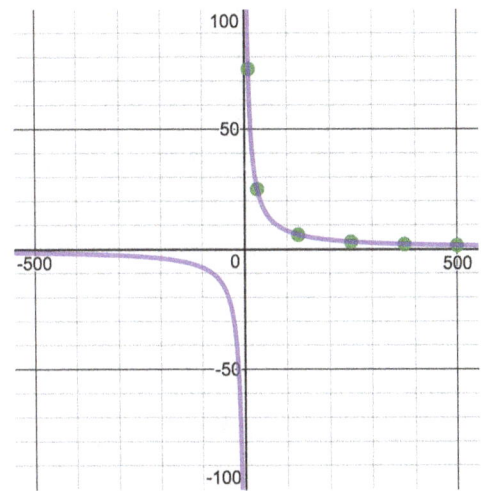

Teacher: *Tell me about this graph and how it relates to the graph of $y = \dfrac{1}{x}$. I've put the graph up here in the same window. What's going on?*

Student: *That looks like it's just the axes.*

Student: *So, if the 750 is a vertical stretch and we are looking at a pretty big window ... I am wondering what a smaller window looks like. Can you zoom in?*

If students do not suggest a vertical stretch and zooming in, wonder aloud about what is happening around the origin.

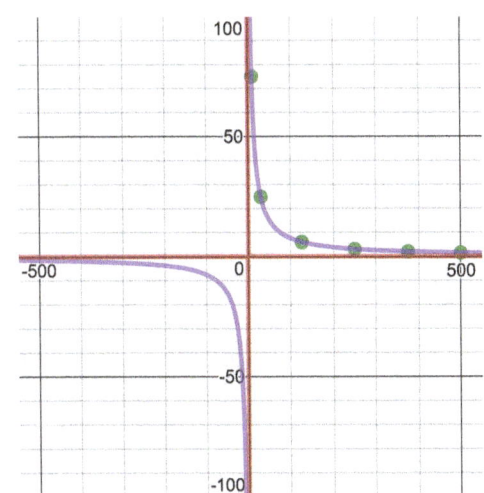

(continued)

Student: *Right, it's just way in there. The 750 is a vertical stretch. Cool.*

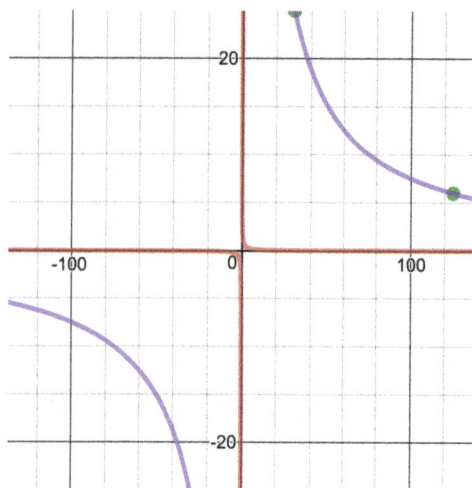

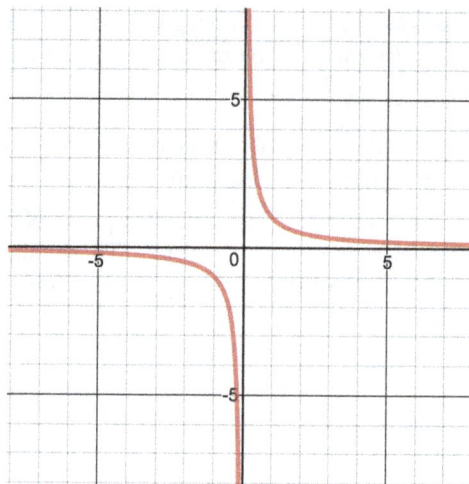

Teacher: *How would you summarize some of the things that came up in this string today?*

Elicit the following:

- *We can model inverse variation situations as dilations of the rational parent function,* $y = \dfrac{1}{x}$.
- *If* $x \cdot y = k$, *then* $y = \dfrac{k}{x}$ *and k is the vertical stretch factor.*
- *A function represents what is happening for all x's in its domain. That may include values that do not make sense in a given situation.*

Sample Final Display

Your display could look like this at the end of the problem string:

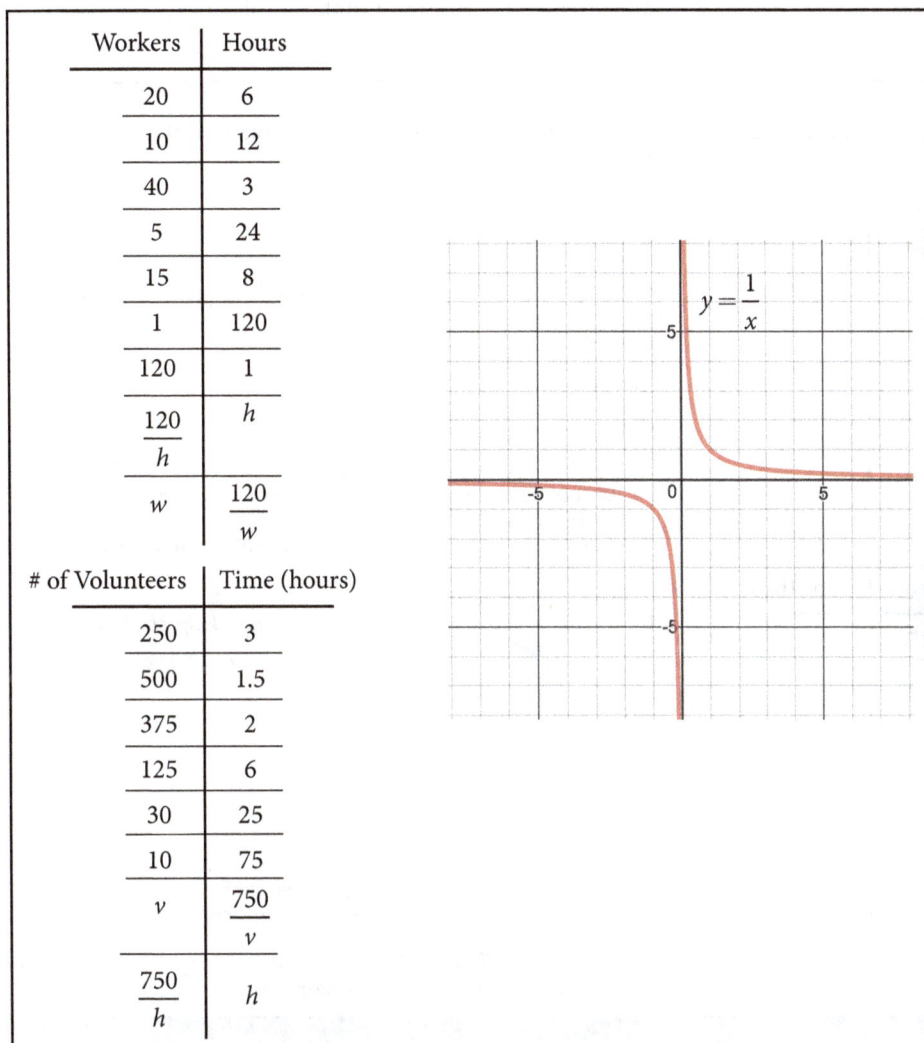

Workers	Hours
20	6
10	12
40	3
5	24
15	8
1	120
120	1
$\dfrac{120}{h}$	h
w	$\dfrac{120}{w}$

# of Volunteers	Time (hours)
250	3
500	1.5
375	2
125	6
30	25
10	75
v	$\dfrac{750}{v}$
$\dfrac{750}{h}$	h

$$y = \frac{1}{x}$$

(continued)

Facilitation Notes

This version of the problem string lists short notes for important teacher moves during the string. After you've done the string yourself and studied the relationships involved, you might make similar notes for the things you want a reminder of or deem important.

Workers	Hours	
20	6	Remind of context, 20 workers can get the job done in 6 hours. Put in table. What if we only have 10 workers? Halve/halve or halve/double? Why?
10		
40		Repeat.
	24	What if we need to get the job done in 24 hours? How many workers?
15		Repeat.
1		Repeat.
	h	Generalize.
w		Generalize. What would a graph look like? Graph points then y=120/x on display grapher in first quadrant. What do you notice? Extends? What if we look further out? Display all four quadrants.

# of Volunteers	Time (hours)	
250	3	Repeat for this scenario. Don't spend too much time figuring. Just help students get the feel of inverse variation.
500		
	2	
	6	
	25	
10		
	h	Generalize. What would a graph look like? Graph points then y=750/x on display grapher in first quadrant. What do you notice? Extends? What if we look further out? Display all four quadrants. How does this relate to the parent function y=1/x? Zoom in.
v		

Algebra Problem Strings
©2017 Kendall Hunt Publishing

8.1 | Squaring and Unsquaring

At a Glance	Objectives

At a Glance

$(6 \times 6)^*$

$(-6 \times -6)^*$

$(12 \times 12)^*$

$(-12 \times -12)^*$

$x^2 = 64$

$x^2 - 17 = 64$

$x^2 + 15 = 64$

$(x + 3)^2 = 49$

$(x - 7)^2 = 25$

$(x + 2)^2 + 1 = 82$

*optional problems

Objectives

The goal of this string is to bridge students' earlier work finding the square and square roots of values with current work solving quadratic functions by using inverse operations—a strategy also known as "solving by inspection."

Placement

You can use this problem string as students begin to solve quadratic equations.

Students had previous experiences with the squaring function (lesson 7.3) and will be asked to solve quadratic functions symbolically in Chapter 8 of the textbook. You can use this string at the beginning of the chapter to help prepare students to use inverse operations with flexibility and ease.

Guiding the Problem String

As you are leading the string be mindful of going a bit deeper than simply soliciting solutions. This means probing students' understandings and giving them the chance to articulate to the class how they know something is true. This also means creating a bit of puzzlement so that students are really invited to express what they know and what makes sense. To be clear, this does not mean (a) pretending to not know the mathematics yourself or (b) having students trying to convince you.

For example, instead of the question "Does it work?," think about asking "Does it make sense?," followed by "Why?" or "Why not?" In this way we assure kids that just because the algebra of quadratic functions may feel different and potentially more challenging than linear functions, the role of sense-making in mathematics still matters.

Another way to keep students thinking critically about the meaning of squaring is to consistently ask, "Could it be anything else?" Early in the string students may wonder why you are asking this, but later it will become an important question, especially to those students who are satisfied by only one solution. More importantly, you are offering students an important question for them to consider when doing all kinds of mathematics.

The structure of the string is designed to move slowly, but deliberately, from friendly values towards more complex functions that resemble the kinds of quadratic functions they will soon encounter.

About the Mathematics

When solving quadratic functions using inverse operations, students will encounter the need to undo the squaring operation. Without a strong sense of the meaning of the square root symbol and absolute value symbol, they are likely to make errors or not consider multiple values that could make the function true. A simple example of this is when students assume that when $x^2 = 16$ then x must be 4, ignoring the value of -4 as a valid solution.

(continued)

Sample Interactions

Use the following as you plan how to elicit and model student strategies. This is not meant as a script, but as a view into the relationships involved and the intent of the problem string.

Teacher: *Okay, everyone. Let's get our brains warmed up with a problem string. You might not need paper and pencil, but be sure you are sitting near your math partner so you can talk together. I'll put a problem on the board and ask that you think about the answer or answers and be able to explain how you know. Here we go! The first problem is x squared is 64. What is x?*	$x^2 = 64$
Student: *So, this is 8 because you are asking what number squared is 64 and that is 8.* **Teacher:** *What do other people think?* **Student:** *Same idea but squaring is like what number multiplied by itself is 64. I agree, it's got to be 8.*	$x^2 = 64$ $8^2 = 64$ $x \cdot x = 64$ $8 \cdot 8 = 64$
Teacher: *Seems like we agree. Could it have another solution? Is there anything else x could be?* **Student:** *Ahhh, it could be negative. Negative 8 also works.* **Teacher:** *So this is also true?* **Students:** *Yes!*	$x^2 = 64$ $(-8)^2 = 64$ $-8 \cdot -8 = 64$ $(-8)^2$ or -8^2 ?

Teacher: *Turn and talk to your partner for 30 seconds about how I wrote that. Why did I use parentheses here? Is there a difference between $(-8)^2$ and -8^2?*

Student: *So my partner and I discussed that the difference might be about order of operations, in the first expression it's clear from what you wrote that we are supposed to square negative 8, and in the second one it's a little murky. Maybe that means square the 8, then make the answer negative which would not give us 64, but −64.*

Teacher: *What do other people think?*

Student: *The first notation is just clearer so I get why you used it. And we don't have to use it with positive 8 because it's not necessary.*

Teacher: *In fact, mathematicians agree with you that this notation could be murky and so they have decided to use parentheses to mean negative 8 times negative 8 and when there are no parentheses, it means 8 squared and that is negative, so −64. We'll use these order of operations from now on.*	$(-8)^2 = -8 \cdot -8 = 64$ $-8^2 = -(64) = -64$
Teacher: *And, so we are all clear going forward, I'm going to use this symbol to capture the idea of a positive and negative solution. Everyone okay with that and understand what it means? We mean positive and negative 8 when we write this, but sometimes mathematicians would say, "plus or minus eight" when they use this symbol.*	$x = \pm 8$

Algebra Problem Strings
©2017 Kendall Hunt Publishing

Teacher: *Okay, now I'm going to tinker with things a little bit. Be thinking about how this solution, or these solutions, will relate to the positive and negative eight.*	$x^2 - 17 = 64$
Teacher: *What is x?* **Student:** *I got 9.* **Student:** *I got both plus and minus 9.* **Teacher:** *How did you find x?* **Student:** *I wasn't sure about this one, so I added 17 to both sides and got x squared equals 81 and then it was way easier.* **Teacher:** *Can you say why that made sense to you?* **Student:** *Well, the way it is written I couldn't figure it out. I wanted just a variable on one side and a number on the other, like the last one.* **Student:** *I thought that if something minus 17 is 64, than that number must be 17 more than 64.* **Teacher:** *Nice. Can someone finish this thinking?* **Student:** *Sure, so based on the last one, x has to be 9 or −9.* **Teacher:** *Sometimes people call this solving by inspection or using inverse operations.*	$x = -9, 9$ $x^2 - 17 = 64$ $+17+17$ $x^2 = 81$ $x = \pm 9$
Teacher: *Now what? What if this time we add 15 to x squared to get 64. How will that relate? What is x now?* **Student:** *If something plus 15 is 64, the number has to be 15 less than 64.* **Student:** *I first subtracted 15 from both sides to isolate the x^2 and I was really happy that I got 49. Then I knew it would be 7 and −7.* **Teacher:** *Okay. Let me capture your strategy. Who understands why Jeffrey was excited about the 49?* **Student:** *It's a perfect square and we know how to find the square roots without a calculator. Makes life easier.*	$x^2 + 15 = 64$ $x^2 + 15 = 64$ $-15-15$ $x^2 = 49$ $x = \pm 7$
Teacher: *How would you solve this one? Just notice what's the same and what's different here. This time it's the sum of x and 3 squared to get 49. What is x?* Students work. The teacher circulates, looking for students who found only one solution, $x = 4$ and students who found both solutions. After most students have a solution, the teacher asks students to share their thinking with their partner.	$(x + 3)^2 = 49$

(continued)

Teacher: *Who can tell us how their partner solved this one?* Asking students to explain their partner's thinking can help encourage students to pay attention to other strategies and make sense of other's thinking. **Student:** *My partner thought about this as "what number plus three, when you square it, would give you 49." Since that number is 7, we knew x was 4 so 4 plus 3 is 7. Square it and you get 49.* **Teacher:** *Thoughts on this strategy?*	$(x+3)^2 = 49$ $(\_+3)^2 = 49, \quad (7)^2 = 49$ so $x = 4$
Student: *I like it, but I feel like there's another answer. In all of the previous ones there were two answers not one. So I'm just wondering if we mean plus and minus 4?* **Teacher:** *Interesting. What do we think?* **Student:** *Uhhh, −4 won't work, but I like the idea of checking for a negative solution. What about −10?* **Teacher:** *Say more about that.* **Student:** *In the previous problems when a number was squared we had two solutions, one negative and one positive. That is true here, but we are kinda thinking about how to get x plus 3 to equal 7 and how to get x plus 3 to equal −7.* **Teacher:** *Raise your hand if you are convinced by this idea. Anybody not convinced?*	Since $(x+3)^2 = 49$ $x+3 = \pm 7$ $x+3 = 7 \qquad$ or $\qquad x+3 = -7$ $x = 4 \qquad\qquad\qquad x = -10$
Teacher: *Using the ideas we've talked about so far, how would you solve this one?* Students work. The teacher circulates. **Teacher:** *Turn and tell your partner how you are thinking about this one?* Teacher records conversation like the last one, being sure to wonder about a second solution if only one solution is mentioned.	$(x-7)^2 = 25$ $x-7 = \pm 5 \qquad x-7 = 5$ or $x-7 = -5$ $x = 12 \qquad\qquad x = 2$
Teacher: *Alright my friends, let's end with this one. Try to make it friendlier to solve, like we always do. This time it's the sum of x and 2, that quantity squared, then add 1 to get 82. What's x?* Students work. The teacher circulates. **Teacher:** *Go ahead and hear how your partner solved this one.* **Teacher:** *Luis, will you share your strategy with all of us?* **Student:** *Sure. We talked about how getting rid of the one made sense—so now we have x plus two squared equals 81. Then we saw the perfect square and knew we were looking for plus or minus nine.* **Teacher:** *Am I capturing your thinking here?* **Student:** *Yeah, exactly. So then it's just a linear equation. Where x will equal 7 or −11.*	$(x+2)^2 + 1 = 82$ $(x+2)^2 = 81$ $(x+2) = \pm 9$ $x+2 = 9 \quad$ or $\quad x+2 = -9$ $x = 7 \qquad\qquad x = -11$ $x = -11, 7$

Algebra Problem Strings
©2017 Kendall Hunt Publishing

Teacher: *How would you summarize some of the things that came up in this string today?*

Elicit the following:

- *When you are looking for what squared is a number, you have to think of both the positive and negative possibilities.*

- *You can often find one value by guessing and checking but the other one might need some more work to find.*

- *Just like you can add/subtract/multiply/divide to both sides of an equation, you can "unsquare" or take the square root of both sides of an equation.*

Sample Final Display

Your display could look like this at the end of the problem string:

$6 \times 6 = 36$
$-6 \times -6 = 36$

$x^2 = 64$	$x^2 = 64$	$x^2 = 64$	
$x = -8, 8$	$8^2 = 64$	$(-8)^2 = 64$	$(-8)^2 = -8 \cdot -8 = 64$
$x = \pm 8$	$x \cdot x = 64$	$-8 \cdot -8 = 64$	$-8^2 = -(64) = -64$
	$8 \cdot 8 = 64$	$(-8)^2$ or -8^2?	

$x^2 - 17 = 64$ 　　　 $x^2 - 17 = 64$
　　$x = \pm 9$ 　　　　　　　 $+17 \ +17$
　　　　　　　　　　　　　　 $x^2 = 81$
　　　　　　　　　　　　　　 $x = \pm 9$

$x^2 + 15 = 64$ 　　　 $x^2 + 15 = 64$
　　$x = \pm 7$ 　　　　　　　 $-15 \ -15$
　　　　　　　　　　　 $x^2 = 49$
　　　　　　　　　　　 $x = \pm 7$

$(x+3)^2 = 49$ 　　　 $(\_+3)^2 = 49, \ (7)^2 = 49$ 　　 Since $(x+3)^2 = 49$
　　$x = -10, 4$ 　　 so $x = 4$ 　　　　　　　　 $x + 3 = \pm 7$
　　　　　　　　　　　　　　　　　　　　　　 $x + 3 = 7$ 　　 or 　　 $x + 3 = -7$
　　　　　　　　　　　　　　　　　　　　　　 $x = 4$ 　　　　　　 $x = -10$

$(x-7)^2 = 25$ 　　　 $x - 7 = \pm 5$ 　 $x - 7 = 5$ *or* $x - 7 = -5$
　　$x = 2, 12$ 　　　　　　　　　　 $x = 12$ 　　　　 $x = 2$

$(x+2)^2 + 1 = 82$ 　　　　　 $(x+2)^2 = 81$
　　$x = -11, 9$ 　　　　　　 $(x+2) = \pm 9$
　　　　　　　　　　　　 $x + 2 = 9$ 　 or 　 $x + 2 = -9$
　　　　　　　　　　　　 $x = 7$ 　　　　　　 $x = -11$
　　　　　　　　　　 $x = -11, 7$

(continued)

Facilitation Notes

This version of the problem string lists short notes for important teacher moves during the string. After you've done the string yourself and studied the relationships involved, you might make similar notes for the things you want a reminder of or deem important.

$(6 \times 6)^*$	Use these only if we need them.
$(-6 \times -6)^*$	Will students consider both possibilities when solving?
$(12 \times 12)^*$	
$(-12 \times -12)^*$	
$x^2 = 64$	Friendly problem. Could there be another solution? Pause to clarify notation here: $(-8)^2$ or -8^2 Name and use this symbol: ± 8
$x^2 - 17 = 64$	Tinkering with the structure. Could there be another solution?
$x^2 + 15 = 64$	Keeping structure the same—but subtracting instead of adding. Could there be another solution?
$(x + 3)^2 = 49$	New structure—now what? Sum of x and 3, that quantity squared is 49, what's x?
$(x - 7)^2 = 25$	Related structure. Difference between x and 7, that quantity squared is 25, what's x? Be sure to have students articulate why two solutions and how to find them.
$(x + 2)^2 + 1 = 82$	Challenge problem, but friendly enough to solve mentally.

Algebra Problem Strings
©2017 Kendall Hunt Publishing

8.2 Solving and Graphing

At a Glance
$y = 3x + 6$
$0 = 3x + 6$
$y = x^2$
$x^2 = 0$
$x^2 - 16 = 0$
$x^2 - 36 = 0$
$x^2 - 2 = 34$
$x^2 - 72 = 0$
$x^2 - 10 = 62$
$x^2 + 72 = 0^*$

*optional problem

Objectives
This string builds on the previous string of squaring and "unsquaring" by inviting students to solve equations that involve a slightly different structure, are less friendly, or are in fact functions instead of equations to solve. They will also have the chance to predict and then study the graphs of these as functions—exploring the emergent relationships between the solutions (roots) of an equation and the x-intercepts of its graph.

Placement
You can use this string to support the zero product property and to support students making connections between quadratic functions, solving quadratic equations, and key features of the graph.

This problem string can happen before, during, or after textbook Lesson 8.2 Finding the Roots and the Vertex.

Guiding the Problem String
This string starts with a friendly linear function to remind students of the difference between solving an equation and representing a function. Later, setting a function equal to zero is a deliberate move in this string, highlighting the value of finding the roots of quadratic functions.

Acknowledge to students that they have seen some problems like this before. Let them know that today you will be graphing these functions. Wonder with them, encourage them to think about how the solutions to the equations might be related to features of the graph. This is meant to be initial or emergent work. That is, you are suggesting that perhaps you can predict some things about the shape of the graph from an equation.

Specifically question and wonder about the solution(s) of a quadratic equation and the x-intercept(s) of the graph and how to symbolically solve these equations, even when perfect squares are not present.

Let your students do the work here—do not feel the need to explain to students what to look for. Instead see what they are able to notice on their own. Your questioning, not your telling, will allow this to happen. Work to help students realize that equations have equivalent forms and that these forms give us insight into how they might be solved.

About the Mathematics
Functions represent sets of ordered pairs, such as $y = x^2 - 72$, and equations in one variable represent sets of values, such as $x^2 = 72$. Equations show equivalence of expressions and functions show relationships of inputs and outputs.

The zeros of a function $f(x)$ are the x-intercepts of the graph of $y = f(x)$ and the roots of the equation $f(x) = 0$.

The zero product property states that if the product of factors is zero, then either of the factors is zero or both factors are zero.

(continued)

Sample Interactions

Use the following as you plan how to elicit and model student strategies. This is not meant as a script, but as a view of how a teacher might elevate the conversation from simply solving towards making sense of the strategies and big ideas.

Teacher: *Today we'll start the string with something we saw earlier in the year. And my questions for you are: Can you solve it? Why or why not?* Quick think time. **Teacher:** *Turn and tell your partner what you are thinking.*	$y = 3x + 6$

Teacher: *Let's have a couple of you share what you are thinking.*

Student: *So you could graph it, it's linear, but you can't solve it.*

Teacher: *But I see a variable, can't we solve for x or solve for y?*

Student: *Well, it's already solved for y.*

Student: *You have an infinite set of solutions here. Like a whole bunch of points, or (x, y) values, that would work. If you want to solve it for a number you have to replace the x or the y with a value.*

Teacher: *Reflect on your own about what you think.*

Teacher: *Okay, so what happens when I do this? Why do you think I chose zero for y?* *Can you solve this? What does the solution tell us?*	$0 = 3x + 6$
Student: *If you wanted to find the x-intercept, this would be a good move. You replaced the y with zero, so your solution is where this crosses the x-axis.*	$0 = 3x + 6$
Teacher: *And where is that and how do you know?*	$-6 = 3x$
Student: *I solved it and got −2.*	$x = -2$
Teacher: *So, x is −2.*	

Teacher: *Okay, so I'm going to use the original function, y = 3x + 6, and let's graph this and check out what it looks like. Verify this claim that the x-intercept is −2.*	$y = 3x + 6$

Algebra Problem Strings
©2017 Kendall Hunt Publishing

Teacher: *I am going to present the next two problems at the same time: $y = x^2$ and $0 = x^2$. Are these functions that represent lots of points or equations we need to solve?*

Student: *The first one is the same idea except we know it's a parabola and can graph it, but finding a single solution isn't possible.*

Teacher: *Are you convinced by what she said? What about the second one, $0 = x^2$?*

Student: *I can solve that.*

Teacher: *Back to $y = x^2$. Let's imagine the x-intercept here. How would we find it? What will the graph look like? Turn and talk with your partner and then we'll talk together.*

Student: *So, my partner and I did the same thing as in the first two problems. We made y a zero and then solved it and I think we got x equals zero.*

Teacher: *What does that tell us exactly?*

Student: *So this is going to be a parabola that faces up and crosses the x-axis at (0, 0).*

Teacher: *Hold on, how does Jessica know that this faces up?*

Student: *She looked at the x squared term and because that was positive it's going to go up, not down.*

Teacher: *Let's graph it.*

$$y = x^2$$
$$0 = x^2$$
$$x = 0$$

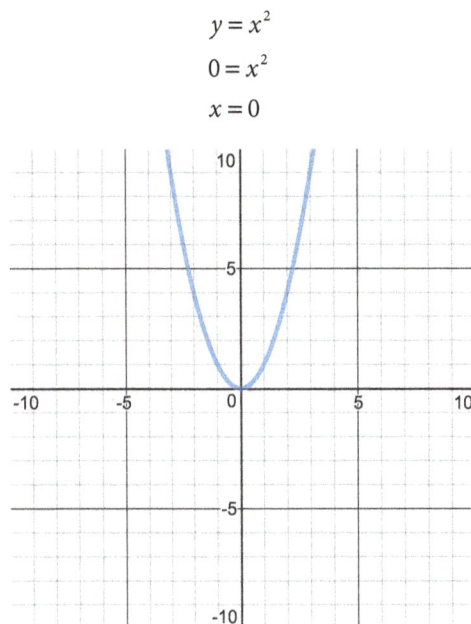

Teacher: *Let's keep going. Here are the next two problems. Since you are all convinced that the first one is graphable, but not something to solve, solve the second one and then tell your partner three things:*

- *How you solved it.*
- *What the solution tells us.*
- *What you think the graph will look like.*

Students work.

Teacher: *Ready to share? Go!*

Elicit the equivalent equation $x^2 = 16$, the solutions of $x = \pm 4$, the graph, and the x-intercepts at $(-4, 0)$ and $(4, 0)$.

Pause and compare the last two graphs. It's important to leave all the representations up throughout the entire string so that students begin to construct relationships.

You might ask about the shape of the graphs and how that compares. Likely a student will notice the y-intercept or vertex at $(0, -16)$ and you can begin to wonder aloud with students about whether this is related to or predictable from the equation.

Is there a way to find the vertex?

Do not be too pointed here so that students who are new to these ideas will begin to be curious with you.

$$y = x^2 - 16 \qquad x^2 = 16$$
$$0 = x^2 - 16 \qquad x = \pm 4$$

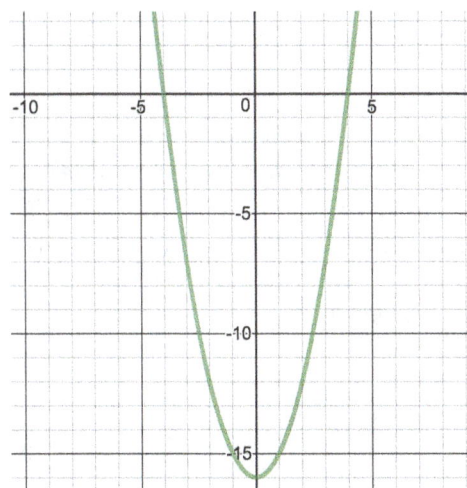

(continued)

Teacher: *Let's keep going. How would you solve this one?* **Student:** *That's just 6 and –6. Subtract 36 from both sides and then you have x squared equals 36, which we can take the square root of both sides.*	$x^2 - 36 = 0$ $x^2 = 36$ $\sqrt{x^2} = \sqrt{36}$ $x = -6, 6$
Teacher: *Tell your partner quickly what you think the graph of $y = x^2 - 36$ will look like.* Elicit some of the key features of the graph: • roots at 6 and –6 • opens up • crossing the *y*-axis at –36 In each case, ask students to explain what makes them think this. After this conversation, present the graph.	
Teacher: *How about this one, $x^2 - 2 = 34$?* Students work. **Teacher:** *All right, who can get us started here?* **Student:** *Well I subtracted two from both sides and then it looked like the last one. So I think the solutions will be 6 and –6 again.* **Teacher:** *Seems like a lot of you agree. Could I graph this? What function would I graph?* **Student:** *I think in order to graph it you would need one side equal to zero, so this is equivalent to $x^2 = 36$. And we know from the last problem that this is equivalent to $x^2 - 36 = 0$. So I think we would just graph $y = x^2 - 36$.* **Teacher:** *Turn and tell your partner what we just said to see if you are convinced by it. I'm wondering if we even need to graph it.*	$x^2 - 2 = 34$ $x^2 = 36$ $\sqrt{x^2} = \sqrt{36}$ $x = \pm 6$ $x^2 - 2 = 34$ $x^2 = 36$ $x^2 - 36 = 0$ $y = x^2 - 36$

Algebra Problem Strings
©2017 Kendall Hunt Publishing

Teacher: *Just a few more today. Think about this one:* $x^2 - 72 = 0$.

Teacher: *I heard a lot of groans. Why was that?*

Student: *The 72 made it yucky. I don't know the square root of 72 off the top of my head so I stopped there.*

Teacher: *How about other folks? Is this what made you groan? Okay, so without a calculator, someone remind us how we could estimate the solution. I think you are saying that x is the square root of 72, though, yes?*

Student: *So, it's probably between 8 and 9 because 8 squared is 64 and 9 squared is 81 and 72 is right in the middle.*

Student: *But aren't there going to be two solutions. Something between 8 and 9 and something between −8 and −9?*

$$x^2 - 72 = 0$$
$$x^2 = 72$$
$$\sqrt{x^2} = \sqrt{72}$$
$$x = \pm\sqrt{72}$$

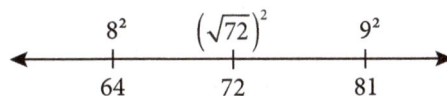

Teacher: *What do we think, everybody? Does this seem reasonable? How else could we get a good estimate here?*

Student: *We could graph it. The x-intercepts would be between 8 and 9 and between −8 and −9.*

Teacher: *Should we try it? Anyone think it might be something else? Okay. Let's look at the graph. Notice where it crosses the x-axis.*

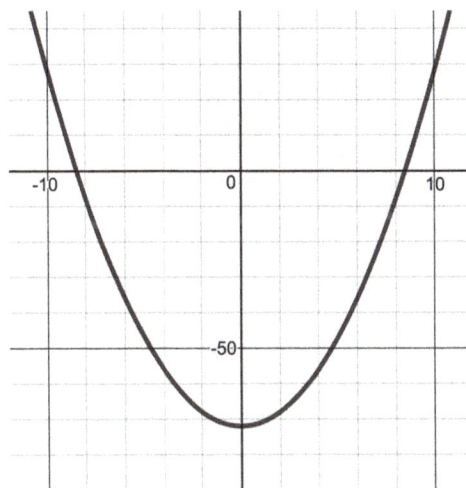

Teacher: *Let's keep going. How would you solve this one? What have we done previously to help you think about this one?*

Here, students will realize that it is equivalent to the previous problem, so there is no need to re-graph it once this is the consensus of the group.

$$x^2 - 10 = 62$$

(continued)

Teacher: *All right, crew. Last one. What will you do with this one?*

Student: *I love it.*

Teacher: *You love it? Say more…*

Student: *You cannot do this one. It's totally impossible because you would get x squared equals −72 and there is no number you can square and get a negative, so I think you were trying to trick us.*

Teacher: *Definitely not trying to trick you, maybe make you think a little bit. Who agrees with what she is saying?*

Student: *I agree with that. There isn't a number that would make this true so I'm going to guess that it's not a real thing— kind of an impossible situation like what she says.*

Teacher: *Interesting. In this class, we are going to leave that answer as "not a real number." Did you know that mathematicians wondered what would happen if you could find the square root of a negative number? You'll look at those ideas in future courses. Okay, so maybe we'll leave it there today and keep thinking about this.*

$$x^2 + 72 = 0$$
$$x^2 = -72$$
$$\sqrt{x^2} = \sqrt{-72}$$
$$x = \sqrt{-72} \quad \text{not a real number}$$

Algebra Problem Strings
©2017 Kendall Hunt Publishing

Sample Final Display

Your display could look like this at the end of the problem string:

$y = 3x + 6$ $0 = 3x + 6$

$0 = 3x + 6$ $-6 = 3x$

     $x = -2$ $x = -2$

$y = x^2$

$x^2 = 0$

     $x = 0$

$y = x^2 - 16$

             $x^2 = 16$

$0 = x^2 - 16$ $x = \pm 4$

     $x = -4, 4$

             $x^2 - 36 = 0$

$x^2 - 36 = 0$ $x^2 = 36$

     $x = -6, 6$ $\sqrt{x^2} = \sqrt{36}$

             $x = -6, 6$

         $x^2 - 2 = 34$ $x^2 - 2 = 34$

$x^2 - 2 = 34$ $x^2 = 36$ $x^2 = 36$

     $x = -6, 6$ $\sqrt{x^2} = \sqrt{36}$ $x^2 - 36 = 0$

             $x = \pm 6$ $y = x^2 - 36$

$x^2 - 72 = 0$ $x^2 - 72 = 0$

    $x = \pm\sqrt{72}$ $x^2 = 72$

$x^2 - 10 = 63$ $\sqrt{x^2} = \sqrt{72}$

             $x = \pm\sqrt{72}$

$x^2 + 72 = 0$ $x^2 = -72$ no real solution

             $x = \pm\sqrt{-72}$

$y = x^2 - 72$

Number line:

8^2 $\left(\sqrt{72}\right)^2$ 9^2

64 72 81

Facilitation Notes

This version of the problem string lists short notes for important teacher moves during the string. After you've done the string yourself and studied the relationships involved, you might make similar notes for the things you want a reminder of or deem important.

$y = 3x + 6$	Start with friendly linear—Can you solve it? Is this a function or an equation to solve?
$0 = 3x + 6$	Can you solve it now? What does the solution tell us? Graph it.
$y = x^2$	Is it a function that represents lots of points? Or an equation to solve?
$x^2 = 0$	What are x-intercepts? What will the graph look like?
$x^2 - 16 = 0$	Solve the second to help you think about the first. Turn and talk: how you solved it, what the solution tells us, what you think the graph will look like.
$x^2 - 36 = 0$	Same type—a chance to relate the equation to the graph. Compare to last problem—What is the same? What is different? Speculate on why.
$x^2 - 2 = 34$	Same as last one, when re-arranged. No need to graph.
$x^2 - 72 = 0$	Not so friendly—Conversation about estimating square roots. Estimate first, then show graph to verify.
$x^2 - 10 = 62$	Same as last one, when re-arranged. No need to graph.
$x^2 + 72 = 0*$	Optional: Is this possible? Why or why not? Solution is not a real number.

Algebra Problem Strings
©2017 Kendall Hunt Publishing

8.3 Factor Puzzles 1

At a Glance

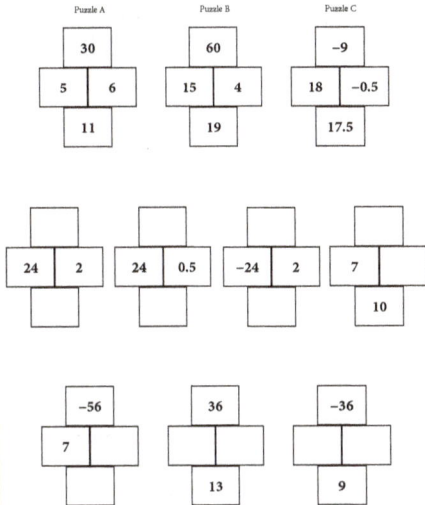

Objectives

The goal of this string is to support students in using relational thinking to find the products, sums, factors, and addends of various quantities.

Placement

This is the first in a series of two problem strings. These puzzles can be introduced at any time, but are designed to support students' work with quadratic functions, specifically, factoring trinomials.

You can introduce this string anytime in textbook Chapter 8, but it should precede or accompany the work in textbook Lesson 8.4 Factored Form.

Guiding the Problem String

This string requires some work on the teacher's part to establish the notation as well as the relationships between the values of the products and sums. Once the notation and the relationships between the values are established, the remaining time can focus on generating the missing parts of the puzzle.

It is often difficult for students to factor quadratic trinomials. These puzzles give students experience with the sum and product relationships involved in factoring quadratic trinomials without the variables. It is not important now to articulate this relationship. Focus instead on developing efficient strategies using the sums and the products to find the base numbers. How are some students using the structure of the numbers to generate possibilities? In what ways are students using the factors and addends of a number to find the right pair of base numbers? This is the thinking to be looking for and highlighting for the class during the string.

As you are leading the string, watch for students who have systematic strategies for finding the values and draw out those strategies. This is the act of making students' thinking visible for others to use and is an important way to move students beyond a "guess and check" strategy.

Though the string is designed as a puzzle, the emphasis is not solving the puzzle quickly or getting it "right." Instead, have students do the work of sharing their strategies and convincing their classmates that they have solved the puzzle. If students turn to you and ask, "Well, is it right?," you can question them, "Well, he just showed us his thinking and we seemed pretty convinced. Could there be another solution to this puzzle?"

About the Mathematics

The puzzles have this structure that relates to the factors of quadratic trinomials:

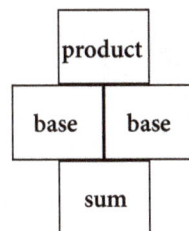

$$(x+a)(x+b) = x^2 + (a+b)x + ab$$

sum of a, b   product of a, b

(continued)

Sample Interactions

Use the following as you plan for ideas on how to elicit and model student strategies. This is not meant as a script, but as a view into the relationships involved and the intent of the problem string.

Teacher: *I'm going to put up three puzzles that we haven't talked about before and I want you to take 30 seconds to study them. Then, I'll give you a chance to talk to your partner about what is going on here. With your partner discuss:* • *How are these puzzles designed?* • *Are these values related, and if so, how?* *Any questions? Okay, here we go.* Pause while students work. **Teacher:** *Okay, let's make a list of your ideas so far. Who wants to get us started? I'm going to chart your ideas to see if we all agree.* **Student:** *It seems like the number on top is like the product.* **Teacher:** *Can you say more about that? The product of what?* **Student:** *I guess those numbers on the side, like the second row.*	Puzzle A — top 30, middle 5 and 6, bottom 11 Puzzle B — top 60, middle 15 and 4, bottom 19 Puzzle C — top −9, middle 18 and −0.5, bottom 17.5
Teacher: *What do we think of this idea? Does it seem to fit with the puzzles we have? Okay, so let's call this top box the product, and just so we are using the same language, let's call these middle boxes the base numbers or base values.* *And you are saying that the base numbers, when multiplied give us this value right here, or the product. Is that right? Okay, what else?* **Student:** *The number on the bottom is the sum. The sum of the base numbers.* **Teacher:** *What do we think of this idea? Yes? Okay, so I'll record that so we can keep track of these relationships. Can anyone give use a nice one sentence description of the entire puzzle, maybe for someone who missed today's class?* **Student:** *There are two numbers and you multiply them and put that number on top, the product, and then you add them and put that number on the bottom, the sum.*	Puzzle A — top box 30 labeled *product (×)*, middle boxes 5 and 6 labeled *base values*, bottom box 11 labeled *sum(+)*
Teacher: *So what I just showed you we will call a Factor Puzzle and they are really going to help us with quadratics, but don't worry about that right now. Let's play with this a bit. Find some missing parts of a few puzzles.* *Like all of our strings, I'll put a puzzle up one at a time and ask you to think to yourself about how you would complete it. Be ready to share your thinking with the rest of us. Here's the first one.*	Puzzle — middle boxes 24 and 2, top and bottom empty

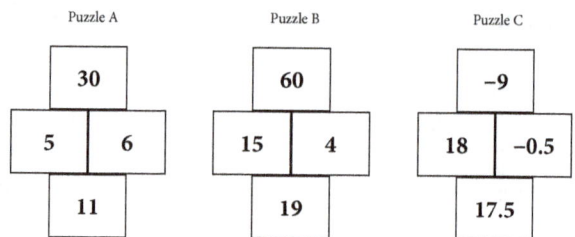

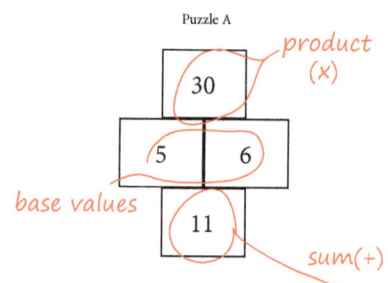

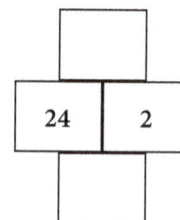

Algebra Problem Strings
©2017 Kendall Hunt Publishing

Student: *The top is 48 and the bottom is 26.* **Teacher:** *So the product, when they are multiplied, is 48 and the sum, when they are added, is 26.* The teacher repeats with the next two puzzles, helping students verbalize the meaning of multiplying by 0.5 (dividing by 2, half of) and multiplying integers.	Puzzle 1: top 48; middle 24 and 2; bottom 26. Puzzle 2: top 12; middle 24 and 0.5; bottom 24.5. Puzzle 3: top −48; middle −24 and 2; bottom −22.
Teacher: *Okay, now that we are warmed up, how would you do this one? Turn and tell your partner how you solved it.*	top (blank); middle 7 and (blank); bottom 10.
Teacher: *I overheard student A and student B talking about this as their solution. Could either of you explain your strategy?* **Student:** *I knew 7 plus something must be 10. So the other base number is 3. Then 7 times 3 is 21.* **Teacher:** *So it sounds like using inverse operations, working backwards, made this one solvable.*	top 21; middle 7 and 3; bottom 10.
Teacher: *How about this one? Will this idea of inverse operations help us here? Why or why not?* **Student:** *Yeah, so I thought about what times 7 would give me −56 and it's −8 so that's the other base number. Then I used −8 and 7 to get −1, which is the sum at the bottom.* **Teacher:** *What do we think? So you found the missing factor. Anyone get a different answer or just solve it differently?*	$7 \times \underline{} = -56$ $7 \times -8 = -56$ top −56; middle 7 and −8; bottom −1.
Teacher: *Okay, so let's do a few more. How would you use what you know so far to complete this puzzle? What relationships are you drawing upon?* Allow students time to work. *Turn and talk to your partner about what is helping you to solve this one.* *Who can start us off? What were you thinking?*	top 36; middle (blank) and (blank); bottom 13.

(continued)

Student: *I got 9 and 4. I thought about what times what is 36. I tried a few until I thought of 9 and 4.* **Teacher:** *So you were thinking about the factors of 36. Did anyone else think about factors of 36 and then see if you could find a factor pair that added to 13? A few of you? Did anyone try something different first? Did anyone think about what adds to 13 first?* **Student:** *Yeah, I did. I thought about 10 and 3, but that's 30. Then I thought about 9 and 4 and 9 times 4 is 36.* **Teacher:** *So, you could think about factors first or addends first.*	$4 \times 9 = 36$ $9 + 4 = 13$ 36 9 | 4 13
Teacher: *The last problem of our string today looks similar. How would you complete this puzzle?* Students work. The teacher ends the string by raising some important questions. **Teacher:** *Did anyone try the addends first? Tell me about that.* **Student:** *I looked for things that added to 9, like 4 and 5, 3 and 6, but those don't multiply to −36. Since it's a negative product, I need a positive and negative number. That's a lot more options, −1 and 10, −2 and 11. It keeps going on.* **Teacher:** *That is a lot of options. Did anyone try the factors first? Tell me about that.* **Student:** *Since the product is −36, I tried 6 times −6, 4 times −9, −4 times 9 and finally 12 times −3.* **Teacher:** *I wonder if one of those, addends or factors, is more efficient? We'll talk more about that in our next string.*	$12 \times -3 = -36$ $12 + (-3) = 9$ -36 12 | -3 9

Algebra Problem Strings
©2017 Kendall Hunt Publishing

Sample Final Display

Your display could look like this at the end of the problem string:

Puzzle A

30 — product (×)

5 | 6 — base values

11 — sum (+)

Puzzle B

60

15 | 4

19

Puzzle C

−9

18 | −0.5

17.5

48

24 | 2

26

12

24 | 0.5

24.5

−48

−24 | 2

−22

21

7 | 3

10

7 × ___ = −56
7 × −8 = −56

−56

7 | −8

−1

4 × 9 = 36
9 + 4 = 13

36

9 | 4

13

12 × −3 = −36
12 + (−3) = 9

−36

12 | −3

9

(continued)

Facilitation Notes

This version of the problem string lists short notes for important teacher moves during the string. After you've done the string yourself and studied the relationships involved, you might make similar notes for the things you want a reminder of or deem important.

Present these three to get at the structure of the puzzles.
Turn and talk. Label product, sum, base values.

Puzzle A

30

5	6

11

Puzzle B

60

15	4

19

Puzzle C

−9

18	−0.5

17.5

Quick—just getting used to the puzzles

24	2

Quick

24	0.5

Quick
Integers

−24	2

Tag inverse operations or working backwards or missing addend.

7	
	10

Same 7 but now negatives

−56

7	

Ah-ha! Now what?
More time here.
Elicit: factors or addends?

36

13

Who used addends?
Factors?
I wonder which is more efficient?

−36

9

Algebra Problem Strings
©2017 Kendall Hunt Publishing

For Display

Puzzle A

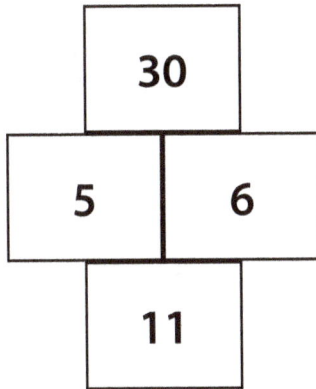

Puzzle B

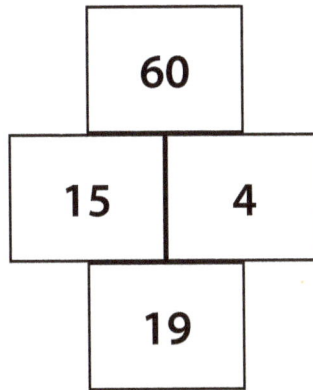

Puzzle C

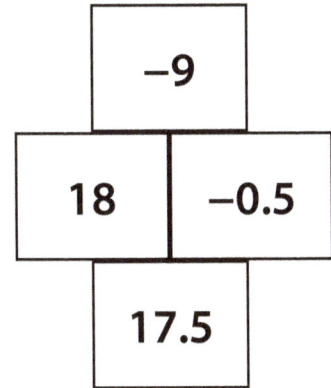

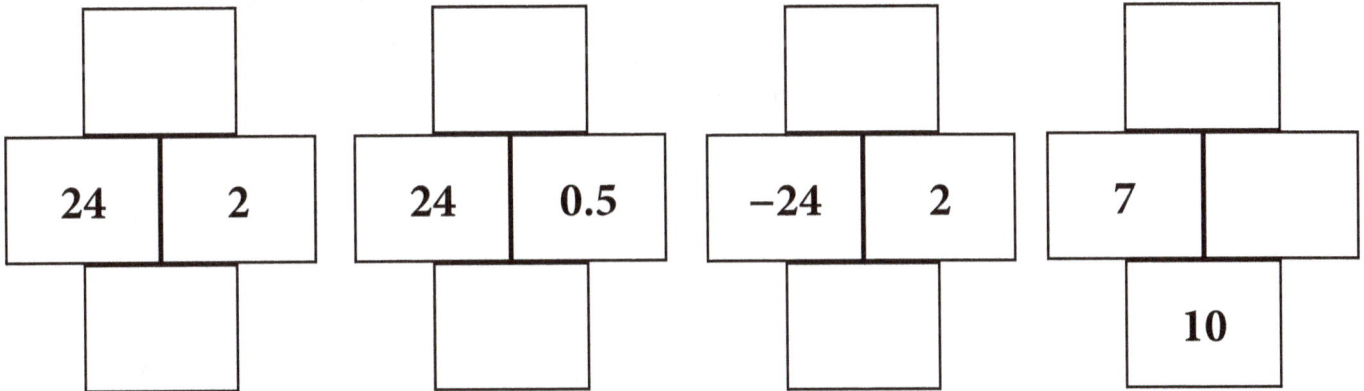

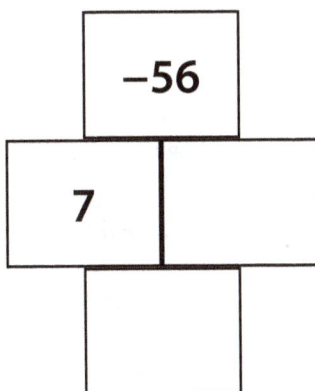

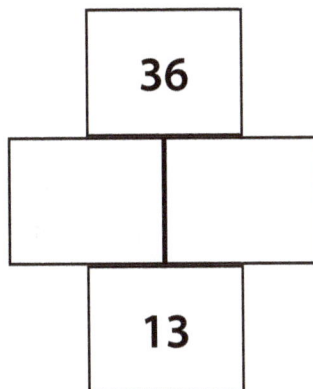

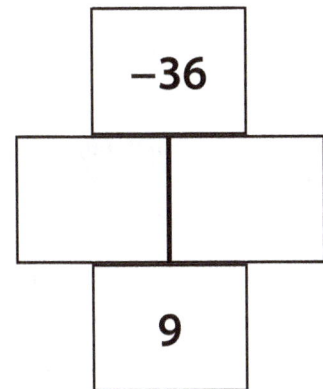

8.4 | Factor Puzzles 2

At a Glance

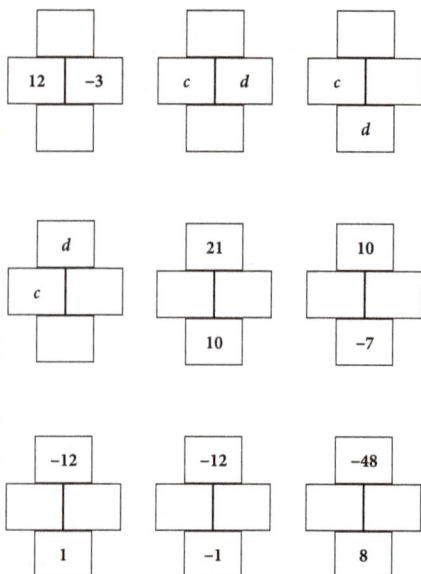

Objectives

The goal of this string is to support students as they use relational thinking to find the base values when given the product and sum of a factor puzzle. These strategies will support students to be able to factor quadratic trinomials.

Placement

These puzzles can be introduced at any time, but are designed to support students' work with quadratic functions, specifically factoring trinomials.

You can introduce this string anytime in textbook Chapter 8, but ideally it should precede and accompany the work in Lesson 8.4 Factored Form.

Guiding the Problem String

The string begins with a warm-up problem to crystallize the key relationships in the puzzle and uses variables as a way to codify these relationships.

Then, the string moves into puzzles where the product and sum are given and students are asked to provide the base numbers. The sequence of puzzles are increasingly more complex, some involving negative sums and products so students must consider integers. Because some strategies have come up in string 8.3, you can draw upon those strategies here, discussing which puzzles are easy to solve and which are harder. What makes a puzzle harder to solve? What about the given values makes the puzzle more or less challenging?

Guide the conversation so that students realize that considering the factors first is more efficient than the addends first. Help students make organized lists if necessary, especially when dealing with negative products.

About the Mathematics

One strategy is to consider the addends first and then try to find which ones create the correct product. The list of possible addends is infinite, thus not efficient. The list of possible factors is much smaller and is therefore a more efficient place to start.

Important Questions

See the previous string for ideas on leading this problem string.

Use the following as you plan how to elicit and model student strategies.

- *How do the numbers in the puzzles relate to each other?*
- *Let's get a little general by using variables. If you have any number c and any number d, what is their product? Their sum? How can we represent the product and sum using variables?*
- *For the fifth puzzle, what if we only have the product and the sum? How many possibilities are there?*
- *Could you find all of the possible addends?*
- *Could you find all of the possible factors?*
- *What happens when one of the numbers is negative? How does that affect the other numbers?*
- *Would it help to make a list?*

- *How did you organize your list?*
- *What are you looking for in your list? Why? How does that help you?*
- *Are there more possible addends or factors?*
- *Which is more efficient, looking for possible addends first or possible factors first? Why?*
- *Which puzzles are easy to solve and which are harder?*
- *What makes a puzzle harder to solve? What about the given values makes the puzzle more or less challenging?*

Sample Final Display

Your display could look like this at the end of the problem string:

(continued)

Facilitation Notes

This version of the problem string lists short notes for important teacher moves during the string. After you've done the string yourself and studied the relationships involved, you might make similar notes for the things you want a reminder of or deem important.

Super quick.

Puzzle 1 (middle row): **12** | **−3**

Present the next three problems together. Look, then turn and talk.

Second, third, fourth—getting a little general.

Puzzle 2 (middle row): *c* | *d*

Puzzle 3 (middle row): *c* | (blank); bottom: *d*

Puzzle 4: top: *d*; middle row: *c* | (blank)

This one's a puzzle!

Puzzle 5: top: **21**; bottom: **10**

Now 10 is the product. And a negative sum.

Puzzle 6: top: **10**; bottom: **−7**

Negative product, how does that work?
Talk about strategy, factors or addends?

Puzzle 7: top: **−12**; bottom: **1**

Same −12, but now −1. How can the last problem help think about this one?

Puzzle 8: top: **−12**; bottom: **−1**

Expose and describe strategies: addends too inefficient. Listing factors, do you need to list them all?

Puzzle 9: top: **−48**; bottom: **8**

Algebra Problem Strings
©2017 Kendall Hunt Publishing

For Display

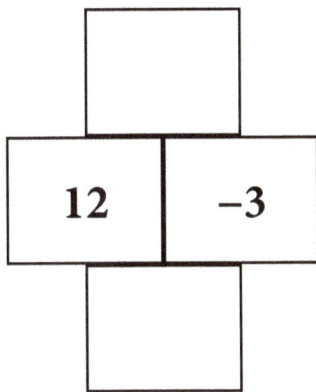

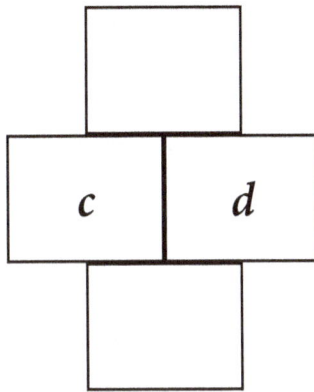

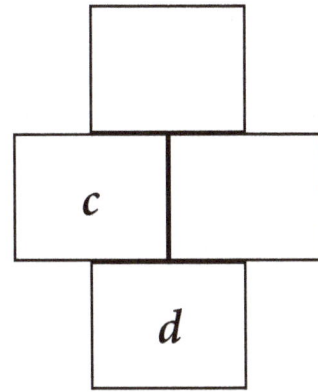

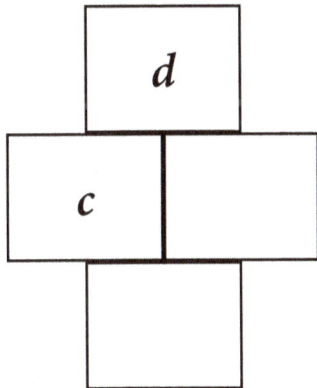

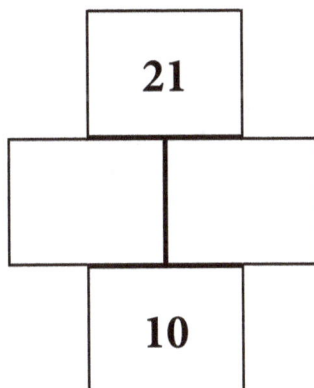

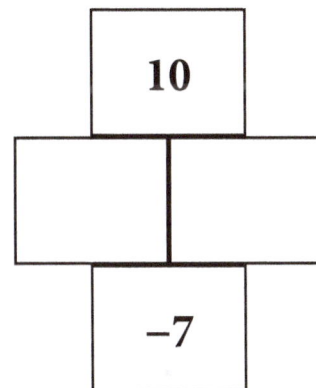

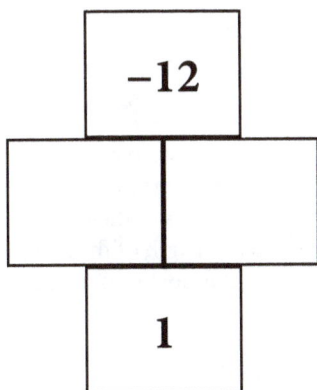

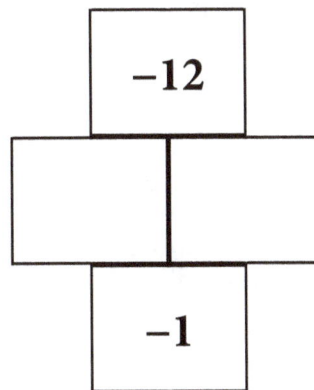

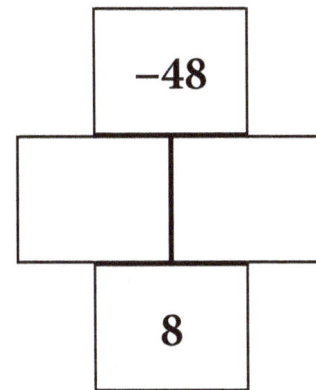

8.5

Expanding and Envisioning

At a Glance

10×10

10×12

13×10

13×12

$(10+3)(10+2)$

$(x+3)(x+2)$

$5(10+2)$

$5(x+2)$

$(x+2) \times 5$

$(x+6)(x+8)$

$(x-7)(x+8)$

Objectives

The purpose of this problem string is to help students change algebraic expressions to different but equivalent forms. The rectangular diagram gives students a way to envision how binomial expressions and their products are related.

Placement

This string is designed to follow the factor puzzle work, so students have some familiarity with factoring and multiplying before they encounter it in the rectangular diagram.

You can use this problem string to introduce textbook Lesson 8.5 Projectile Motion.

Guiding the Problem String

The mathematics of this string should be quite accessible to students without paper and pencil, so encourage students to try to solve the problems with mental visualization. The focus is on envisioning the expressions in relation to their products, all captured in one rectangular diagram. Later, we will use this same diagram to factor and solve.

About the Mathematics

The string uses the open array, a model students have likely encountered in elementary or middle school with multiplication and division. Recall that when multiplying (or dividing) using the array the factors are thought of as the length and width of the array, and the product as the area.

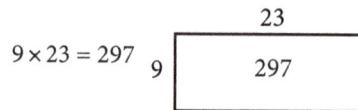

$9 \times 23 = 297$

The open array does not need to be drawn with absolute precision (it is not a ruler) because it is meant to capture students' thinking and be a model for the relationships between values. It should, however, be roughly to scale. That is, a side length of 12, for example, should be drawn somewhat longer than a side length of 4.

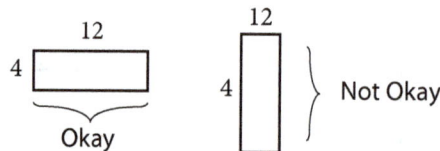

A similar model is a version of the array called a rectangular diagram, meant to capture the general relationship between factors and their product. In a rectangular diagram the factors (here monomial and binomial expressions) are analogous to the lengths and widths of the rectangle and the product to the area, but there is no real attention to the relative length of the sides. In this way, the rectangular diagram can be considered more of a computational tool than a model.

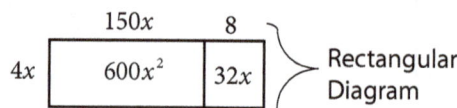

Algebra Problem Strings
©2017 Kendall Hunt Publishing

Sample Interactions

Use the following as you plan how to elicit and model student strategies. This is not meant as a script, but as a view into the relationships involved and the intent of the problem string.

Teacher: *I'm going to put a very friendly multiplication problem on the board. I know you know it, so today be thinking about what it would look like as an array. We're going to use what you did in elementary or middle school to help us with multiplying expressions.* *So, what is 10 times 10?* *What does it look like?* *Where are the tens and where is the 100?* *Does this make sense?* *Are we convinced?*	10×10 $10 \times 10 = 100$
Teacher: *Okay, can you use the last problem to make the array of this one?* *Turn and tell your partner:* *What changed?* *What stayed the same?* *What's the product?* *What will this look like?*	10×12 $10 \times 12 = 120$
Teacher: *How about this one, 13 times 10?* *What's the product?* *Who can share what this one looks like?* *Could you have used a previous problem to build the array of this one? How?*	13×10 $13 \times 10 = 130$

(continued)

Teacher: *Let's try this, 13 times 12. If it's helpful to make smaller arrays and put them together to make this one, go ahead.* *Tell your partner two things:* • *What does the array look like?* • *How can you use the array to solve this problem, without paper and pencil?*	13×12 $13 \times 12 = 156$ 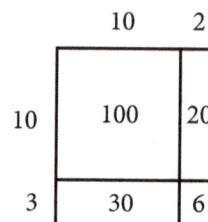 $100 + 20 + 30 + 6 = 156$
Teacher: *All right, what's going on with this one, 10 plus 3 all times the quantity 10 plus 2?* *Have we seen anything like this before? How is it different from previous problems?*	$(10+3)(10+2)$ $(10+3)(10+2) = 156$ 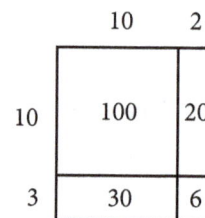
Teacher: *And if I give you this to think about, some x plus 3 all times the quantity of that same x plus 2. Here some of the side lengths are unknown. How do you envision it?* *Can you still multiply these factors?* *And if so, what does your answer tell you?* *Think time first. Then, turn and talk.* *[Optional] Make a sketch with your partner as you discuss.* Mention that you are now moving from an open array that represents dimensions and area to a rectangular diagram that only represents factors and products. Because the value of x is unknown, there are infinite possibilities. So the rectangular diagram is a way to capture a generalized version without the relative magnitude of the terms.	$(x + 3)(x + 2)$ $(x+3)(x+2) = x^2 + 5x + 6$ 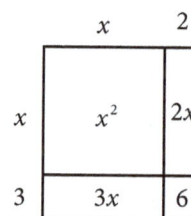

Algebra Problem Strings
©2017 Kendall Hunt Publishing

Teacher: *Okay, how about this one, 5 times the sum of 10 and 2?*	$5(10+2)$ $5(10+2)=(5\times10)+(5\times2)=60$ *[area model: width 5; length split into 10 and 2; regions 50 and 10; total 60]*
Teacher: *The next problem is 5 times the sum of some x and 2?* *What are you picturing here?*	$5(x+2)$ $5(x+2)=5x+10$ *[area model: width 5; length split into x and 2; regions 5x and 10]*
Teacher: *Alright, let's do this, the sum of x and 2, all times 5.* *Can I use the same model as before? Why or why not? Turn and talk.*	$(x+2)\cdot5$ $(x+2)\cdot5=5x+10$ *[area model: width 5; height split into x and 2; regions 5x and 10]*
Teacher: *So, let's try this one, the sum of x and 6 all times the sum of x and 8.* *What exactly is going on here? Think time first. Then, turn and talk.* *[Optional] Make a sketch with your partner as you discuss.*	$(x+6)(x+8)$ $(x+6)(x+8)=x^2+14x+48$ *[area model: top split into x and 8; side split into x and 6; regions x^2, $8x$, $6x$, 48]* $x^2+6x+8x+48$ $x^2+14x+48$

(continued)

Teacher: *Last one friends, the difference of x and 7 times the sum of x and 8.*

What do you picture?

How would you use the diagram to find the product?

What's different about this one?

Does the rectangular diagram still help us?

$$(x - 7)(x + 8)$$

$$(x-7)(x+8) = x^2 + x - 56$$

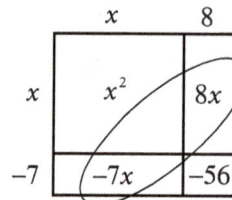

$$8x + (-7x) = x$$

$$x^2 + x - 56$$

Teacher: *How would you summarize some of the things that came up in this string today?*

Elicit the following:

- *We can use an open array to represent multiplication of numbers.*

- *We can use a rectangular diagram to represent the multiplication of expressions with variables.*

- *We can add the pieces together to find the total product.*

- *Some of these were related to the numbers in the factor puzzles.*

Algebra Problem Strings
©2017 Kendall Hunt Publishing

Sample Final Display

Your display could look like this at the end of the problem string:

$10 \times 10 = 100$

$10 \times 12 = 120$

$13 \times 10 = 130$

$13 \times 12 = 156$

$(10+3)(10+2) = 156 = 13 \times 12$ $100+20+30+6 = 156$

$(x+3)(x+2) = x^2 + 5x + 6$

$5(10+2) = (5 \times 10) + (5 \times 2) = 60$

$5(x+2) = 5x + 10$

$(x+2) \cdot 5 = 5x + 10$

$(x+6)(x+8) = x^2 + 14x + 48$

$x^2 + 6x + 8x + 48$
$x^2 + 14x + 48$

$(x-7)(x+8) = x^2 + x - 56$

$8x + (-7x) = x$

$x^2 + x - 56$

(continued)

Facilitation Notes

This version of the problem string lists short notes for important teacher moves during the string. After you've done the string yourself and studied the relationships involved, you might make similar notes for the things you want a reminder of or deem important.

10×10	Really friendly. What does it look like on an open array?
10×12	Envision an array. What changed? What's the product?
13×10	Another easy one.
13×12	What is the product? Would smaller arrays help?
$(10+3)(10+2)$	How does this relate? Same thing!
$(x+3)(x+2)$	Some of the lengths are unknown. How do you envision it? Shift from open array to rectangular diagram language.
$5(10+2)$	Back to numbers. Quick.
$5(x+2)$	What are you picturing here?
$(x+2) \times 5$	Slightly new structure—now what?
$(x+6)(x+8)$	What exactly is going on here? Turn and talk. Sketch w/ partner.
$(x-7)(x+8)$	Challenge problem w/ negative term. Connect to factor puzzle work.

Algebra Problem Strings
©2017 Kendall Hunt Publishing

At a Glance

$$x(x+4)=5$$
$$x(x-1)=12$$
$$x(x-2)-8=0$$
$$x^2+2x=15$$

Follow-up string:

$$x(x+3)=18$$
$$x(x-7)=-10$$
$$x(x+6)+9=0$$
$$x^2-5x=24$$

Objectives

The goal of this problem string is to help students develop the "factor pair" strategy and the "factoring then using the zero product property" strategy when solving quadratic equations.

Placement

This string could come after students have worked with factor puzzles, multiplying linear binomials, factoring quadratic trinomials, and have begun to work with solving quadratic equations using the zero product property.

You could deliver this string after textbook Lesson 8.4 Factored Form and before Lesson 8.7 The Quadratic Formula.

Guiding the Problem String

The first two problems are given in the format of $x(x-b)=c$ to nudge some students to consider the relationships in the factor pair strategy. Other students will solve by factoring then using the zero product property: expand, move everything to one side of the equation, factor and use the zero product property. Model both strategies. Compare. The second problem can suggest consecutive integer solutions x and $x-1$, which can make it easier for students to experiment with factor pairs. The third and forth problems change the format and might nudge students to choose factoring then using the zero product property. This can also help students learn to change to a format that is more efficient to solve. Ask students to share both strategies and compare.

By juxtaposing the two strategies, students have the opportunity to grapple with the relationships, parse out the similarities and differences, and build confidence in choosing an efficient strategy. Since not all students will grasp the factor pair strategy right away, there is a follow-up string provided to give students more opportunity to build this strategy.

About the Mathematics

The factor pair strategy is a limited, but useful, strategy for solving some quadratic equations using factor pairs of c for the form $x(x-b)=c$. The factor pair strategy works if $x(x-b)=c$ has integer solutions. Often students can find one of the integer solutions by quickly substituting factors of c. To find the other solution, if there are two solutions, it can be helpful to think about the equivalent equation $x^2-bx-c=0$, where it is more obvious that solutions $x=r_1, r_2$ must be factors of $-c$. For example, for $x(x-1)=12$, students often recognize quickly that $x=4$ is a solution because $4(4-1)=12$. The other solution is the partner with the first solution in the factor pair of -12, so with 4 is the second solution -3.

The factoring then using the zero product property strategy is the traditional method of setting the quadratic equation equal to zero, factoring into linear binomials, setting each linear binomial equal to zero, and solving for x.

Both of these strategies are limited to rational solutions, the factor pair strategy is limited further to integer solutions.

(continued)

Sample Interactions

Use the following as you plan how to elicit and model student strategies. This is not meant as a script, but as a view into the relationships involved and the intent of the problem string.

Teacher: *We have been working with quadratic functions and equations. Today's string has some equations for us to solve. To start, no paper and pencil. Just look and see what you can find. The first problem is $x(x+4)=5$. What is x?* Brief think time.	$$x(x+4)=5$$
Teacher: *I see some of you nodding. What are you thinking so far?* **Student:** *Well, I see that two things multiplied together have to be 5.* **Student:** *I see that if we multiplied that out, we'd have the quadratic equation $x^2 + 4x = 5$.* **Teacher:** *Great. Go ahead and solve for x. What is x?* Students work and the teacher circulates, looking for students starting to use a factor pair strategy and students factoring then using the zero product property. The teacher also questions to prompt students who might be stymied.	$$x(x+4)=5$$ $$x^2$$
Teacher: *I saw some of you doing what we were working on yesterday. Will you please share your thinking?* **Student:** *I multiplied it out and moved the 5 over.* **Teacher:** *What did you get when you multiplied x times x + 4?* **Student:** *I got x squared and 4x. Then I brought the 5 over so now it's all equal to 0. Then I factored into x plus 5 times x minus 1. Then x is −5 and 1.* **Teacher:** *So, you factored and since the two factors were multiplying to get 0, you knew that you could set each factor to 0. Does anyone remember what we call that property?* **Student:** *It's on the anchor chart we made yesterday, the zero product property.*	$$x^2+4x-5=0$$ $$(x+5)(x-1)=0$$ $$x+5=0 \quad x-1=0$$ $$x=-5,1$$
Teacher: *Let's take a quick look at that graphically. I'll put it in the class display grapher. Where is this quadratic equal to 0? Where are the y-values 0?* **Student:** *Yeah, the y-vaues are 0 at the x-axis so that agrees with what we got, x is −5 and 1.*	 $$y=x^2+4x-5$$

Teacher: *I'd like to look graphically at the original equation, $x(x+4)=5$ before you got it all equal to 0. To do that, let's graph $y=x(x+4)$ and $y=5$ and see where they inter- sect. What do you see?*

Student: *The red quadratic intersects the x-axis at the same x-values as the blue quadratic intersects 5.*

Student: *I don't know. It looks like the red ones are further out than the blue ones.*

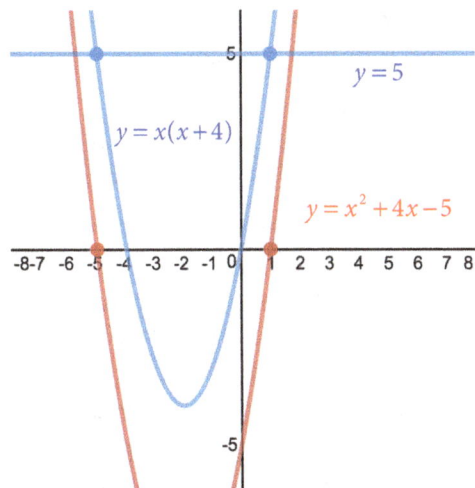

Teacher: *We can graph the vertical lines at $x = -5$ and $x = 1$. What do you see now?*

Student: *Ahhh, right. They do intersect at the same x-values.*

Teacher: *So equivalent equations, $x(x+4)=5$ and $x^2 + 4x - 5 = 0$, have the same solutions. And we can see that in the graphs. Interesting.*

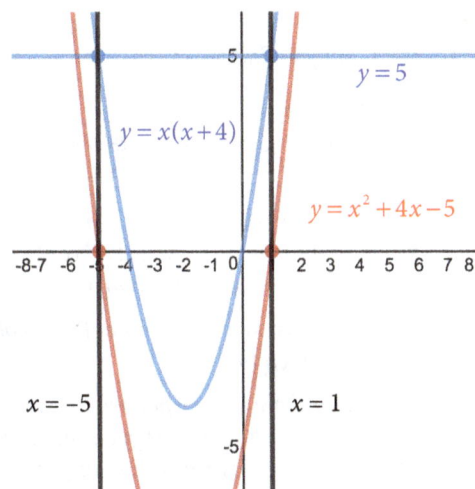

Teacher: *Nice work. But that's not what everyone was doing. Will you please tell us what you were working on?*

Student: *Yeah, I looked at it like there were two numbers times each other that got the answer 5. And so the only things are 1 and 5. So I tried 1 and it worked.*

Teacher: *You thought about the factors of 5, which are 1 and 5. What do you mean 1 worked?*

Student: *Well, if x is 1, then 1 times 5 is 5 because 1 and 4 is 5.*

$$x(x+4)=5$$

$$1(1+4)=5$$

(continued)

Teacher: *Does everyone follow what she was thinking? And does that fit with the solutions the others found? And the graph?*

Student: *It looks like we found one of the same solutions. But there must be another one.*

Student: *Yeah, we should find −5 too.*

Teacher: *Does −5 work if you plug it in? Sure does. Did anyone find the −5 before the others had found it using factoring and the zero product property that we already have on the board?*

Student: *I did. I just guessed all of the factors of 5. You had said 1 and 5 were the only factors, but so are −1 and −5. When I tried them, −5 worked.*

Teacher: *I wonder if there is a way to find that other factor besides guessing and checking? Something more efficient. Let's do another problem and we'll come back to that idea.*

Teacher: *Okay, here's the second problem, $x(x-1)=12$. What is x?* Students pick up pencils and work and the teacher circulates, looking for students using the factor pair strategy and factoring then using the zero product property.	$x(x-1)=12$
Teacher: *What did you find for x?* **Student:** *I got 4 and −3.* **Teacher:** *I saw some of you messing with the factors of 12. Tell us about that.* **Student:** *Well, like they said in the last problem, you have two things times each other is 12. So I could just see that 4 would work. 4 times 1 less than 4, or 3, is 12.* **Student:** *And then I just kept trying other factors of 12 until I found that −3 worked too.* **Teacher:** *I wonder if there is some pattern to it that we can figure out so we don't have to just guess and check all the time. How would you describe this strategy?* **Student:** *Guessing 'til it works.* **Student:** *Guessing factors.*	$x = 4, -3$ $4(4-1)=12$ $-3(-3-1)=-3(-4)=12$

Algebra Problem Strings
©2017 Kendall Hunt Publishing

Teacher: *How about those of you who multiplied stuff out? Tell us about that.* Students explain expanding and using the zero product property. **Teacher:** *How would you describe this strategy?* **Student:** *Get everything on one side and 0 on the other, then factor.* **Student:** *And because it's two things times each other getting 0, you make them both 0.* **Teacher:** *Right, using the zero product property. Does anyone see a connection between the two strategies?* **Student:** *Well, they both got the same answers.* **Student:** *Yeah, but also they were both thinking about factors of 12.* **Teacher:** *What do we think about that? Were they both thinking about factors of 12? −12? Let's see what happens with this next problem.*	$x(x-1)=12$ $x^2-x-12=0$ $(x-4)(x+3)=0$ $x-4=0 \quad x+3=0$ $x=4,-3$
Teacher: *Here is the third problem,* $x(x-2)-8=0$. *How do you feel like finding x for this problem? Find x and then we'll discuss your thinking.* Students work. The teacher looks for both strategies and nudges students to try the other strategy if they finish early to help them decide which strategy is the most efficient for which problems.	$x(x-2)-8=0$
Teacher: *First, what did you find for x?* **Student:** *I found x is 4 and −2.* **Teacher:** *What were you thinking?* **Student:** *Well, since the problem started with the −8 already on the left side, I multiplied and factored.* The student details the strategy. **Teacher:** *Why might you want to factor and use the zero product property strategy for this problem?* **Student:** *Like he said, the −8 was already over there so it made sense to multiply and factor.*	$x(x-2)-8=0$ $x^2-2x-8=0$ $(x-4)(x+2)=0$ $x-4=0 \quad x+2=0$ $x=4,-2$

(continued)

Teacher: *What about guessing a factor, where you thought about the factor pairs of 8? Did anyone try that?*

Student: *I did. I just added 8 to both sides, so then I was thinking about factors of 8. I noticed that it was a number and that number minus 2, so the numbers have to be two apart. The factors of 8 that are two apart are 2 and 4.*

Teacher: *That's interesting. What do you all think about what she noticed?*

Student: *Wait, what do you mean two apart?*

Student: *Like, 8 and 1 are not two apart but 4 and 2 are. Since it says x times x − 2, the numbers are two apart. So, since 4 worked, then I tried 2 but it didn't work so I tried −2 and it did.*

Teacher: *Was that true in the previous problems? Interesting insight.*

$$x(x-2)=8 \qquad 4(2)=8, \quad -2(-4)=8$$
$$x=4, -2$$

Teacher: *Let's see what you do with the last problem, which is $x^2 + 2x = 15$.*

The teacher repeats modeling both a factor pair strategy and the factoring then using the zero product property strategy. As the strategies are being discussed, the teacher uses the name/descriptors for the strategies. The guessing factors strategy is tagged as the "using factor pairs" strategy.

$$x^2 + 2x = 15 \qquad x^2 + 2x - 15 = 0 \qquad\qquad x(x+2)=15 \qquad 3(5)=15, \quad -5(-3)=15$$
$$x = -5, 3 \qquad (x+5)(x-3)=0 \qquad\qquad x = 3, -5$$
$$x+5=0 \quad x-3=0$$
$$x = -5, 3$$

Teacher: *Let's take a look at the two different equations you solved graphically. Are the x-values of the intersections the same?*

The teacher leads a conversation about the connections between the graphs, the x-intercepts, the intersection points, and the solutions to the equation.

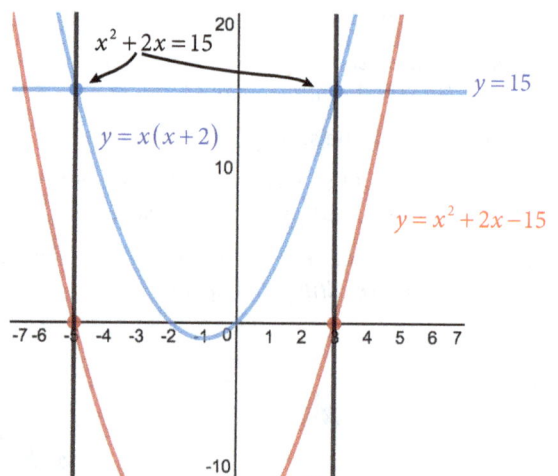

Teacher: *How would you summarize some of the things that came up in this string today?*

Elicit the following:

- *Solving with factor pairs involves finding a factor that works and then finding the other factor. There are patterns in the solutions, they are factor pairs.*

- *It makes sense to use the factor pair strategy when the equation is already set up in the form $x(x-b)=c$.*

Algebra Problem Strings
©2017 Kendall Hunt Publishing

- *Solving by factoring and using the zero product property makes sense when the form is $x^2 - bx - c = 0$.*
- *Equivalent equations have the same solutions. These can be seen on a graph as the same x-values of intersection points.*

Sample Final Display

Your display could look like this at the end of the problem string:

$x(x+4)=5$ $\quad$ $x(x+4)=5$ $\quad$ $x^2+4x-5=0$ $\quad$ $1(1+4)=5$

$\quad\quad x = -5, 1$ $\quad\quad\quad\quad\quad\quad$ $(x+5)(x-1)=0$

$\quad\quad\quad\quad\quad\quad\quad\quad\quad\quad$ $x+5=0 \quad x-1=0$

$\quad\quad\quad\quad\quad\quad\quad\quad\quad\quad$ $x = -5, 1$

$x(x-1)=12$ $\quad$ $4(4-1)=12$ $\quad\quad$ $x(x-1)=12$

$\quad\quad x = -3, 4$ $\quad$ $-3(-3-1)=-3(-4)=12$ $\quad$ $x^2-x-12=0$

$\quad\quad\quad\quad\quad\quad\quad\quad\quad\quad\quad\quad$ $(x-4)(x+3)=0$

$\quad\quad\quad\quad\quad\quad\quad\quad\quad\quad\quad\quad$ $x-4=0 \quad x+3=0$

$\quad\quad\quad\quad\quad\quad\quad\quad\quad\quad\quad\quad$ $x = 4, -3$

$x(x-2)-8=0$ $\quad$ $x^2-2x-8=0$ $\quad$ $x(x-2)=8$ $\quad$ $4(2)=8, \quad -2(-4)=8$

$\quad\quad x = -2, 4$ $\quad$ $(x-4)(x+2)=0$ $\quad$ $x=4, -2$

$\quad\quad\quad\quad\quad\quad\quad\quad$ $x-4=0 \quad x+2=0$

$\quad\quad\quad\quad\quad\quad\quad\quad$ $x = 4, -2$

$x^2+2x=15$ $\quad$ $x^2+2x-15=0$ $\quad$ $x(x+2)=15$ $\quad$ $3(5)=15, \quad -5(-3)=15$

$\quad\quad x = -5, 3$ $\quad$ $(x+5)(x-3)=0$ $\quad$ $x=3, -5$

$\quad\quad\quad\quad\quad\quad\quad\quad$ $x+5=0 \quad x-3=0$ $\quad\quad\quad$ Use factor pairs.

$\quad\quad\quad\quad\quad\quad\quad\quad$ $x = -5, 3$

$\quad\quad\quad\quad$ Factor, use the zero product property

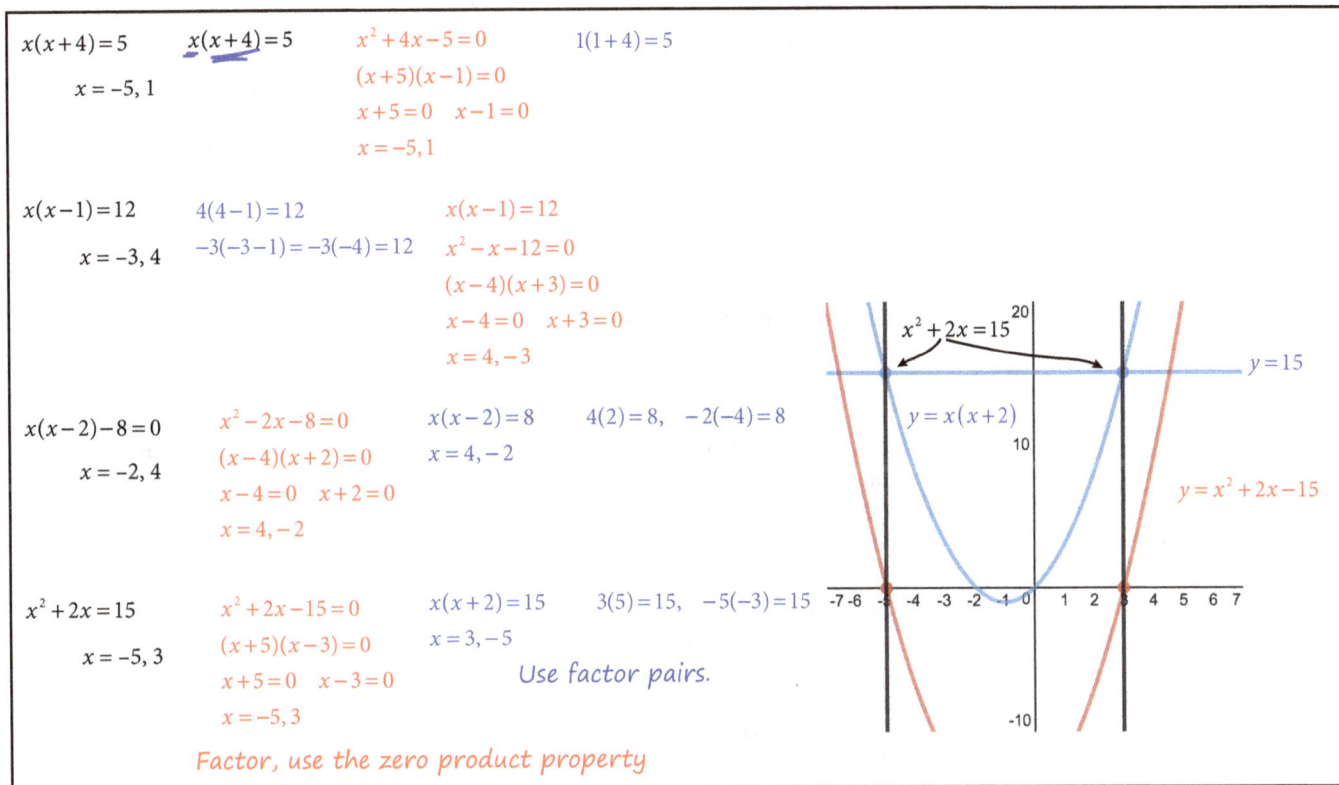

Facilitation Notes

This version of the problem string lists short notes for important teacher moves during the string. After you've done the string yourself and studied the relationships involved, you might make similar notes for the things you want a reminder of or deem important.

$x(x+4)=5$	First, what do you think the solutions might be? Solve for x. Find factor pairs, factor with zero product property. Graph given and general form. Make connections.
$x(x-1)=12$	Now what's x? Start describing strategies. Might notice that numbers are 1 apart. What do the strategies have in common?
$x(x-2)-8=0$	Different structure. What about the structure influences your strategy choice? Notice that numbers are 2 apart. Which is more efficient?
$x^2+2x=15$	Different structure. How do solutions relate to each other? Factor pairs? Tag the strategy name. Add to anchor chart. Notice patterns when using factor pairs. Graph.

8.7 How Would You Solve and Why?

At a Glance

$$5x^2 - 45 = 0$$

$$x^2 + 5x = 24$$

$$4x^2 + x - 3 = 0$$

$$14 + 2x^2 = 0$$

$$x^2 + 6x - 1 = 0$$

$$(x - 4)(x + 7) = 0$$

Objectives

This string is designed to help students choose a strategy for solving quadratic equations before introducing the next strategy, the quadratic formula.

Placement

Use this problem string after students have wrestled with a variety of strategies and types of quadratic equations. Ideally they have had the chance to solve by factoring, factor pairs, graphing, using inverse operations (solving by inspection), and completing the square.

You can use this problem string to support students to solve equations before they learn the general solution tool in textbook Lesson 8.7 The Quadratic Formula.

Guiding the Problem String

There are two "levels" of this string. On one level you could present the problems one at a time and ask students, without paper or pencil, to describe how they would solve and justify their choices. What did you pay attention to that made you think solving by inspection was a good choice here? Why wouldn't graphing be a smart strategy on this one? In this way you are making visible the kinds of choices that mathematicians make when solving these equations. You want this knowledge to be shared by the entire class. Consider making an anchor chart.

Depending on where your students are, you might also take the conversation to a second level—inviting students to describe their strategies in greater detail (e.g., "Walk us through your thinking…"), while you as the teacher are modeling for the class. This is also the place to try to generalize key insights that some students have but others do not. For example, students' ability to predict how many or what kind of solutions will result, or students' recognition of common factors or the presence of perfect squares.

Unlike most problem strings that are designed so that the problems are closely related and scaffolded to both support and challenge students, the problems in this string are less closely related so that students can talk more about strategy and less about relatedness to the previous problem. Students might use previous problems to highlight differences, and if so, we want to connect their observation of differences to their choice of strategy. This string could be considered a "strategy talk" where we are not introducing new mathematical ideas, but pausing to solidify some shared understandings as a class.

The mathematical work is coming to consensus as a class about what features of an equation help to choose smart strategies. Let the students do this work. If they are not sure, invite them to play with a few strategies to see what feels efficient, elegant, or clever. This is not the place to insert yourself as a mathematician or mathematical authority, but create space for students to share their reasoning and try to convince each other of the merits.

About the Mathematics

While the quadratic formula is the strategy to solve any quadratic equation, the other strategies in this session can be much more efficient given the numbers in the problem as many of the problems in this string demonstrate.

Sample Interactions

Use the following as you plan how to elicit and model student strategies. This is not meant as a script, but as a view into the relationships involved and the intent of the problem string.

Teacher: *Let's warm up with a problem string today that gets us really thinking about strategy. Someone name the ways that we know how to solve a quadratic equation so far.*

Student: *We could graph it, factor it, use factor pairs, complete the square, and use inverse operations.*

Teacher: *Today I'm going to put a few equations on the board, but I want us to focus on what strategy you would use and why that strategy feels like a smart choice—better than the others, perhaps.*

We may not even get to solutions today and that's fine. Our goal is to really know when to use one strategy over another and why. Does that make sense? Okay, here we go.

Teacher: *How would you solve this one? Maybe think about what your first instinct is, what's your opening move?* **Student:** *I guess I would first factor out the fives and then it would be pretty easy to solve.* **Teacher:** *Who understands why this was Hector's first move? Why does this make sense?*	$5x^2 - 45 = 0$ $5(x^2 - 9) = 0$ $x^2 - 9 = 0$
Student: *It makes sense because in a way you are always looking for a common factor first, so that you can make the equation look simpler. I thought of it differently. I saw that everything could be divided by 5 and that would give you x squared minus 9 equals 0, which I can do.* **Teacher:** *I could model that this way.*	$\dfrac{5x^2}{5} - \dfrac{45}{5} = \dfrac{0}{5}$ $x^2 - 9 = 0$ $x^2 = 9$ $x = \pm 3$

Teacher: *Two ideas just came up. One, is the idea that a good first move is to look across all terms for a common factor. Two, is the question of whether factoring out a 5 is the same as dividing by 5. Turn and talk to your partner about these two ideas.*

Start a separate poster or section of the board "Choosing a Smart Strategy" where ideas that are shared can be recorded, such as "Look for common factors across the terms."

Teacher: *Sounds like we agree that we've really got two different looks for using the same relationships, factoring or dividing out the common factor. Looking for a common factor is a good first move, which makes this one something we can solve easily, yes?*

(continued)

Teacher: *Okay, how about this one? What do you pay attention to so you know how to start? Turn and tell your partner what strategy you would use.*	$x^2 + 5x = 24$

The teacher listens for factoring using the zero product property and using factor pairs.

Teacher: *I'm hearing several of you using factoring as your strategy. Who can say a little bit about factoring as a strategy? Why this one? Why now?*

Student: *This one is kind of set up for you to factor. I mean all those factor puzzles we did, or you could represent it in a rectangular diagram, too. I don't know off the top of my head if it is factorable, but I would try that. These numbers look friendly.*

Student: *Yes, but I think you could just leave the 24, then factor out the x, and look for two factors of 8 that are 5 apart. That's easy for me to see that one of them is 3.*

Teacher: *I see people nodding—seems like we are now saying, "Look for a common factor first. Then see if it's worth factoring or using factor pairs next." Do I have that right? Would anyone have graphed it?*

Student: *You could. Depends on if you are doing it by hand or with a calculator. But it's easy enough to factor, that you don't need to.*

$$x^2 + 5x - 24 = 0$$

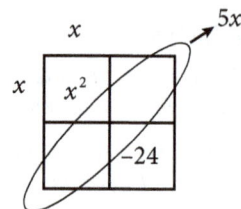

$$x(x+5) = 24$$

$$x = -8, 3$$

Teacher: *Okay, let's keep thinking about where the other strategies might be as good, or maybe even better. Here's another problem to think about.*

$$4x^2 + x - 3 = 0$$

Student: *Based on what we've said so far, there isn't a common factor, so next I would think about factoring.*

Teacher: *Other folks?*

Student: *I agree with Janice. Those numbers are small enough that you aren't playing with tons of options—you are going to have either 4 and 1 or 2 and 2 when factoring the $4x^2$, and you are really only dealing with −3 and 1 or −1 and 3 to factor the last term. I would try factoring and if that didn't work, I would then try completing the square.*

Student: *I don't like to factor with that 4 coefficient. So, for this one, I'd probably graph it on my calculator. If I didn't have a calculator, I agree that I would try factoring or completing the square next.*

The teacher chooses whether to have students solve this one with a partner or move on to the next problem.

Algebra Problem Strings
©2017 Kendall Hunt Publishing

Teacher: *Well, what about now? Use what we've said so far to find the best strategy here. Be prepared to convince all of us.*	$14 + 2x^2 = 0$
Student: *So, I would take out a 2. You guys know what I mean by that, right? Factor out a 2 or divide everything by 2. Then I'm left with 7 + x² = 0. I'm not loving it.*	$2(7 + x^2) = 0$ $7 + x^2 = 0$
Teacher: *Why is Daniel not loving this?*	$x^2 = -7$
Student: *Because he realizes that this can't be solved. Because x squared equals −7 is impossible. That actually makes it easy because we are done.*	no real solutions
Teacher: *I agree that the square root of −7 is not a real number. How might a graph support that there are no real solutions to this equation? What does a graph of y = 14 + 2x² look like?*	
Student: *Right, it would totally be above the x-axis. It wouldn't intersect at all. So, yep, no real solutions.*	
Teacher: *All right, how about this one?*	$x^2 + 6x - 1 = 0$
Think time.	
Teacher: *Turn and tell your partner what strategy you would use and why?*	$x^2 + 6x + 9 = 1 + 9$ $x^2 + 6x + 9 = 10$
Teacher: *Who changed their thinking, based on what their partner said? You were open to new ideas and had a convincing partner. Go ahead.*	$(x + 3)^2 = 10$ $x + 3 = \pm\sqrt{10}$
Student: *We both started by trying to factor. She thought about factor pairs and there's no way to have factor pairs that are 6 apart multiply to 1. I tried to factor and couldn't get that to work either. When we realized it wasn't factorable, I was going to graph it and my partner said, "complete the square." Both would have worked, but I didn't think about completing the square and now it makes sense to me here.*	$x = 3 \pm \sqrt{10}$
Teacher: *If I record your thinking, can you walk us through it?*	
Student: *Sure, so the idea is to get a perfect square trinomial, which this is not. Kick that −1 to the other side and bring in a 9. I mean, add 9 to both sides. Now we are good because x² + 6x + 9 is factorable—it's (x + 3)².*	
Teacher: *Let me pause you. When you say "kick," you mean...?*	
Student: *Add it to both sides.*	
Teacher: *Who can take over from here?*	
Student: *I got it. Take the square root of both sides. Now let's kick the 3 to the other side. So our solutions are 3 plus the square root of 10 and 3 minus the square root of 10. Voila!*	

(continued)

Teacher: *Was completing the square our best option here? Or maybe I should say, why was completing the square a nice option here?*

Student: *I think the way it was written—the form, I guess—made it pretty nice for completing the square. It didn't have a leading coefficient and the 6 was even and pretty small. That made it friendly to work with.*

Teacher: *Let's return to the idea of graphing here. Could we have graphed it?*

Student: *We could have, but the answers would have been approximations.*

Teacher: *Nice, so for exact answers, completing the square makes sense.*

Teacher: *Last one, friends. What's your strategy for solving here?*	$(x-4)(x+7)=0$
Student: *It is solved! So easy.*	
Student: *No, not really, but you basically can do it mentally.*	$x-4=0 \quad x+7=0$
Teacher: *Someone else say what these two are talking about—that I gave you such an easy one here. What about this makes it really friendly?*	$x=4 \qquad x=-7$
Students will talk about the factored form and the coefficients that make this friendly to solve.	

Teacher: *Okay, so to return to our strategy chart. Take a moment and read through our ideas so far. This chart will stay up and is meant to help all of us make good strategy choices and be more efficient and clever with our choices. Also, we'll keep adding to it as we go.*

Sample Anchor Chart

Your anchor chart could look like this at the end of the problem string:

Choosing a Smart Strategy

Look for a common factor across all terms—if you find one, factor it out (divide by it)	$5x^2-45=0$ $5(x^2-9)=0$
Notice what form it is in Question we're still working on—how does the form help you choose a strategy?	$x^2+5x=24$ $50=-16x^2+48x$ $0=x^2+10x+25$ $0=(x+4)(x-12)$
Some forms are easy to factor because the coefficient of the x's is one and the whole thing is equal to zero	$(x-4)(x+7)=0$
If you can factor it, you should try. Use a rectangular diagram if that helps you.	$4x^2+x-3=0$
If it is not factorable, but it's close to general form, you could complete the square.	$x^2+6x-1=0$
Graphing is probably good if the numbers are messy or if you have your calculator.	

Algebra Problem Strings
©2017 Kendall Hunt Publishing

Sample Final Display

Your display could look like this at the end of the problem string:

$5x^2 - 45 = 0$

$x = \pm 3$

$5x^2 - 45 = 0$

$5(x^2 - 9) = 0$

$x^2 - 9 = 0$

$\dfrac{5x^2}{5} - \dfrac{45}{5} = \dfrac{0}{5}$

$x^2 - 9 = 0$

$x^2 = 9$

$x = \pm 3$

$x^2 + 5x = 24$

$x = -8, 3$

$x^2 + 5x - 24 = 0$

$x(x+5) = 24$

$4x^2 + x - 3 = 0$

$x = -1, 0.75$

$(4x - 3)(x + 1) = 0$

$x = 0.75, -1$

$14 + 2x^2 = 0$

no real solutions

$2(7 + x^2) = 0$

$7 + x^2 = 0$

$x^2 = -7$

no real solutions

$x^2 + 6x - 1 = 0$

$x = 3 \pm \sqrt{10}$

$x^2 + 6x + 9 = 1 + 9$

$x^2 + 6x + 9 = 10$

$(x + 3)^2 = 10$

$x + 3 = \pm\sqrt{10}$

$x = 3 \pm \sqrt{10}$

$(x - 4)(x + 7) = 0$

$x = -7, 4$

$x - 4 = 0 \quad x + 7 = 0$

$x = 4 \qquad x = -7$

Facilitation Notes

This version of the problem string lists short notes for important teacher moves during the string. After you've done the string yourself and studied the relationships involved, you might make similar notes for the things you want a reminder of or deem important.

$5x^2 - 45 = 0$	Today we're focusing on strategy choice. What strategies do we know so far? List. How would you solve? What is your first move? Common factors sound important. Start anchor chart "Choosing a Smart Strategy."
$x^2 + 5x = 24$	What do you pay attention to with this problem? Why factoring?
$4x^2 + x - 3 = 0$	Factor, graph?
$14 + 2x^2 = 0$	No real solutions. Does graph support the solution?
$x^2 + 6x - 1 = 0$	Won't factor? What else do we know? Complete the square. Does graph support the solution?
$(x - 4)(x + 7) = 0$	Easy! Why? What makes it so friendly?

Perfect Squares and Square Roots

At a Glance

$$\sqrt{16}$$

$$\sqrt{4}$$

$$\sqrt{16} \times \sqrt{4}$$

$$\sqrt{16 \times 4}$$

$$\sqrt{9 \times 100}$$

$$\sqrt{9} \times \sqrt{100}$$

$$\sqrt{4} \times \sqrt{36}$$

$$\sqrt{4 \times 36}$$

$$\sqrt{9} + \sqrt{9}$$

$$\sqrt{5} + \sqrt{5}$$

$$\sqrt{4} \times \sqrt{5}$$

$$\sqrt{4 \times 5}$$

$$\sqrt{200}$$

Objectives

This string is designed to support students as they compose, decompose, and reason about quantities that contain square roots. This helps students more confidently work with and reason about these values when various operations and grouping symbols are used, like in the quadratic formula.

Placement

When students use the quadratic formula to solve quadratics, their fluency with square root calculations must be strong. For this reason the string might precede, accompany, or even follow the introduction of the quadratic formula, depending on what your students seem to need. The need to simplify square root values is inevitable.

This string can precede, accompany, or follow textbook Lesson 8.7 The Quadratic Formula.

Guiding the Problem String

The values in this string, mostly perfect squares, are chosen purposefully to support students to reason mentally and not depend on estimation or a calculator. When you encounter a non-perfect square (e.g., 5) let the students acknowledge that it isn't a friendly number and reason about it without calculating its approximate value. Use calculators to get decimal approximations to check for equivalence.

As with the design of many problem strings, we are attempting to keep one idea or quantity the same while something else changes. This design is intentional because it nudges students to pay attention to small details and decide how the changes effect the outcome—here, the equivalence.

We intend for students to reason about the values here, without calculating them exactly, unless they are perfect squares as many of them are. The string also allows you to find out how students think about square roots and what, if any, strategies they have for estimating when the values are not friendly.

About the Mathematics

Given that students will need to be able to simplify and sometimes calculate the values within the discriminant of the quadratic formula, it's crucial that they have fluency with square roots.

The quadratic formula:

$$x = \frac{-b \pm \sqrt{b^2 - 4ac}}{2a}$$

The discriminant:

$$b^2 - 4ac$$

Sample Interactions

Use the following as you plan how to elicit and model student strategies. This is not meant as a script, but as a view into the relationships involved and the intent of the problem string.

Teacher: *Today we're going to warm up with some square roots which I know you've seen before. This will give us a chance to think about how to operate on them and how we might compose or decompose them to make them easier to work with.* *I'm going to put a very friendly square root on the board and just ask you to rename or simplify it when you can. And since these values will be pretty friendly, I'll ask that you pay attention to how the problems are related and what we might be able to generalize from the expressions here.*	
Teacher: *What is the square root of 16?* *How do you know?* *How can we write 16 in terms of 4?*	$\sqrt{16}$ $\sqrt{16} = 4$ $4 \times 4 = 16$ $4^2 = 16$
Teacher: *How about this one, the square root of 4? No problem, right?*	$\sqrt{4}$ $\sqrt{4} = 2$ $2 \times 2 = 4$ $2^2 = 4$
Teacher: *And what about the square root of 16 times the square root of 4?* *If you know the square root of 16 and 4, then you can just multiply them? Everyone agree?*	$\sqrt{16} \times \sqrt{4}$ $\sqrt{16} \times \sqrt{4}$ 4×2 8
Teacher: *What if I do this? What if it's the square root of the product of 16 times 4?* Make sure students note that this and the previous problem are equivalent. Use the equal sign to reinforce the idea of this equivalence. Then wonder with students. **Teacher:** *Will this always be true? Does it make sense?*	$\sqrt{16 \times 4}$ $\sqrt{16 \times 4}$ $\sqrt{64}$ 8 $\sqrt{16 \times 4} = \sqrt{16} \times \sqrt{4}$
Teacher: *How would you simplify this expression: Square root of the product of 9 times 100?* Share finding the product first.	$\sqrt{9 \times 100}$ $\sqrt{9 \times 100}$ $\sqrt{900}$ 30

(continued)

Teacher: *Anyone want to guess what the next problem is? What about the square root of 9 times the square root of 100?* Share finding the square roots of the factors first. Make note of the equivalence between this and the previous problem. **Teacher:** *Will this always be true? Does it make sense?*	$\sqrt{9} \times \sqrt{100}$ $\sqrt{9} \times \sqrt{100}$ 3×10 30
Teacher: *Let's look at these next two problems at the same time. Turn and tell your partner what you notice. What's the same? What's different? Since they are equivalent, which one is easier for you to simplify?* Students may say that the first problem is easier to simplify mentally because it doesn't involve finding the product of 4 and 36. The focus should be on being more and more certain that these are two equivalent forms—this means that if one doesn't feel friendly, students can use the other one.	$\sqrt{4} \times \sqrt{36} \quad \sqrt{4 \times 36}$ $\sqrt{4} \times \sqrt{36} \qquad \sqrt{4 \times 36}$ $2 \times 6 \qquad\quad \sqrt{144}$ $12 \qquad\qquad 12$ $\sqrt{4 \times 36} = \sqrt{4} \times \sqrt{36}$
Teacher: *What about this problem, the square root of 9 plus the square root of 9? What are you thinking?* Student may attempt to combine the expressions as shown. Solicit all possible answers and then let the class decide which one or ones makes sense. Compute approximations with a calculator to set the equal and not equal symbols correctly.	$\sqrt{9} + \sqrt{9}$ $\sqrt{9} + \sqrt{9} = 3 + 3 = 6$ $\sqrt{9} + \sqrt{9} \neq \sqrt{9 + 9}$ $\sqrt{9} + \sqrt{9} \neq \sqrt{18}$ $\sqrt{9} + \sqrt{9} \neq \sqrt{81}$ $\sqrt{9} + \sqrt{9} = 2\sqrt{9} = 2 \cdot 3 = 6$
Teacher: *How does that influence you for the square root of 5 plus the square root of 5? Perfect square?* You might hear that there isn't a way to simplify, because 5 is not a perfect square or that at best, we could rewrite it as $2\sqrt{5}$. Compute approximations with a calculator as necessary. **Teacher:** *So, what do we think so far? When can we simplify and when can we not?*	$\sqrt{5} + \sqrt{5}$ $\sqrt{5} + \sqrt{5} \neq \sqrt{10}$ $\sqrt{5} + \sqrt{5} \neq \sqrt{5 + 5}$ $\sqrt{5} + \sqrt{5} \neq \sqrt{25}$ $\sqrt{5} + \sqrt{5} = 2\sqrt{5}$
Teacher: *Let's use what we are thinking to make sense out of the square root of 4 times the square root of 5. What do you think?* Compute approximations with a calculator to verify.	$\sqrt{4} \times \sqrt{5}$ $\sqrt{4} \times \sqrt{5}$ $2\sqrt{5}$ or $\sqrt{5} + \sqrt{5}$

Algebra Problem Strings
©2017 Kendall Hunt Publishing

Teacher: *You knew I was going to ask this one. Can you simplify the square root of the product of 4 and 5?* By this point, we hope to hear students talking about how looking for square roots as factors is an important step in simplifying.	$\sqrt{4\times5}$ $\begin{array}{ll} \sqrt{4\times5} & \sqrt{4\times5} \\ \sqrt{20} & \sqrt{4}\times\sqrt{5} \\ 2\sqrt{5} & 2\sqrt{5} \end{array}$
Teacher: *And finally, how would you simplify the square root of 200? Turn and talk to your partner about what you would do and why?* This problem gives you the chance to compare simplifying strategies, and think about equivalence and what it means to simplify. You want to encourage a few approaches so that these issues come up. **Teacher:** *Of all of these, is one more simplified than the other? Why do you think so?*	$\sqrt{200}$ $\begin{array}{ll} \sqrt{4\times50} & \sqrt{100\times2} \\ \sqrt{4}\times\sqrt{50} & 10\sqrt{2} \\ 2\sqrt{50} & \end{array}$ $2\sqrt{50}=10\sqrt{2}$

Teacher: *How would you summarize some of the things that came up in this string today?*

Elicit the following:

- When you are multiplying within a square root, you can split up the factors and find the square roots of the factors.

- When you are multiplying square roots, you can multiply the numbers first and then take the square root.

- When you have a non-perfect square within a square root, a decimal is only an approximation of the value.

(continued)

Sample Final Display

Your display could look like this at the end of the problem string:

$$\sqrt{16} = 4 \qquad 4 \times 4 = 16$$

$$4^2 = 16$$

$$\sqrt{4} = 2 \qquad 2 \times 2 = 4$$

$$2^2 = 4$$

$$\sqrt{16} \times \sqrt{4} = 8 \qquad 4 \times 2$$

$$8$$

$$\sqrt{16 \times 4} = 8 \qquad \sqrt{64} \qquad \sqrt{16 \times 4} = \sqrt{16} \times \sqrt{4}$$

$$8$$

$$\sqrt{9 \times 100} = 30 \qquad \sqrt{900}$$

$$30$$

$$\sqrt{9} \times \sqrt{100} = 30 \qquad 3 \times 10$$

$$30$$

$$\sqrt{4} \times \sqrt{36} = 12 \qquad \sqrt{4} \times \sqrt{36} \qquad \sqrt{4 \times 36}$$

$$2 \times 6 \qquad \sqrt{144}$$

$$12 \qquad\qquad 12$$

$$\sqrt{4 \times 36} = 12 \qquad\qquad \sqrt{4 \times 36} = \sqrt{4} \times \sqrt{36}$$

$$\sqrt{9} + \sqrt{9} \qquad \sqrt{9} + \sqrt{9} = 3 + 3 = 6$$

$$\sqrt{9} + \sqrt{9} \neq \sqrt{9+9}$$

$$\sqrt{9} + \sqrt{9} \neq \sqrt{18}$$

$$\sqrt{9} + \sqrt{9} \neq \sqrt{81}$$

$$\sqrt{9} + \sqrt{9} = 2\sqrt{9} = 2 \cdot 3 = 6$$

$$\sqrt{5} + \sqrt{5} = 2\sqrt{5} \qquad \sqrt{5} + \sqrt{5} \neq \sqrt{10}$$

$$\sqrt{5} + \sqrt{5} \neq \sqrt{5+5}$$

$$\sqrt{5} + \sqrt{5} \neq \sqrt{25}$$

$$\sqrt{5} + \sqrt{5} = 2\sqrt{5}$$

$$\sqrt{4} \times \sqrt{5} = 2\sqrt{5} \qquad 2\sqrt{5} \text{ or } \sqrt{5} + \sqrt{5}$$

$$\sqrt{4 \times 5} = 2\sqrt{5} \qquad \sqrt{4 \times 5} \qquad \sqrt{4 \times 5}$$

$$\sqrt{20} \qquad \sqrt{4} \times \sqrt{5}$$

$$2\sqrt{5} \qquad 2\sqrt{5}$$

$$\sqrt{200} = 2\sqrt{50} = 10\sqrt{2} \qquad \sqrt{4 \times 50} \qquad \sqrt{100 \times 2}$$

$$\sqrt{4} \times \sqrt{50} \qquad 10\sqrt{2}$$

$$2\sqrt{50}$$

$$2\sqrt{50} = 10\sqrt{2}$$

Algebra Problem Strings
©2017 Kendall Hunt Publishing

Facilitation Notes

This version of the problem string lists short notes for important teacher moves during the string. After you've done the string yourself and studied the relationships involved, you might make similar notes for the things you want a reminder of or deem important.

$\sqrt{16}$	What is the square root of 16? Quick.
$\sqrt{4}$	Quick.
$\sqrt{16} \times \sqrt{4}$	What is the square root of 16 times the square root of 4? Don't model square root of product yet.
$\sqrt{16 \times 4}$	What is the square root of the product of 16 and 4? Are the two problems equivalent? Will this always be true?
$\sqrt{9 \times 100}$ $\sqrt{9} \times \sqrt{100}$	Repeat, one problem at a time.
$\sqrt{4} \times \sqrt{36}$ $\sqrt{4 \times 36}$	What about both of these? Since they are equivalent, which is easier for you to simplify?
$\sqrt{9} + \sqrt{9}$	What about this? Pull out calculators to support. Decimal approximations.
$\sqrt{5} + \sqrt{5}$	How does the previous help? This time, not perfect squares.
$\sqrt{4} \times \sqrt{5}$	How do you make sense of the square root of 4 times the square root of 5? Use calculators to support.
$\sqrt{4 \times 5}$	You knew I was going to ask this one.
$\sqrt{200}$	Challenge problem, but friendly enough to solve mentally.

8.7 | Quick Image Squares and Square Roots

At a Glance

Show images quickly so students can make a mental picture or actual drawing of what they saw and then reason about the relationship between area and side length.

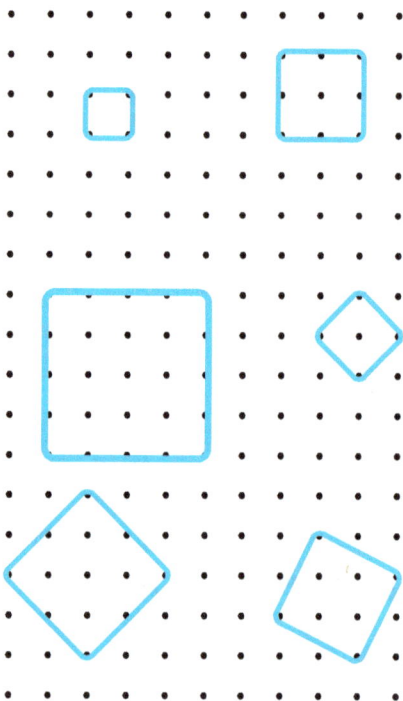

Objectives

This string is designed to encourage students to consider the relationship between square numbers and their square root in a geometric model—that is between the area of a square and its side length.

Placement

This string could be used to precede or accompany other work in which students are asked to simplify or operate on values that include squares and square roots. One logical place for this string would be just before the introduction of the quadratic formula—whose design requires students to wrestle with potentially unfriendly square root values.

You could use this problem string to introduce or help students deal with square roots in textbook Lesson 8.7 The Quadratic Formula.

Guiding the Problem String

The images in this string are designed to start with friendly perfect squares—as seen in the upright squares—and then move towards less friendly numbers in the tilted squares. Showing the images quickly can support a number of things including:

- a heightened sense of engagement and anticipation by students,
- a reason to talk to a partner, or as a class, to verify what we think we saw, and
- a way for students to describe in their own mathematical language what square they saw and how it was oriented or arranged.

The use of the quick images is not designed to trick students or to emphasize speed in a math classroom, but just to increase the overall classroom level of energy and interest in the shared work. If students need to see the image again, you can certainly show it again. The emphasis should not be on getting the square "right" but on reasoning about the side length and area of the square that the whole class saw.

About the Mathematics

The mathematical idea at the heart of this string is the relationship between the area of a square and its side length, between a value and its square root. Too often, this relationship taught without any mention of the geometric underpinnings. When you ask algebra students to describe why 4 is a perfect square and 5 is not, they often do not know or are not grounded in the context of geometry.

This sequence of quick images is designed to help students to construct (or re-construct) the idea that prefect square numbers relate to the area of "upright squares" and other squares are possible, but might be envisioned as "tilted squares." Framed differently, 4 square tiles or 9 square tiles or 16 square tiles would allow us to make a square (2×2, 3×3, or 4×4) but 18 square tiles would not. Students are familiar with finding the area of a square as $s^2 = a$, but can be less developed with the equivalent relationship, $\sqrt{a} = s$. Using areas of perfect squares first can help students recognize and prepare them to work with non-whole number square roots.

Sample Interactions

Use the following as you plan how to elicit and model student strategies. This is not meant as a script, but as a view into the relationships involved and the intent of the problem string.

Teacher: *Today, I'm going to show you an image and give you only a few seconds to take a look at it. Your job is to try to make a mental picture of it, so that you could describe it to your partner, describe it to all of us, and be able to tell us some important features of the image.* *Does that make sense? It will get clearer as we go.*	

Teacher: *Okay, get ready. Here we go.* Show image briefly. **Teacher:** *Tell your partner what you saw and what you can say about it mathematically.* Students will likely say that they saw a square with an area of one and side length of one. You might prompt with questions such as: What did you see? How did you see it?	length: 1 area: 1
Teacher *How about this one? Don't miss it, I'm going to show it quickly.* We recommend that you re-draw each new image as students try to verify what they saw. Next to each new image record students' claims about the area and side length of the image. If students notice perimeter you can acknowledge this, but do not record because the focus is on area and side length.	length: 2 area: 4
Teacher: *Okay, here comes image number three. Be thinking about what you think I'm going to show you. Ready? Go!* Brief wait time. *What did you see? What can you say about it?*	length: 4 area: 16
Teacher: *How about this one? Are you ready? Okay, here it is.* Here you may need to put up the image again that students can wrestle with how to find the area.	

(continued)

The shift from upright to tilted squares is a deliberate one, so that you as a class can reason about non-perfect squares and think about area as the composition of sub-areas. You will likely hear a splitting and re-arranging strategy based on right triangles with base and height of one. Then you can ask about how to find the side length once the area is determined.

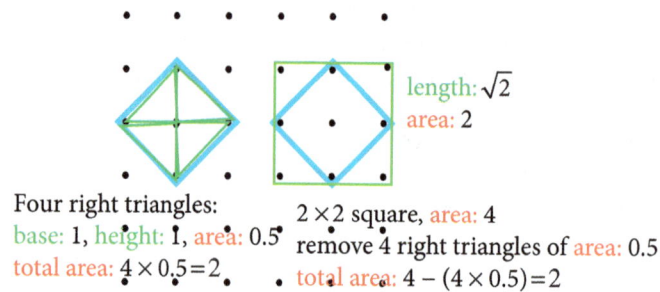

Four right triangles:
base: 1, height: 1, area: 0.5
total area: $4 \times 0.5 = 2$

length: $\sqrt{2}$
area: 2

2×2 square, area: 4
remove 4 right triangles of area: 0.5
total area: $4 - (4 \times 0.5) = 2$

Teacher: *Here's one more and I'll leave it up for a few seconds so that you and your partner can make a mental picture of it and then decide if you agree. Ready?*

At this point in the string students may debate what they saw. You can decide to take some guesses and record those and then reveal the original image. You don't want the conversation to turn into a guessing game. Instead, move the conversation from "guess my square" to "here's my square, what can you say about it?"

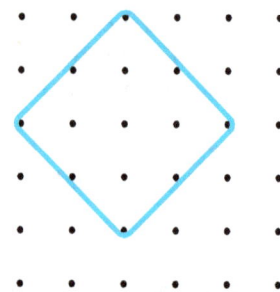

Two primary strategies for finding the area will probably emerge—cutting the area into upright squares and isosceles triangles and surrounding the tilted square in a large upright square and removing the extra parts (right triangles).

Elicit the relationship between the last two squares so that students see that one was scaled by a factor of four to make the other. Encourage them to make use of this relationship to find the side length as $\sqrt{2} + \sqrt{2} = 2\sqrt{2} = \sqrt{8}$.

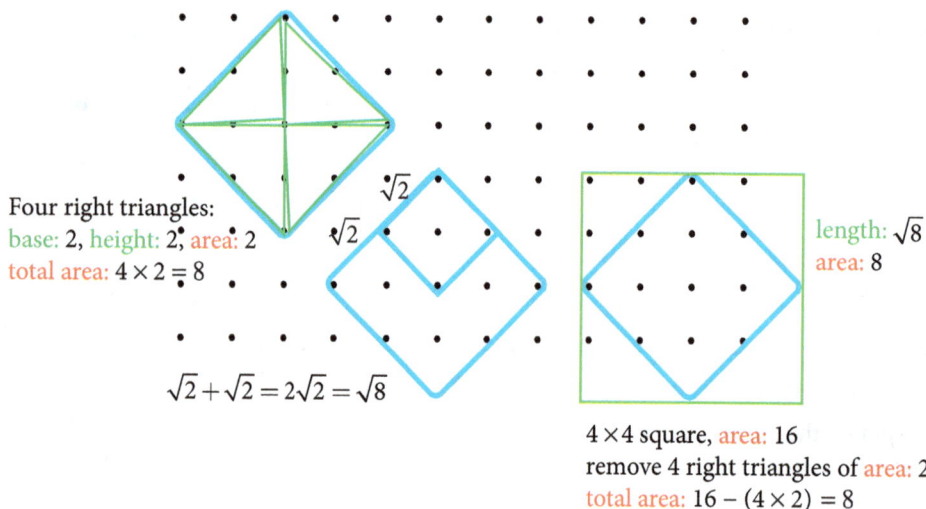

Four right triangles:
base: 2, height: 2, area: 2
total area: $4 \times 2 = 8$

$\sqrt{2}$
$\sqrt{2}$

$\sqrt{2} + \sqrt{2} = 2\sqrt{2} = \sqrt{8}$

length: $\sqrt{8}$
area: 8

4×4 square, area: 16
remove 4 right triangles of area: 2
total area: $16 - (4 \times 2) = 8$

Algebra Problem Strings
©2017 Kendall Hunt Publishing

Teacher: *Okay, last one. I'm going to leave this image up for a few more seconds, just so you have a way to describe it. And then, if as a class you want to see it again, I will show it to you in a slightly different way.*

Expect that students will need to see this image a second time and perhaps have the image left up so that they can determine its area.

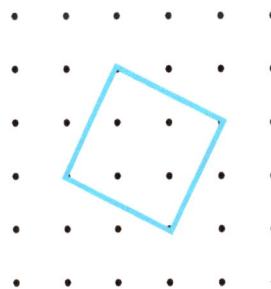

Keep images displayed, giving students a variety of strategies for finding the area, and then the side length.

Be sure that a student names the relationship clearly—once we found the area of any square, taking the square root of that number gave us the side length.

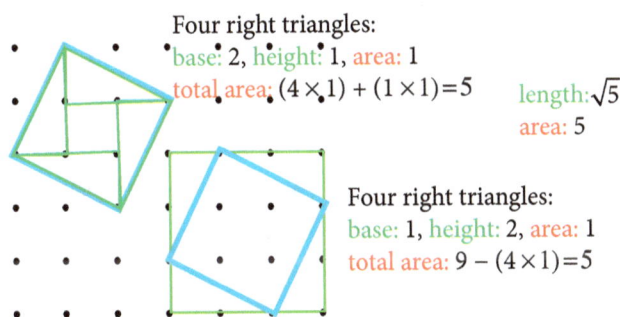

Four right triangles:
base: 2, height: 1, area: 1
total area: $(4 \times 1) + (1 \times 1) = 5$
length: $\sqrt{5}$
area: 5

Four right triangles:
base: 1, height: 2, area: 1
total area: $9 - (4 \times 1) = 5$

Teacher: *How would you summarize some of the things that came up in this string today?*

Elicit the following:

- *Squares have equal sides, length s. The product of s times s, s^2, is the area of the square.*

- *When the square is tilted, we can use triangles inside the square to help us find the area or we can use a square that surrounds the tilted square and remove the extra triangles' areas.*

- *When the square is tilted, we can use the Pythagorean theorem to help find the side length.*

- *We can find equivalencies by using smaller squares to find the length of bigger squares, such as $\sqrt{2} + \sqrt{2} = 2\sqrt{2} = \sqrt{8}$.*

(continued)

Sample Final Display

Your display could look like this at the end of the problem string:

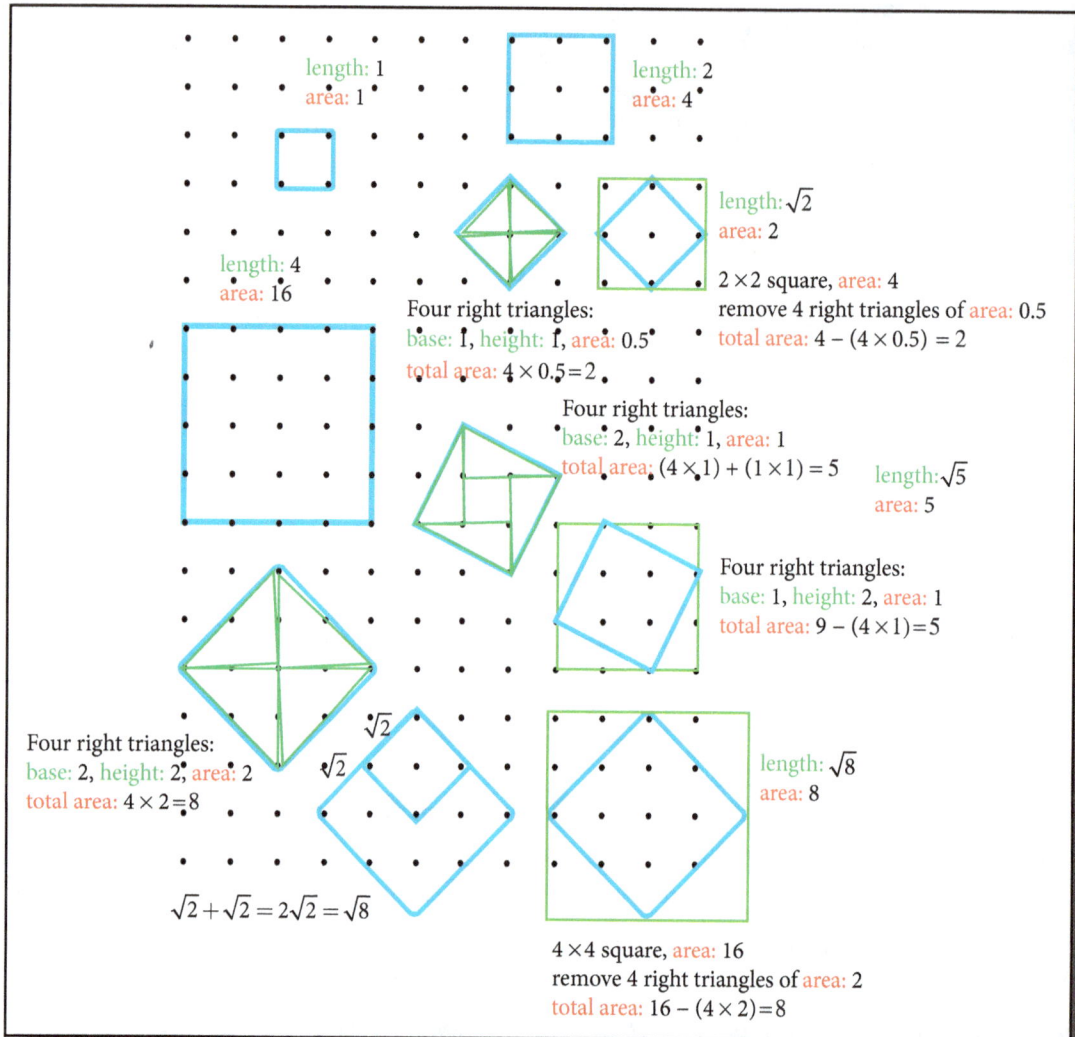

length: 1
area: 1

length: 2
area: 4

length: $\sqrt{2}$
area: 2

length: 4
area: 16

Four right triangles:
base: 1, height: 1, area: 0.5
total area: $4 \times 0.5 = 2$

2×2 square, area: 4
remove 4 right triangles of area: 0.5
total area: $4 - (4 \times 0.5) = 2$

Four right triangles:
base: 2, height: 1, area: 1
total area: $(4 \times 1) + (1 \times 1) = 5$

length: $\sqrt{5}$
area: 5

Four right triangles:
base: 1, height: 2, area: 1
total area: $9 - (4 \times 1) = 5$

Four right triangles:
base: 2, height: 2, area: 2
total area: $4 \times 2 = 8$

$\sqrt{2}$

$\sqrt{2}$

length: $\sqrt{8}$
area: 8

$\sqrt{2} + \sqrt{2} = 2\sqrt{2} = \sqrt{8}$

4×4 square, area: 16
remove 4 right triangles of area: 2
total area: $16 - (4 \times 2) = 8$

Algebra Problem Strings
©2017 Kendall Hunt Publishing

Facilitation Notes

This version of the problem string lists short notes for important teacher moves during the string. After you've done the string yourself and studied the relationships involved, you might make similar notes for the things you want a reminder of or deem important.

Today we'll look at some images.
Just a few seconds to make a mental picture, so you can describe.
What did you see? How did you see it?
What is the area? What is the side length?
Quick.

Repeat.
Quick.

Repeat.
Quick.

Repeat. Linger.
What is this shape? What is the area? How can you find it?
What is the side length?
Elicit 4 small right triangles.
Elicit area of large, surrounding square – area of 4 right triangles.

Repeat. Linger.
What is this shape? What is the area? How can you find it?
What is the side length?
Elicit 4 small right triangles.
Elicit area of large, surrounding square – area of 4 right triangles.
Elicit connection between previous problem, 4 of them fit,
so $\sqrt{2} + \sqrt{2} = 2\sqrt{2} = \sqrt{8}$.

Repeat. Linger.
What is this shape? What is the area? How can you find it?
What is the side length?
Elicit 4 small right triangles.
Elicit area of large, surrounding square – area of 4 right triangles.
How is the square root of a number related to the area and side length of a square?

Problem Strings at a Glance

Lesson 1.1

Subtraction as Difference
Page 1

92 − 60

91 − 59

90 − 58

190 − 158

1090 − 1058

80 − 16

380 − 316

170 − 119

104 − 99.5

600 − 489

Follow up problem string

52 − 4

61 − 57

202 − 18

581 − 47

2081 − 47

71 − 37

73 − 39

121 − 87

Lesson 1.2

Finding the Mean
Page 8

Find the mean:

10, 10, 10, 10, 10

9, 10, 10, 10, 11

5, 5, 15, 15

6, 7, 9, 10

16, 14, 10, 8

12, 8, 10, 12, 8

24, 24, 20, 16, 16

12, 12, 12, 10, 10

12, 10, 12, 10, 12

Lesson 1.3

Mean, Median, Mode
Page 15

Find the mean, median and mode:

20, 20, 20, 20, 20

18, 20, 20, 20, 22

18, 18 20, 22, 22

18, 18, 18, 22, 22

18, 18, 22, 22, 22

18, 18, 18, 18, 22

Lesson 1.4

Making Sense of Histograms
Page 22

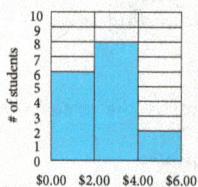

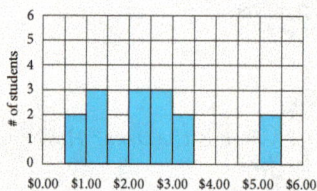

Lesson 1.5

Percents of 360
Page 27

100% of 360°

50% of 360°

25% of 360°

75% of 360°

10% of 360°

5% of 360°

15% of 360°

35% of 360°

144° is what percent of 360°

Lesson 1.7

Graphs: What's the Story?
Page 33

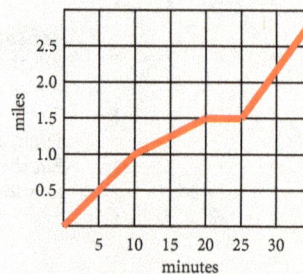

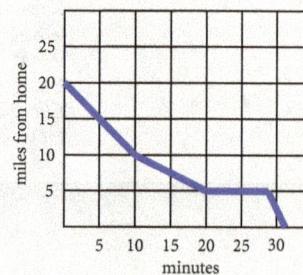

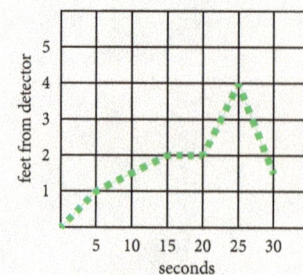

Algebra Problem Strings
©2017 Kendall Hunt Publishing

Lesson 1.8

Playing the Game "Guess It!"
Page 40

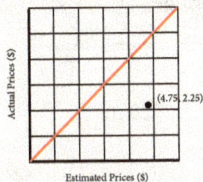

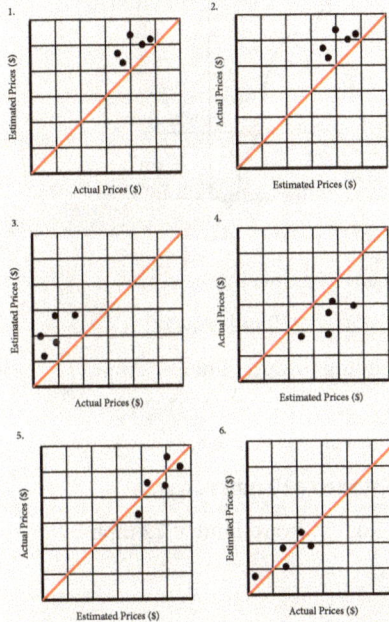

Lesson 2.0

Proportional Reasoning with Ratio Tables 1
Page 47

# of Packs	# of Sticks
1	17
2	
4	
20	
5	
15	
	153

Lesson 2.1

Proportional Reasoning with Ratio Tables 2
Page 54

Time in Car (hr)	Distance (mi)
0.75	36
1.5	
3	
4.5	
1	
	108

Lesson 2.2

Reasoning About Percents
Page 59

Hours left	% of battery
21	70%
	35%
	10%
	45%
	5%
	56%
	100%

Lesson 2.3

Conversions
Page 65

Number of Bleeps	Number of Meeps
32	24
48	
	18
1	
	1

Create a ruler.

Lesson 2.4

Using Ratio Tables versus Dimensional Analysis
Page 70

Convert:

40 feet to inches

52 centimeters to meters

60 inches to centimeters

70 mph to miles per minute

5 meters per second to miles per hour

Lesson 2.5

Reasoning About Inverse Variation 1
Page 75

Workers	Hours
20	6
10	
40	
	24
15	
1	
	1

Lesson 2.6

Reasoning About Inverse Variation 2
Page 81

# of Volunteers	Time (hours)
250	3
500	
	2
	6
	25
10	
v	h

Lesson 2.7

Solving Equations 1

Page 86

$x = 3$

$x = -2$

$-x = 5$

$-x = -4$

$x - 4 = 6$

$x + 4 = -6$

$x - 4 = -10$

Follow-up String:

$x = -6$

$-x = -2$

$2x = 12$

$2x = -12$

$-2x = 10$

$-2x = -8$

$\frac{1}{2}x = 3$

$\frac{1}{3}x = -2$

$-\frac{1}{4}x = -5$

Lesson 2.8

Solving Equations 2

Page 91

$3x + 7 = 22$

Where is $3x + 17$?

Where is $3x$?

Where is $3x - 2$?

Where is x?

$\frac{x}{6} - 20 = -19$

Where is 0?

Where is $\frac{x}{6}$?

Where is x?

Follow-up string:

$\frac{x+9}{3} - 1 = 4$

Where is $\frac{x+9}{3} + 5$?

Where is $\frac{x+9}{3} - 7$?

Where is $\frac{x+9}{3}$?

Where is $x + 9$?

Where is x?

$3(x - 5) + 7 = -14$

Where is $3(x - 5)$?

Where is $x - 5$?

Where is x?

Lesson 3.1

Arithmetic Sequences

Page 96

1, 3, 5, 7

−2, −5, −8

1, ___, 13

3, ___, ___, 15

17, ___, ___, ___, ___, 32

−10, ___, ___, 23

17, ___, ___, ___, 5

last term

↓

−15   6   ___ ← first term

Lesson 3.2

Linear Plots

Page 102

Mackinac Bridge

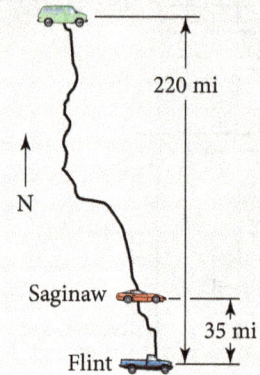

220 mi

N

Saginaw

35 mi

Flint

Green minivan,

started 220 miles away,

going toward Flint at 72 mph

Purple motorcycle,

started 110 miles away,

going toward Flint at 72 mph

Pink SUV,

started 220 miles away,

going toward Flint at 60 mph

Red sports car,

started 35 miles away,

going away from Flint at 48 mph

Light blue bicycle,

started 35 miles away,

going away from Flint at 20 mph

Green bus,

started 50 miles away,

going away from Flint at 48 mph

Lesson 3.3

Time-Distance Rates
Page 106

Time (s)	Dist (ft)
0	5
3	11
6	17

Time (s)	Dist (ft)
0	10
2	6
4	2

Time (s)	Dist (ft)
2	12
4	6
6	0

Time (s)	Dist (ft)
3	30
6	45
9	60

Time (s)	Dist (ft)
0	5
4	7
8	9

Time (s)	Dist (ft)
2	9
6	6
12	3

Lesson 3.4

Linear Equations 1
Page 115

$y = 215 + 3.8x$

$y = 150 + 3.8x$

$y = 3.8x$

$y = -100 + 3.8x$

$y = 215$

$y = 215 + 4.5x$

$y = 215 + 2x$

$y = 350 + 2.5x$

Lesson 3.5

Linear Equations 2
Page 122

Start at 2 feet,
walk away at 1.5
feet per second

$y = -2 + 1.5x$

$y = 10 - x$

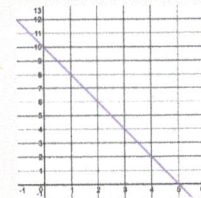

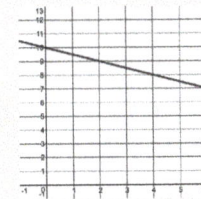

Time (s)	Dist (ft)
0	12
1	11

Time (s)	Dist (ft)
0	5
2	3

$y = 3 + 2.5x$

Lesson 3.6

Solving Equations Using Relational Thinking
Page 125

$$\frac{2x + 3}{}$$
$$17$$

$2x - 3 = 17$

$$\frac{2x + 3}{}$$
$$-17$$

$2x - 3 = -17$

$2 + 4x = x + 8$

Follow-up String

$14x = 28$

$14x + 10 = 66$

$-14x + 10 = -46$

$2x + 4 = -22$

$2(x + 4) = -18$

$2x + 4 = 3x - 5$

Lesson 3.7

Rate of Change
Page 134

Time (s)	Dist (ft)
0	4
2	10

Time (s)	Dist (ft)
0	4
2	7

Time (s)	Dist (ft)
2	5
6	8

Time (s)	Dist (ft)
−2	9
1	7

Time (s)	Dist (ft)
−3	17
0	13

Time (s)	Dist (ft)
0	10
5	8

Time (s)	Dist (ft)
0	10
5	2

Lesson 4.0

Relations and Functions 1
Page 138

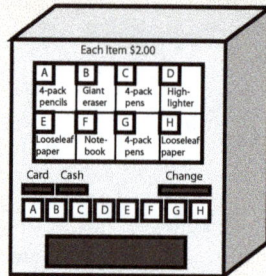

Button (x)	Item (y)
7	folder
2	graph paper
8	mechanical pencil
3	paper
4	folder
1	hand-sanitizer
3	eraser
5	mechanical pencil
6	paper
4	folder

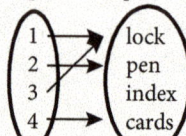

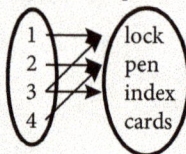

Lesson 4.1

Relations and Functions 2
Page 146

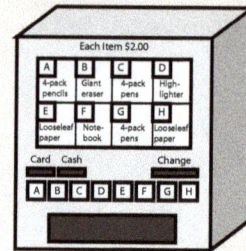

Button (x)	Item (y)
7	folder
2	graph paper
8	mechanical pencil
3	paper
4	folder
1	hand-sanitizer
3	eraser
5	mechanical pencil
6	paper
4	folder

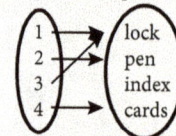

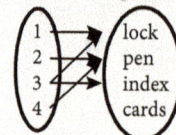

Lesson 4.2

Functions and Graphs
Page 153

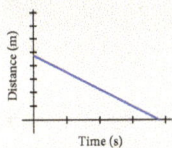
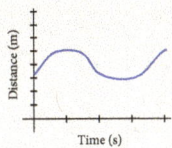

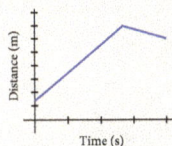
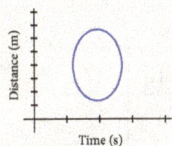

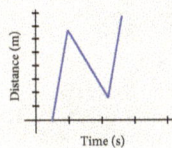
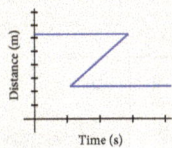

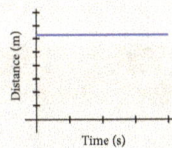
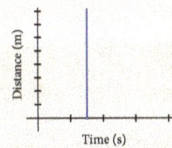

Lesson 4.3

Reading Graphs
Page 158

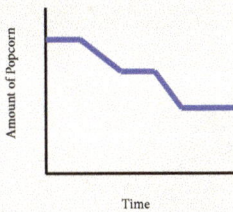

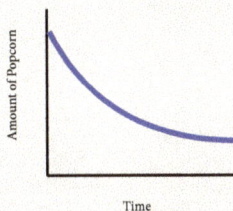

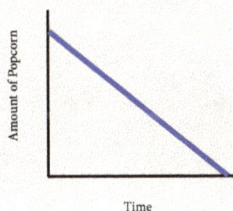

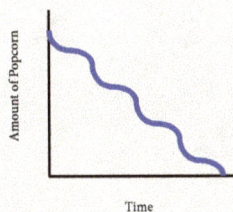

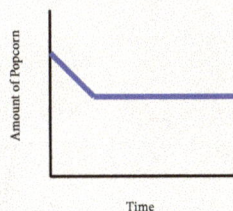

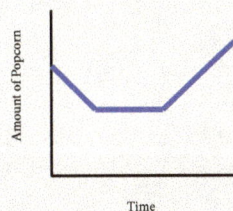

Lesson 4.4

Writing Linear Equations 1
Page 165

rate of 2 m/s, start at 0

rate of 2 m/s, (3, 7)

rate of 3 m/s, (4, 10)

Lesson 4.5

Writing Linear Equations 2
Page 175

rate of 2 m/s, (30, 55)

rate of −5 m/s, (2, 4)

rate of −1.5 m/s, (4, −5)

Lesson 4.6

Equivalent Expressions
Page 181

12×8

12×18

$2(x+4)$

$(x+3) \cdot 7$

$-3(x-5)$

$-4(x+3)$

$$\begin{array}{|c|c|c|} \hline 2 & 2x & -8 \\ \hline \end{array}$$

$$\begin{array}{|c|c|c|} \hline & 3x & -9 \\ \hline \end{array}$$

$$\begin{array}{|c|c|c|} \hline -4 & \underline{} & 20 \\ \hline \end{array}$$

$-3(\underline{}+4) = -3x + \underline{}$

$2(x + \underline{}) = \underline{} - 10$

Lesson 4.7

Writing Linear Equations 3
Page 188

Write the equation of the line that contains:

(5, 10) (3, 0)

(−3, −5) (0, 4)

(−4, −1) (−6, −½)

(−2, 10) (10, −2)

Lesson 4.8

Writing Linear Equations 4
Page 192

x	y
−5	10
3	2
6	−1

x	y
4	2
10	8
15	13

x	y
−15	−15
−8	−8
20	20

x	y
−15	12
−7	10
5	−2

Lesson 4.9

Writing Linear Equations 5
Page 199

x	y
48	52
52	48
99	1

x	y
−5	−8
5	4
11	11.2

x	y
3	−97
99	−1
105	5

x	y
−13	13
−7	7
21	−21

Lesson 4.10

Writing Linear Equations 6
Page 204

x	y
−3	10
3	4
7	0

x	y
−15	27
0	22
8	−18

x	y
−8	28
8	−12
12	−22

Lesson 5.0

Division of a Sum
Page 207

$$816 \div 8 = \frac{816}{8}$$

$$792 \div 8 = \frac{792}{8}$$

$$\frac{4x}{4}$$

$$\frac{4x+8}{4}$$

$$\frac{3x}{3}$$

$$\frac{3x-9}{3}$$

$$\frac{5x}{2}$$

$$\frac{5x+3}{2}$$

$$\frac{4x-3}{8}$$

$$\frac{10-2x}{-5}$$

Algebra Problem Strings
©2017 Kendall Hunt Publishing

Lesson 5.1

Systems of Linear Equations
Page 213

5 sandwiches & 4 cookies cost $62

4 sandwiches & 5 cookies cost $55

6 sandwiches & 3 cookies cost $69

3 sandwiches & 6 cookies cost $48

10 sandwiches & 8 cookies cost $124

8 sandwiches & 10 cookies cost $110

2 sandwiches & 7 cookies cost $41

1 sandwich & 8 cookies cost $34

How much is a cookie?

How much is a sandwich?

Lesson 5.2

Parallel and Perpendicular Slopes
Page 224

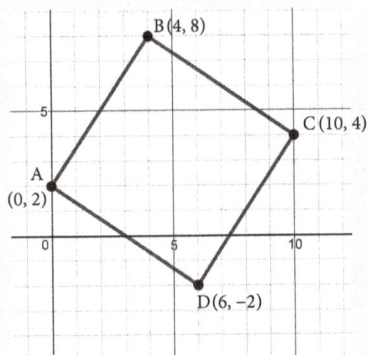

Find the equation of $\overline{AD}$.

Find the equation of $\overline{BC}$.

Find the equation of $\overline{AB}$.

Find the equation of $\overline{CD}$.

Lesson 5.3

Solving for One Variable
Page 230

Solve for x:

$x + 3y = 8$

$x - 3y = 2$

Solve for y:

$2x + y = 3$

$3x - y = 9$

Lesson 5.4

Choose a Strategy
Page 236

Choose a strategy to solve each system.

$8x + 2y = 60$
$-8x - 5y = 72$

$-4x + 3y = -20$
$y = 5x + 8$

$2x + 4y = 72$
$3x + 5y = 80$

$y = 2x - 12$
$y = 3x - 12$

Lesson 5.5

Inequalities in One Variable
Page 241

$x = 3$

$x \geq 3$

$-x = 3$

$-x \geq 3$

$-x < -5$

$3x = -6$

$3x \leq -6$

$-3x \leq -6$

$-4x \geq 12$

Lesson 5.6

Graphing Standard Form
Page 246

$2x + y = 4$

x	y
0	4
2	0

$3x - 5y = -15$

$(0, \underline{\quad})$ $(\underline{\quad}, 0)$

$4x + 3y = -12$

x	y
0	
	0

$5x - 6y = 30$

$2x + y = -2$

Problem Strings at a Glance (continued)

Lesson 5.7

What's Your Solution?
Page 255

$x = -3$

$x \geq -3$

$x < -3$

$2x - 6 = 0$

$y = 2x - 6$

$y > 2x - 6$

$y \leq 2x - 6$

$y = 2x - 6$
$y = \frac{1}{2}x$

$y > 2x - 6$
$y \leq \frac{1}{2}x$

Lesson 6.0

Exponents 1: Definition
Page 264

4^3

5^3

$5 \cdot 5^2$

$4^2 \cdot 4$

$2^3 \cdot 2^2$

$2^2 \cdot 2^4$

x^m

Lesson 6.1

Division as Ratio
Page 269

$4 \times \underline{\quad} = 12$

$4 \times \underline{\quad} = 14$

$4 \times \underline{\quad} = 13$

$4 \times \underline{\quad} = 15$

$\overset{\times\underline{\quad}}{\frown}$
5, 30, $\underline{\quad}$

$\overset{\times\underline{\quad}}{\frown}$
5, 32.5, $\underline{\quad}$

Term #	Term
1	8
2	4

$\big) \times \underline{\quad}$

Term #	Term
1	8
2	6

$\big) \times \underline{\quad}$

Lesson 6.2

Exponents 2: Multiplying Like Bases
Page 275

3^3

$3^2 \cdot 3$

$3^2 \cdot 3^2$

$2^3 \cdot 5^2$

$3^2 \cdot 5^2$

$2^2 \cdot 2^3$

$2^a \cdot 2^b$

$x^m \cdot x^n$

$ab^2 \cdot 3ab$

$2a^{15}b \cdot 3a^3b^{10}$

Lesson 6.3

Exponents 3: Multiplication
Page 282

$(3 \cdot 3)^2$

$(2^2)^3$

$(2^3)^2$

$(a^4)^3$

$(a^6)^2$

$(3^a)^3$

$(3^2)^b$

$(x^a)^b$

$(2a^2)^3$

$(-3ab^{10})^2$

Lesson 6.4

Exponents 4: Equivalence
Page 288

$(3x^2y)^3 = 9x^5y^4$

$(2a^3b^4)(4a^3b^2) = 2^3a^9b^8$

$(4a^3b^6)^4 = (8^2a^6b^{12})^2$

$(-2x^{20}y^3)^3 = (-4x^{20}y^4)(2x^3y^2)$

Algebra Problem Strings
©2017 Kendall Hunt Publishing

Problem Strings at a Glance (continued)

Lesson 6.5

Exponents 5: Division Property
Page 291

$$\frac{8}{4}$$

$$\frac{32}{4}$$

$$\frac{2 \cdot 2 \cdot 2 \cdot 2 \cdot 2}{2 \cdot 2}$$

$$\frac{3^4}{3}$$

$$\frac{2^m}{2^n}$$

$$\frac{x^m}{x^n}$$

$$\frac{2^5 \cdot 3}{2^2}$$

$$\frac{2^4 3^3}{12}$$

Lesson 6.6

Exponents 6: Negative Exponents
Page 298

$$\frac{3}{3 \cdot 3 \cdot 3}$$

$$\frac{2^2}{2^3}$$

$$\frac{4^5}{4^3}$$

$$\frac{3^3}{3^5}$$

$$\frac{3^2 \cdot 2^2}{6^3}$$

$$\frac{x^m}{x^n}$$

$$m^{-a}$$

$$\frac{1}{m^{-a}}$$

Lesson 6.7

Fitting Exponential Models to Data
Page 302

$$y = 201(1 - 0.198)^x$$

$$y = 100(1 - 0.198)^x$$

$$y = 250(1 - 0.198)^x$$

$$y = 201(1 - 0.6)^x$$

$$y = 201(1 - 0.1)^x$$

"Years" elapsed	"Atoms" remaining
0	100
1	88
2	77

Lesson 6.8

Simply Simplifying
Page 305

$$\left(\frac{y^5}{y}\right)^{-3}$$

$$\left(x^{-3}\right)^{-4}$$

$$\left(\frac{x^{-2}}{x^{-3}}\right)^{-5}$$

$$\left(\frac{x^{-3}}{x}\right)^{-4}$$

$$\left(\frac{x^2 y^{-5}}{x^{-3} y^{-4}}\right)^{-1}$$

Lesson 7.1

Function Notation
Page 311

$$(2, 7)$$

$$f(-1) = 1$$

$$f(-4) = -5$$

$$(\tfrac{1}{2}, 4)$$

$$f(1) = \underline{\quad}$$

$$f(\underline{\quad}) = 0$$

$$f(x) = \underline{\quad\quad\quad}$$

Lesson 7.2

Lines and Absolute Value
Page 318

$$y = x$$

$$y = |x|$$

$$y = x + 3$$

$$y = |x| + 3$$

$$y = |x + 3|$$

$$y = 2x - 4$$

$$y = |2x - 4|$$

$$y = |2x| - 4$$

$$y = |x - 2|$$

$$y = |x| - 2$$

Algebra Problem Strings
©2017 Kendall Hunt Publishing

431

Lesson 7.3

Connecting Linear, Absolute Value, and Quadratic Equations

Page 324

$x - 3 = 2$

$|x - 3| = 2$

$(x - 3)^2 = 4$

$x + 4 = 3$

$|x + 4| = 3$

$(x + 4)^2 = 9$

Lesson 7.4

Rate of Change

Page 331

$(0, 0)$ $(1, 1)$

$(1, 1)$ $(4, 2)$

$(4, 2)$ $(9, 3)$

Lesson 7.5

Transforming Functions 1: Translating

Page 336

$y = x^2$

$y = |x|$

$y = (x - 250)^2$

$y = |x| - 750$

$y = |x + 3000| - 1200$

$y = (x - 120)^2 + 425$

Lesson 7.6

Transforming Functions 2: Reflecting

Page 344

$y = 2^x$

$y = -2^x$

$y = 2^{-x}$

$y = -2^{-x}$

$y = -2^x + 100$

$y = -2^{x+100}$

$y = -x^2$

$y = -|x|$

$y = -|x - 400| - 600$

$y = -(x + 2000)^2 + 5000$

Lesson 7.7

Transforming Functions 3: Stretching and Compressing

Page 347

$y = x^2$

$y = 0.01x^2$

$y = 1000x^2$

$y = 10(x - 250)^2$

$y = \frac{1}{4}x^2 - 500$

$y = -20x^2 - 10$

$y = -\frac{1}{5}(x - 30)^2 + 150$

Lesson 7.8

Transforming Functions 4: Combinations

Page 355

$y = -0.1|x - 35| - 200$

$y = 50(x + 400)^2 + 87$

$y = 5\sqrt{-x} + 15$

Lesson 7.9

The Rational Parent Function

Page 358

Workers	Hours
20	6
10	
40	
	24
15	
1	
	h
w	

# of Volunteers	Time (hours)
250	3
500	
	2
	6
	25
10	
	h
v	

Algebra Problem Strings
©2017 Kendall Hunt Publishing

Problem Strings at a Glance (continued)

Lesson 8.1

Squaring and Unsquaring
Page 365

(6×6)*

(-6×-6)*

(12×12)*

(-12×-12)*

$x^2 = 64$

$x^2 - 17 = 64$

$x^2 + 15 = 64$

$(x+3)^2 = 49$

$(x-7)^2 = 25$

$(x+2)^2 + 1 = 82$

*optional problems

Lesson 8.2

Solving and Graphing
Page 371

$y = 3x + 6$

$0 = 3x + 6$

$y = x^2$

$x^2 = 0$

$x^2 - 16 = 0$

$x^2 - 36 = 0$

$x^2 - 2 = 34$

$x^2 - 72 = 0$

$x^2 - 10 = 62$

$x^2 + 72 = 0$*

*optional problem

Lesson 8.3

Factor Puzzles 1
Page 379

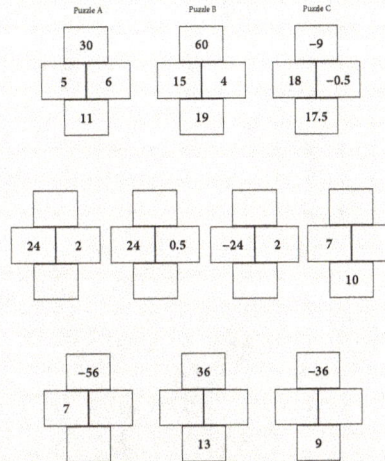

Lesson 8.4

Factor Puzzles 2
Page 386

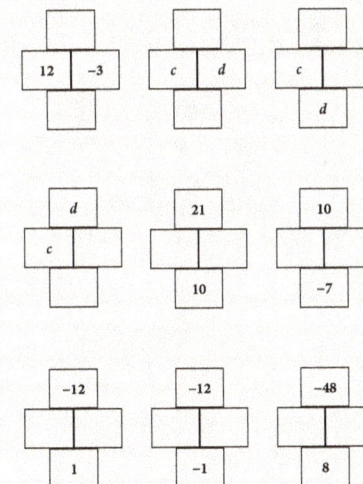

Lesson 8.5

Expanding and Envisioning
Page 390

10×10

10×12

13×10

13×12

$(10+3)(10+2)$

$(x+3)(x+2)$

$5(10+2)$

$5(x+2)$

$(x+2) \times 5$

$(x+6)(x+8)$

$(x-7)(x+8)$

Lesson 8.6

Factors and Factoring
Page 397

$x(x+4) = 5$

$x(x-1) = 12$

$x(x-2) - 8 = 0$

$x^2 + 2x = 15$

Follow-up string:

$x(x+3) = 18$

$x(x-7) = -10$

$x(x+6) + 9 = 0$

$x^2 - 5x = 24$

How Would You Solve and Why?

Page 404

$5x^2 - 45 = 0$

$x^2 + 5x = 24$

$4x^2 + x - 3 = 0$

$14 + 2x^2 = 0$

$x^2 + 6x - 1 = 0$

$(x-4)(x+7) = 0$

Perfect Squares and Square Roots

Page 410

$\sqrt{16}$

$\sqrt{4}$

$\sqrt{16} \times \sqrt{4}$

$\sqrt{16 \times 4}$

$\sqrt{9 \times 100}$

$\sqrt{9} \times \sqrt{100}$

$\sqrt{4} \times \sqrt{36}$

$\sqrt{4 \times 36}$

$\sqrt{9} + \sqrt{9}$

$\sqrt{5} + \sqrt{5}$

$\sqrt{4} \times \sqrt{5}$

$\sqrt{4 \times 5}$

$\sqrt{200}$

Quick Image Squares and Square Roots

Page 416

Show images quickly so students can make a mental picture or actual drawing of what they saw and then reason about the relationship between area and side length.

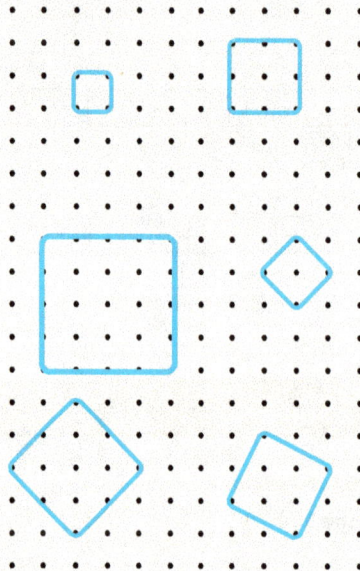

Algebra Problem Strings
©2017 Kendall Hunt Publishing

www.ingramcontent.com/pod-product-compliance
Lightning Source LLC
Chambersburg PA
CBHW082124210326
41599CB00031B/5861